NANOSTRUCTURED POLYMER BLENDS AND COMPOSITES IN TEXTILES

Edited by
Mihai Ciocoiu, PhD
Seghir Maamir, PhD

Reviewer and Editorial Board Member
A. K. Haghi, PhD

AAP | APPLE ACADEMIC PRESS

Apple Academic Press Inc.
3333 Mistwell Crescent
Oakville, ON L6L 0A2
Canada

Apple Academic Press Inc.
9 Spinnaker Way
Waretown, NJ 08758
USA

Exclusive worldwide distribution by CRC Press, a member of Taylor & Francis Group

International Standard Book Number-13: 978-1-77188-143-2 (hbk)

Typeset by Accent Premedia Services (www.accentpremedia.com)

Library and Archives Canada Cataloguing in Publication

Nanostructured polymer blends and composites in textiles / edited by Mihai Ciocoiu, PhD, Seghir Maamir, PhD ; reviewer and editorial board member, A.K. Haghi, PhD.

Includes bibliographical references and index.
Issued in print and electronic formats.
ISBN 978-1-77188-143-2 (hardcover).--ISBN 978-1-77188-287-3 (pdf)
1. Textile fabrics. 2. Nanocomposites (Materials). 3. Polymers.
I. Ciocoiu, Mihai, editor II. Maamir, Seghir, editor III. Haghi, A. K., author, editor

TS1765.N35 2015 677 C2015-906476-7 C2015-906477-5

Library of Congress Cataloging-in-Publication Data

Names: Ciocoiu, Mihai (Textiles professor) | Maamir, Seghir. | Haghi, A. K.
Title: Nanostructured polymer blends and composites in textiles / [edited by] Mihai Ciocoiu, PhD, Seghir Maamir, PhD ; reviewer and editorial board member, A.K. Haghi, PhD.

Description: Toronto : Apple Academic Press, 2015. | Includes index.
Identifiers: LCCN 2015037693 | ISBN 9781771881432 (alk. paper)
Subjects: LCSH: Fibrous composites. | Textile fabrics. | Polymers. | Nanostructures.
Classification: LCC TA418.9.C6 N334 2015 | DDC 677/.02835--dc23
LC record available at http://lccn.loc.gov/2015037693

Apple Academic Press also publishes its books in a variety of electronic formats. Some content that appears in print may not be available in electronic format. For information about Apple Academic Press products, visit our website at **www.appleacademicpress.com** and the CRC Press website at **www.crcpress.com**

CONTENTS

LIST OF CONTRIBUTORS

A. Afzali
University of Guilan, Rasht, Iran

A. A. Berlin
N. Semenov Institute of Chemical Physics, RAS, 4 Kosygin Str., Moscow, 119991, Russian Federation

D. S. Davtyan
State Engineering University of Armenia, 105 Teryana Str., Yerevan, 375009, Armenia

S. P. Davtyan
State Engineering University of Armenia, 105 Teryana Str., Yerevan, 375009, Armenia

A. Grumezescu
Department of Science and Engineering of Oxide Materials and Nanomaterials, Faculty of Applied Chemistry and Materials Science, University Politehnica of Bucharest, 1–7 Polizu Street, 011061 Bucharest, Romania

A. K. Haghi
University of Guilan, Rasht, Iran

A. L. Iordanskii
N. Semenov Institute of Chemical Physics, RAS, 4 Kosygin Str., Moscow, 119991, Russian Federation; E-mail: aljordan08@gmail.com

S. G. Karpova
N.M. Emanuel Institute of Biochemical Physics, RAS, 4 Kosygin Str., Moscow, 119334, Russian Federation

A. V. Khvatov
N.M. Emanuel Institute of Biochemical Physics, RAS, 4 Kosygin Str., Moscow, 119334, Russian Federation

G. V. Kozlov
Kh.M. Berbekov Kabardino-Balkarian State University, Chernyshevsky St., 173, Nalchik, 360004, Russian Federation

Sh. Maghsoodlou
University of Guilan, Rasht, Iran

A. K. Mikitaev
Kh.M. Berbekov Kabardino-Balkarian State University, Chernyshevsky St., 173, Nalchik, 360004, Russian Federation

A. A. Olkhov
G. Plekhanov Russian University of Economics, 9 Stremyannoy per. Moscow, 117997, Russian Federation

O. V. Staroverova
N. Semenov Institute of Chemical Physics, RAS, 4 Kosygin Str., Moscow, 119991, Russian Federation

A. O. Tonoyan
State Engineering University of Armenia, 105 Teryana Str., Yerevan, 375009, Armenia, E-mail: atonoyan@mail.ru

A. Z. Varderesyan
State Engineering University of Armenia, 105 Teryana Str., Yerevan, 375009, Armenia

G. E. Zaikov
N.M. Emanuel Institute of Biochemical Physics of Russian Academy of Sciences, Kosygin St., 4, Moscow-119334, Russian Federation; E-mail: gezaikov@yahoo.com

LIST OF ABBREVIATIONS

AAM	acrylamide
ACS	adsorbate cross sectional area
AFM	atomic force microscopy
AIBN	azoisobutyronitryle
ATO	nanoantimony-doped tin oxide
BET	brunauer, emmett and teller
BSA	bovine serum albumin
CA	cellulose acetate
CB	conduction band
CFD	computational fluid dynamics
CLC	critical liquid content
CNTs	carbon nanotubes
COFs	covalent organic frameworks
CPCs	conductive polymer composites
CPU	central processing unit
CRDCSC	Canadian Research and Development Center of Sciences and Cultures
CVD	chemical vapor deposition
DCPC	dicyclohexylperoxydicarbonate
DSA	drop shape analysis system
ECM	extracellular matrix
EM	electron microscope
ESCs	embryonic stem cells
FCC	face-centered cubic
GAG	glycosaminoglycan
GNC	globular nanocarbon
GNF	graphite nanofibers
GPS	global positioning system
HCP	hexagonal close-packed
HDPE	high density polyethylene
HHI	heat-humidity index

HVAC	heating, ventilating, and air conditioning
IUPAC	The International Union of Pure and Applied Chemistry
MDI	methylenebis(phenyl diisocyanate)
MF	microfiltration
MFI	melt flow index
MMA	methyl-methacrylate
MOFs	metal organic frameworks
MOPs	microporous organic polymers
MWCNTs	multi-walled carbon nanotubes
MWCO	molecular weight cut-off
NF	nanofiltration
NFM	nanofiber mat
NIBIB	The National Institute of Biomedical Imaging and Bioengineering
PA	polyacetylene
PAN	polyacrylonitrile
PAn	polyaniline
PANCMPC	polyacrylonitriles-2-methacryloyloxyethyl phosphoryl choline
PCL	polycaprolactone
PCMs	phase change materials
PES	polyethersulfone
PET	poly(ethylene terephthalate)
PEVA	poly[ethylene-co-(vinyl acetate)]
PGA	poly(glycolic acid)
PHB	poly(3-hydroxybutyrate)
PIMs	polymers of intrinsic microporosity
PLA	poly(l-lactic acid)
PMMA	polymethylmethacrylate
PNA	polyacrylonitrile
POF	plastic optical fiber
PP	polypropylene
PPy	polypyrrole
PRINT	particle repulsion in non-wetting templates
PSD	pore size distribution
PTh	polythiophene

PU	polyurethane
PUCs	polyurethane cationomer
PVC	polyvinylchloride
PVDF	poly(vinylidene fluoride)
PVP	polyvinylpyrrolidone
RO	reverse osmosis
SANS	small angle neutron scattering
SAXS	small angle x-ray scattering
SMAs	shape memory alloys
SMC	shape memory ceramic
SMMs	shape memory materials
SMPs	shape memory polymers
SMPU	shape memory polyurethane
STM	scanning tunneling microscope
SWNTs	single-walled carbon nanotubes
TEMPO	tetramethylpyperidin-1-oxil
TIPS	thermally induced phase separation
TPU	thermoplastic polyurethane
TUFT	tubes by fiber template process method
UF	ultrafiltration
UPF	UV protective factor
VB	valence band
WVP	water vapor permeability

LIST OF SYMBOLS

CHILL	total efferent shivering command
COLDS	cold signal
d	thickness
DILAT	total efferent skin vasodilation command
dL/dt	velocity of front propagation
ERROR	error signal
g	gravity constant
I	intercept
K	geometric factor
k	permeability
L	front position
l	rising height
M	molecular weight of adsorbate
M	shivering metabolic rate
md	mass dry strips
mw	mass of the wet and dry strips
N	Avogadro's number
P	mesh porosity
P0/P	relative pressure
Pa	ambient pressure
P_c	capillary pressure
R	inner radius of the measuring tube
R	mean radius of capillary
r	radius of capillary
r_l	density of gas
r_p	average pore radius
RSW	sweat rate
S	slope
S	specific surface area
S_t	total surface area

STRICT	total efferent skin vasoconstriction command
SWEAT	total efferent sweat command
T	ambient temperature
V_{ads}	volume of gas adsorbed
V_{bl}	blood flow rate
V_{liq}	volume of liquid
V_m	molar vol. of liquid adsorbate
V_T	volume of the sample
W	weight of gas adsorbed
W	weight of penetrating liquid
WARMS	warm signal
W_m	weight of adsorbate as monolayer

Greek Symbols

η	viscosity of liquid
θ	advancing contact angle
ρ	density of measuring liquid
$\rho 0$	density of the polymer
σ	surface tension of liquid
ω	relative porosity

PREFACE

Understanding nanostructured polymer blends and composites in textiles and their properties and behavior is fundamental to modern textile design, engineering, and materials science. Written for professionals and students of chemical science and textile engineering and materials science and design, this book describes the procedures for nanomaterial selection and design in order to ensure that the most suitable materials for a given textile application are identified from the full range of materials, chemicals, and section shapes available.

Several case studies have been developed to further illustrate procedures and to add to the practical implementation of the text.

This new volume reviews recent academic and technological developments behind new-engineered modified nano-textile materials. The book is intended for researchers and those interested in future developments in mechanical and physico-chemical characteristics of modified nano-textile materials and polymer blends. Several innovative applications for different materials are described in considerable detail with emphasis on the experimental data that supports these new applications. From nano-fibers to chemical materials, creative modifications concerning new nano-composites are described that could one day become commonplace. In this book the world's leading experts describe their most recent research in their areas of expertise. The book will also be a useful tool for students and researchers, providing helpful insights into new evolving research areas in nanostructured polymer blends and composites in textiles.

This new book covers a broad range of polymeric materials and textiles and provides industry professionals and researchers in polymer science and technology with important information involved in the functional materials production chain.

This volume presents the latest developments and trends in advanced polymer and textile materials and structures. It discusses the developments of advanced polymers and respective tools to characterize and predict the material properties and behavior.

This book has an important role in advancing polymer materials and textiles in nanoscale. Its aim is to provide original, theoretical, and important experimental results that use non-routine methodologies. It also includes chapters on novel applications of more familiar experimental techniques and analyzes of composite problems that indicate the need for new experimental approaches.

ABOUT THE EDITORS

Mihai Ciocoiu, PhD
Faculty of Textiles-Leather and Industrial Management, Gheorghe Asachi Technical University of Iasi, Iasi, Romania

Mihai Ciocoiu, PhD, is a Professor of Textiles-Leather and Industrial Management at Gheorghe Asachi Technical University of Iasi, Romania. He is the founder and Editor-in-Chief of the *Romanian Textile and Leather Journal*. He is currently a senior consultant, editor, and member of the academic board of the *Polymers Research Journal* and the *International Journal of Chemoinformatics and Chemical Engineering*.

Seghir Maamir, PhD
Associate Professor, Department of Mechanical Engineering, Faculty of Engineer Sciences, University of Boumerdes, Algeria

Seghir Maamir, PhD, is currently Associate Professor in the Department of Mechanical Engineering and Faculty of Engineer Sciences at the University of Boumerdes in Algeria, where he was also an examiner of magister theses in polymers and composites at the University of Boumerdes, Algeria. His fields of interests are fluid mechanics, heat transfer, and mechanical vibrations. He holds a PhD from the University of Franche Comte, France.

A. K. Haghi, PhD
Editor-in-Chief, International Journal of Chemoinformatics and Chemical Engineering; Editor-in-Chief, Polymers Research Journal
Member of the Canadian Research and Development Center of Sciences and Cultures (CRDCSC), Montreal, Quebec, Canada

Dr. A. K. Haghi holds a BSc in urban and environmental engineering from University of North Carolina (USA); a MSc in mechanical engineering from North Carolina A&T State University (USA); a DEA in applied mechanics, acoustics and materials from the Université de Technologie de Compiègne (France); and a PhD in engineering sciences from the Université de Franche-Comté (France). He is the author and editor of 65 books as well as 1000 published papers in various journals and conference proceedings. Dr. Haghi has received several grants, consulted for a number of major corporations, and is a frequent speaker to national and international audiences. Since 1983, he served as a professor at several universities. He is currently Editor-in-Chief of the *International Journal of Chemoinformatics and Chemical Engineering* and *Polymers Research Journal* and is on the editorial boards of many international journals. He is a member of the Canadian Research and Development Center of Sciences and Cultures (CRDCSC), Montreal, Quebec, Canada.

CHAPTER 1

ENGINEERING NANOTEXTILES: DESIGN OF TEXTILE PRODUCTS

A. AFZALI and SH. MAGHSOODLOU

University of Guilan, Rasht, Iran

CONTENTS

ABSTRACT

The study of the history of clothing and textiles traces the availability and use of textiles and other materials and the development of nanotechnology for the making of special applications. Therefore, the first chapter starts by review the textile history and its different types. After that the relationship between nanotechnology and textile industry has been mentioned. In this section, the demonstration of electrospinning process and its variation has been investigated in this field.

1.1 INTRODUCTION

The origin of the textile industry is lost in the past. Fine cotton fabrics have been found in India, dating from 6000 to 7000 years ago, and fine and delicate linen fabrics have been found from 2000 to 3000 years ago, at the height of the Egyptian civilizations. More recent archaeological excavations, among some of Europe's oldest Stone Age sites, have found imprints of textile structures, dating back some 25,000 years, but in the humid conditions obtaining in these more northerly areas, all traces of the actual textiles have long disappeared, unlike those from the dry areas of India and Egypt [1–11].

Until more recent times, the spinning of the yarns and the weaving of the fabrics were generally undertaken by small groups of people, working together – often as a family group. However, during the Roman occupation of England, the Romans established a factory' at Winchester, for the production, on a larger scale, of warm Woolen blankets, to help reduce the impact of the British weather on the soldiers from southern Europe [12].

In the family context, it generally fell to the female side to undertake the spinning, while the weaving was the domain of the men. Spinning was originally done using the distaff to hold the unspun fibers, which were then teased out using the fingers and twisted into the final yarn on the spindle. In the 1530s, in Brunswick, a 'spinning wheel' was invented, with the wheel driven by a foot pedal, giving better control and uniformity to the yarns produced. Often, great skill was developed, as shown by the records of a woman in Norwich, who spun one pound of combed wool into a single yarn measuring 168,000 yards, and from the same weight of cotton, spun a yarn of 203,000 yards. In today's measures this is equivalent to a cotton count of 240, or approximately 25 decitex. Cotton count is the number of hanks of 840 yards (768 meters) giving a total weight of 1 lb (453.6 g). A Tex is a measurement of the linear density of a yarn or cord, being the weight in grams of a 1000 m length; a decitex is the weight in grams of a 10,000 m length [13, 14].

By the eighteenth century, small co-operatives were being formed for the production of textiles, but it was really only with the mechanization of spinning and weaving during the Industrial Revolution, that mass production started.

Up to this time, both spinning and weaving were essentially hand operations. Handlooms were operated by one person, passing the weft (the transverse threads) by hand, and performing all the other stages of weaving manually (*see* Chapter 4 for a description of the weaving process). In 1733, John Kay invented the 'flying shuttle,' which enabled a much faster method for inserting the weft into the fabric at the loom and greatly increased the productivity of the weavers [15].

Until the advent of the flying shuttle, the limiting factor in the production chain for fabrics was the output of the individual weaver, but this now changed and with the more rapid use of the yarns, their production became the limiting factor in the total process. In 1764, this was partly resolved by the invention, by James Hargreaves, of the 'Spinning Jenny,' which was developed further by Sir Richard Arkwright, with his water spinning frame, in 1769, and then in 1779, by Samuel Crompton, with his 'spinning mule' [16].

Alongside these developments in spinning, similar changes were taking place in the weaving field, with the invention of the power loom by Edmund Cartwright, in 1785.

With this increase in mechanization of the whole industry, it was logical to bring the production together, rather than keeping it widely spread throughout the homes of the producers. Accordingly, factories were established [17]. The first of such was in Doncaster in 1787, with many power looms powered by one large steam engine. Unfortunately, this was not a financial success, and the mill only operated for about 3 or 4 years [18].

Meanwhile, other mills were being established, in Glasgow, Dumbarton and Manchester. A large mill was erected at Knott Mill, Manchester, although this burnt down after only about 18 months. The first really successful mill was opened in Glasgow in 1801 [19].

However, this industrialization was not to everyone's liking; many individuals were losing their livelihoods to the mass production starting to come from the increasing number of mills. This led to a backlash from the general public, resulting in the Luddite Riots in 1811–1812, when bands of masked people under the leadership of 'King Ludd' attacked the new factories, smashing all the machinery therein. It was only after very harsh suppression, resulting in the hanging or deportation of convicted Luddites in 1813, that this destruction was virtually stopped. However, there were still some outbreaks of similar actions in 1816, during the depression following the end of the British war with France, and this intermittent action only finally stopped when general prosperity increased again in the 1820s [20].

Following this, the textile industry expanded considerably, particularly in the areas where the raw materials were readily available. For example, the Woolen mills in East Anglia, where there was good grazing for the sheep, and in West Yorkshire and Eastern Lancashire, where either coal was available for powering the new steam engines, or where fast flowing streams existed to provide the energy source for water-powered mills (particularly in central Lancashire). The main Woolen textile production developed in Yorkshire, as it was easier and cheaper to transport the raw wool there, than to carry the large quantities of coal required to power the mills to the wool growing areas [21]. In Lancashire, with the ports of Liverpool and Manchester close by for the importation of cotton from America, the cotton industry grew and flourished. However, in the 1860s, due to the American Civil War, the supply of cotton from America dried up and caused great hardship among the cotton towns of south and east Lancashire [22].

On account of this, and with the great strides being made in chemistry, research was begun to find ways of making artificial yarns and fibers. The first successful artificial yarn was the Chardonnet 'artificial silk,' a cellulosic fiber regenerated from spun nitrocellulose. Further developments lead to the cuprammonium process and then to the viscose process for the production of another cellulosic, rayon [23]. This latter viscose was fully commercialized by Courtaulds in 1904, although it was not widely used in rubber reinforcement until the 1920s, with the development of the balloon type [24].

Research continued into fiber-forming polymers, but the next new fully synthetic yarn was not discovered until the 1930s, when Wallace Hume Carothers, working for DuPont, discovered and developed nylon. This was first commercialized in 1938 and was widely developed during the 1940s to become one of the major yarn types used. Continuing research led to the discovery of polyester in 1941, and over the ensuing decades, polyolefin fibers (although because of their low melting/softening temperatures, these are not used as reinforcing fibers in rubbers) and aramids [25].

As the chemical industry greatly increased the types of yarns available for textile applications, so the machinery used in the industry was being developed. Whereas the basic principles of spinning and weaving have not significantly changed over the millennia, the speed and efficiency of the equipment used for this has been vastly been improved. In weaving, the major changes have been related to the method of weft insertion; the conventional shuttle has been replaced by rapiers, air and water jets, giving far higher speeds of weft insertion [26].

Other methods of fabric formation have similarly been developed, such as the high speed knitting machines and methods for producing fabric webs [27].

The use of nanotechnology in the textile industry made it multifunctional which can produce fibers with variable functions and applications such as UV protection, antiodor, antimicrobial etc. In many cases smaller amounts of the additive are required, for the saving on resources. The success of Nanotechnology and its potential applications in textiles lies in various fields where new methods are combined with multifunctional textile systems, durable etc. without affecting the inherent properties of the textiles including softness, flexibility, washability, etc. Keeping the above factors in consideration, the present review highlighted the use of nanotechnology in textile industry and textile engineering, the types and methods of preparation of different nano composites used in textiles [28, 29].

1.2 TEXTILE TYPES

Textiles can be made from many materials. They are classified on the basis of their component fibers into, animal (wool, silk), plant (cotton, flax, jute), mineral (asbestos, glass fiber), and synthetic (nylon, polyester, acrylic). They are also classified as to their structure or weave, according to the manner in which warp and weft cross each other in the loom. Textiles are made in various strengths and degrees of durability, from the finest gossamer to the sturdiest canvas. The relative thickness of fibers in cloth is measured in deniers. Microfiber refers to fibers made of strands thinner than one denier [30, 31].

1.2.1 ANIMAL TEXTILES

The main animal fiber used for textiles is wool. Animal textiles are commonly made from hair or fur of animals. Silk is another animal fiber produces one of the most luxurious fabrics. Sheep supply most of the wool, but members of the camel family and some goats also furnish wool. Wool, commonly used for warm clothing, refers to the hair of the domestic goat or sheep and it is coated with oil known as lanolin, which is water-proof and dirt-proof making a comfortable fabric for dresses, suits, and sweaters. The term woolen refers to raw wool, while *worsted* refers to the yarn spun from raw wool. Cashmere, the hair of the Indian Cashmere goat, and mohair, the hair of the North African Angora goat, are types of wool known for their softness. Other animal textiles made from hair or fur is alpaca wool, vicuña wool, llama wool, and camel hair. They are generally used in the production of coats, jackets, ponchos, blankets, and other warm coverings. Angora refers to the long, thick, soft hair of the Angora rabbit [32, 33].

Silk is an animal textile made from the fibers of the cocoon of the Chinese silkworm. It is spun into a smooth, shiny fabric prized for its sleek texture. Silk comes from cocoons spun by silkworms. Workers unwind the cocoons to obtain long, natural filaments. Fabrics made from silk fibers have great luster and softness and can be dyed brilliant colors. Silk is especially popular for saree, scarfs and neckties [34].

1.2.2 PLANT TEXTILES

Plants provide more textile fibers than do animals or minerals. Cotton fibers produce soft, absorbent fabrics that are widely used for clothing, sheets, and towels. Fibers of the flax plant are made into linen. The strength and beauty of linen have made it a popular fabric for fine tablecloths, napkins, and handkerchiefs [35].

Grass, rush, hemp, and sisal are all used in making rope. In the first two cases, the entire plant is used for this purpose, while in the latter two;

only fibers from the plant are used. Coir (coconut fiber) is used in making twine, floor mats, door mats, brushes, mattresses, floor tiles, and saking. Straw and bamboo are both used to make hats. Straw, a dried form of grass, is also used for stuffing, as is kapok. Fibers from pulpwood trees, cotton, rice, hemp, and nettle are used in making paper. Cotton, flax, jute, and modal are all used in clothing. Piña (pineapple fiber) and ramie are also fibers used in clothing, generally with a blend of other fabrics such as cotton. Seaweed is sometimes used in the production of textiles. A water soluble fiber known as *alginate* is produced and used as a holding fiber. When the cloth is finished, the alginate is dissolved, leaving an open area [36, 37].

1.2.3 MINERAL TEXTILES

Asbestos and basalt fiber are used for vinyl tiles, sheeting, and adhesives, "transite" panels and siding, acoustical ceilings, stage curtains, and fire blankets. Glass fiber is used in the production of spacesuits, ironing board and mattress covers, ropes and cables, reinforcement fiber for motorized vehicles, insect netting, flame-retardant and protective fabric, soundproof, fireproof, and insulating fibers. Metal fiber, metal foil, and metal wire have a variety of uses, including the production of "cloth-of-gold" and jewelry [38, 39].

1.2.4 SYNTHETIC TEXTILES

Most manufactured fibers are made from wood pulp, cotton linters, or petrochemicals. Petrochemicals are chemicals made from crude oil and natural gas. The chief fibers manufactured from petrochemicals include nylon, polyester, acrylic, and olefin. Nylon has exceptional strength, wears well, and is easy to launder. It is popular for hosiery and other clothing and for carpeting and upholstery. Such products as conveyor belts and fire hoses are also made of nylon. All synthetic textiles are used primarily in the production of clothing [37, 40].

- Polyester fiber is used in all types of clothing, either alone or blended with fibers such as cotton.

- Acrylic is a fiber used to imitate wools, including cashmere, and is often used in place of them.
- Nylon is a fiber used to imitate silk and is tight-fitting; it is widely used in the production of pantyhose.
- Lycra, spandex, and tactel are fibers that stretch easily and are also tight-fitting. They are used to make active wear, bras, and swimsuits.
- Olefin (Polypropylene or Herculon) fiber is a thermal fiber used in active wear, linings, and warm clothing.
- Lurex is a metallic fiber used in clothing embellishment.
- Ingeo is a fiber blended with other fibers such as cotton and used in clothing. It is prized for its ability to wick away perspiration.

1.3 TEXTILE PRODUCTION

1.3.1 PRODUCTION METHODS

Most textiles are produced by twisting fibers into yarns and then knitting or weaving the yarns into a fabric. This method of making cloth has been used for thousands of years. But throughout most of that time, workers did the twisting, knitting, or weaving largely by hand. With today's modern machinery, textile mills can manufacture as much fabric in a few seconds as it once took workers weeks to produce by hand [41]. The production of textiles are done by different methods and some of the common production methods are listed as follows [42]:

(i) *Weaving* (by machine as well as by hand), (ii) *Knitting,* (iii) *Crochet,* (iv) *Felt* (fibers are matted together to produce a cloth, (v) *Braiding, and* (vi) *Knotting. Weaving* is a textile production method that involves interlacing a set of vertical threads (called the warp) with a set of horizontal threads (called the weft). This is done on a machine known as a loom, of which there are a number of types. Some weaving is still done by hand, but a mechanized process is used most often. Tapestry, sometimes classed as embroidery, is a modified form of plain cloth weaving [43].

Knitting and *crocheting* involve interlacing loops of yarn, which are formed either on a knitting needle or crochet hook, together in a line. The two processes differ in that the knitting needle has several active loops at one time waiting to interlock with another loop, while crocheting never has more than one active loop on the needle [44].

Other specially prepared fabrics not woven are *felt* and *bark* (or tapa) cloth, which are beaten or matted together, and a few in which a single thread is looped or plaited, as in crochet and netting work and various laces. *Braiding* or *plaiting* involves twisting threads together into cloth. Knotting involves tying threads together and is used in making macramé [45].

Most textiles are now produced in factories, with highly specialized power looms, but many of the finest velvets, brocades, and table linens are still made by hand. Lace is made by interlocking threads together independently, using a backing and any of the methods described above, to create a fine fabric with open holes in the work. Lace can be made by hand or machine. The weaving of carpet and rugs is a special branch of the textile industry.

Carpets, rugs, velvet, velour, and velveteen are made by interlacing a secondary yarn through woven cloth, creating a tufted layer known as a nap or pile [46].

1.3.2 PRODUCTION OF COTTON CLOTHES

In the 1700s, English textile manufacturers developed machines that made it possible to spin thread and weave cloth into large quantities. Today, the United States, Russia, China and India are major producers of cotton. When cotton arrives at a textile mill, several blenders feed cotton into *cleaning machines*, which mix the cotton, break it into smaller pieces and remove trash. The cotton is sucked through a pipe into *picking machines*. Beaters in these machines strike the cotton repeatedly to knock out dirt and separate lumps of cotton into smaller pieces. Cotton then goes to the *carding machine*, where the fibers are separated. Trash and short fibers are removed. Some cotton goes through a *comber* that removes more short fibers and makes a stronger, more lustrous yarn. This is followed by spinning processes, which do three jobs: *draft* the cotton, or reduce it to smaller structures, straighten and parallel the fibers and lastly, put twist into the yarn. The yarns are then made into cloth by weaving, knitting or other processes [47, 48].

Some of the properties of cotton are discussed as follows: (i) soft and comfortable, (ii) wrinkles easily, (iii) absorbs perspiration quickly, (iv) good color retention and good to print on, and (v) strong and durable [49].

1.3.3 PRODUCTION OF WOOL

The processing of wool involves four major steps. First comes shearing, followed by sorting and grading, making yarn and lastly, making fabric. This is followed by grading and sorting, where workers remove any stained, damaged or inferior wool from each fleece and sort the rest of the wool according to the quality of the fibers. Wool fibers are judged not only on the basis of their strength but also by their *fineness* (diameter), length, *crimp* (waviness) and color [50].

The wool is then scoured with detergents to remove the *yolk* and such impurities as sand and dust. After the wool dries, it is *carded*. The carding process involves passing the wool through rollers that have thin wire teeth. The teeth untangle the fibers and arrange them into a flat sheet called a *web*. The web is then formed into narrow ropes known as *silvers*. After carding, the processes used in making yarn vary slightly, depending on the length of the fibers. Carding length fibers are used to make woolen yarn. Combing length fibers and French combing length fibers are made into *worsted yarn* [51].

Woolen yarn, which feels soft, has a fuzzy surface and it is heavier than worsted. While worsted wool is lighter and highly twisted, it is also smoother, and is not as bulky, thus making it easier to carry or transport about. Making worsted wool requires a greater number of processes, during which the fibers are arranged parallel to each other. The smoother the hard-surface worsted yarns, the smoother the wool it produces, meaning, less fuzziness. Fine worsted wool can be used in the making of athletics attire, because it is not as hot as polyester, and the weave of the fabric allows wool to absorb perspiration, allowing the body to "breathe." Wool manufacturers knit or weave yarn into a variety of fabrics. Wool may also be dyed at various stages of the manufacturing process and undergo finishing processes to give them the desired look and feel [52, 53].

The finishing of fabrics made of woolen yarn begins with *fulling*. This process involves wetting the fabric thoroughly with water and then passing it through the rollers. Fulling makes the fibers interlock and mat together. It shrinks the material and gives it additional strength and thickness. Worsteds go through a process called *crabbing* in which the fabric passes through boiling water and then cold water. This procedure strengthens the fabric.

The exclusive features of cotton fabric are: (i) hard wearing and absorbs moisture, (ii) does not burn over a flame but smolders instead, (iii) light-weight and versatile, (iv) does not wrinkle easily, and (v) resistant to dirt and wear and tear [54].

1.3.4 PRODUCTION OF SILK

Silkworms are cultivated and fed with mulberry leaves. Some of these eggs are hatched by artificial means such as an incubator, and in the olden times, the people carried it close to their bodies so that it would remain warm. Silkworms that feed on smaller, domestic tree leaves produce the finer silk, while the coarser silk is produced by silkworms that have fed on oak leaves. From the time they hatch to the time they start to spin cocoons, they are very carefully tended to. Noise is believed to affect the process, thus the cultivators try not to startle the silkworms. Their cocoons are spun from the tops of loose straw. It will be completed in two to three days' time. The cultivators then gather the cocoons and the chrysales are killed by heating and drying the cocoons. In the olden days, they were packed with leaves and salt in a jar, and then buried in the ground, or else other insects might bite holes in it. Modern machines and modern methods can be used to produce silk but the old-fashioned hand-reels and looms can also produce equally beautiful silk [55, 56].

The properties of silk includes, (i) versatile and very comfortable, (ii) absorbs moisture, (iii) cool to wear in the summer yet warm to wear in winter, (iv) easily dyed, (v) strongest natural fiber and is lustrous, and (vi) poor resistance to sunlight exposure [57].

1.3.5 PRODUCTION OF NYLON MATERIALS

Nylon is made by forcing molten nylon through very small holes in a device called 'spinneret.' The streams of nylon harden into filament once they come in contact with air. They are then wound onto bobbins. These fibers are drawn (stretched) after they cool. Drawing involves unwinding the yarn or filaments and then winding it around another spool. Drawing makes the molecules in each filament fall into parallel lines. This gives the nylon fiber strength and elasticity. After the whole drawing process, the yarn may be twisted a few turns per yard or meters as it is wound onto spools. Further treatment to it can give it a different texture or bulk [58].

The properties of the nylon are as follows: (i) it is strong and elastic, (ii) it is easy to launder, (iii) it dries quickly, (iv) it retains its shape, and (v) it is resilient and responsive to heat setting [59].

1.3.6 PRODUCTION OF POLYESTER

Polyesters are made from chemical substances found mainly in petroleum. Polyesters are manufactured in three basic forms—*fibers*, *films* and *plastics*. Polyester fibers are used to make fabrics. Poly(ethylene terephthalate) (PET) is the most common polyester used for fiber purposes. This is the polymer used for making soft drink bottles. Recycling of PET by re-melting and extruding it as fiber may saves much raw materials as well as energy. PET is made by ethylene glycol with either terephthalic acid or its methyl ester in the presence of an antimony catalyst. In order to achieve high molecular weights needed to form useful fibers, the reaction has to be carried out at high temperature and in a vacuum [60].

1.4 BASIC NANOTECHNOLOGY

Two main approaches are used in nanotechnology that is, the bottom up approach and the top–down approach. In case of the "bottom–up" approach, the different type of materials and the instruments are made up from different types of molecular components, which combine themselves by chemical ways basing on the mechanism of molecular recognition. In case of the "top–down" approaches, various nano-objects are made from various types of components without atomic-level control. Materials reduced to the nanoscale can show different properties compared to what they exhibit on a macro scale, enabling unique applications. The basic premise is that properties can dramatically change when a substance's size is reduced to the nanometer range. For instance, ceramics, which are normally brittle, can be deformable when their size is reduced, opaque substances become transparent (copper); stable materials turn combustible (aluminum); insoluble materials become soluble (gold) [61, 62].

Nanoparticles can be prepared from a variety of materials such as proteins, polysaccharides and synthetic polymers. The selection of materials in mainly depended on factors like size of the nanoparticles required, inherent properties such as aqueous, solubility and stability, surface characteristics, that is, charge and permeability, degree of biodegradability, biocompatibility and toxicity, release of the desired product, antigenicity of the final product, etc. Polymeric nanoparticles have been prepared most frequently be three methods: (i) dispersion of the performed polymers; (ii) polymerization of the monomers; and (iii) ionic gelation of the hydrophobic or hydrophilic polymers. However, techniques like supercritical fluid technology and particle repulsion in non-wetting templates (PRINT) have been also used in modern days [63–65].

1.5 NANOTECHNOLOGY IN TEXTILE INDUSTRY AND TEXTILE ENGINEERING

Of the many applications of nanotechnology, textile industry has been currently added as one of the most benefited sector. Application of nanotechnology in textile industry has tremendously increased the durability of fabrics, increase its comfortness, hygienic properties and have also reduces its production cost. Nanotechnology also offers many advantages as compare to the conventional process in term of economy, energy saving, eco-friendliness, control release of substances, packaging, separating and storing materials on a microscopic scale for later use and release under control condition [66]. The unique and new properties of nanotechnology have attracted scientists and researchers to the textile industry and hence the use of nanotechnology in the textile industry has increased rapidly. This may be due to the reason that textile technology is one of the best areas for development of nanotechnology. The textile fabrics provide best suitable substrates where a large surface area is present for a given weight or a given volume of fabric. The synergy between nanotechnology and textile industry uses this property of large interfacial area and a drastic change in energetic is experienced by various macromolecules or super molecules in the vicinity of a fiber when changing from wet state to a dry state [67].

The application of nanoparticles to textile materials have been the objective of several studies aimed at producing finished fabrics with different functional performances. Nanoparticles can provide high durability for treated fabrics as they posse large surface area and high surface energy that ensure better affinity for fabrics and led to an increase in durability of the desired textile function. The particle size also plays a primary role in determining their adhesion to the fibers. It is reasonable to except that the largest particle agglomerates will be easily removed from the fiber surface, while the smallest particle will penetrate deeper and adhere strongly into the fabric matrix [68, 69]. Thus, decreasing the size of particles to nano-scale dimensional, fundamentally changes the properties of the material and indeed the entire substance.

A whole variety of novel nanotech textiles are already on the market at this moment. Areas where nanotech enhanced textiles are already seeing some applications include sporting industry, skincare, space technology and clothing as well as materials technology for better protection in extreme environments. The use of nanotechnology allows textiles to become multifunctional and produce fabrics with special functions, including antibacterial, UV-protection, easy-clean, water- and stain repellent and anti-odor [68].

1.6 TECHNOLOGICAL USES

Textile materials are materials for the daily use. Besides, textiles are also play a vital role in fashion shows. Technically, they are applied in variety of our life savers including safety belt, and the airbags in the cars, bulletproof vests protect against weapons, used as Implant material in medical applications. Recently, polyurethane (PU) foams that can be combined with Platilon thermoplastic polyurethane (TPU) films are used as excellent material for functional medical wound dressing which are not only help wounds to heal but also allow the wound to breathe by permeate the water-vapor [70, 71].

1.6.1 ADVANCEMENTS IN TEXTILE PRODUCTION

Technological advances during the past decade have opened many new doors for the Textile and Apparel industries, especially in the area of rapid prototyping and related activities. During the past decade, the textile and apparel complex has been scrambling to adjust to a rapidly changing business environment. Textiles and yarns have been around for thousands of years but in the last 50 years, progress in the technology has been most remarkable. The application of textiles and yarns have move beyond clothing and fabrics and they are increasingly used in high value-added applications such as composites, filtration media, gas separation, sensors and biomedical engineering. With the emergence of nanotechnology, the users of textiles and yarns are switching their attention to the production of nanometer diameter fibers [72, 73].

1.6.2 BODY SCANNING

The development of three-dimensional body scanning technologies may have significant potential for use in the apparel industry. First, this technology has the potential of obtaining an unlimited number of linear and non-linear measurements of human bodies (in addition to other objects) in a matter of seconds. Because an image of the body is captured during the scanning process, the location and description of the measurements can be altered as needed in mere seconds, as well. Second, the measurements obtained using this technology has the potential of being more precise and reproducible than measurements obtained through the physical measurement process. Third, with the availability of an infinite number of linear and non-linear measurements the possibility exists for garments to be created to mold to the 3 dimensional shapes of unique human bodies. Finally, the scanning technology allows measurements to be obtained in a digital format that could integrate automatically into apparel CAD systems without the human intervention that takes additional time and can introduce error [74, 75].

1.6.3 APPAREL CAD

Adoption of CAD/CAM technology over the past few decades has increased the speed and accuracy of developing new products, reducing the manpower required to complete the development process. Unfortunately, this technology has also encouraged manufacturers to simplify the design of garments, allowing a more efficient use of materials and making mass production much easier. These systems initially only made an effort to adapt traditional manual methods instead of encouraging innovation in design or fit adaptations. Current developments in the area of information technology help build on the traditional CAD/CAM functions and offer a new way of looking at and using the systems for design and product development [76, 77].

1.6.4 NANOFIBER FABRICATION

As with all new technologies, polymeric nanofibers have brought with it a new beginning to the understanding of polymeric fibers. One apparent advantage of nanofibers is the huge increase in the surface area to volume ratio. Given the huge potential of nanofibers, the key is to use a technique that is able to easily fabricate nanofibers out of most, if not all the different type of polymers. A number of processing techniques such as drawing [78], template synthesis [79], phase separation [80], self-assembly [81], electrospinning [82], etc. have been used to prepare polymer nanofibers in recent years. The drawing is a process similar to dry spinning in fiber industry, which can make one-by-one very long single nanofibers. The template synthesis, as the name suggests, uses a nanoporous membrane as a template to make nanofibers of solid (a fibril) or hollow (a tubule) shape. The phase separation consists of dissolution, gelation, and extraction using a different solvent, freezing, and drying resulting in nanoscale porous foam. The process takes relatively long period of time to transfer the solid polymer into the nano-porous foam. The self-assembly is a process in which individual, preexisting components organize themselves into desired patterns and functions.

1.6.5 ELECTROSPINNING

Electrospinning is a process where continuous fibers with diameters in the sub-micron range are produced through the action of an applied electric field imposed on a polymer solution. Textiles made from these fibers have high surface area and small pore size, making them ideal materials for use in protective clothing. There are currently several applications of electrospinning that are being investigated, including: The fabrication of transparent composites reinforced with nanofibers. The effect of processing conditions on the morphology of polymers has been investigated. The scaling up of current production techniques is one of the main areas of research in which we have ongoing interest is the Adhesives, Permselective membranes, Anti-fouling coatings, Active protective barriers against chemical and biological threats are few examples of nanofibers produced through electrospinning [83].

1.6.5.1 Using Electrospun Nanofibers: Background and Terminology

The three inherent properties of nanofibrous materials that make them very attractive for numerous applications are their high specific surface area (surface area/unit mass), high aspect ratio (length/diameter) and their biomimicking potential. These properties lead to the potential application of electrospun fibers in such diverse fields as high-performance filters, absorbent textiles, fiber-reinforced composites, biomedical textiles for wound dressings, tissue scaffolding and drug-release materials, nano- and microelectronic devices, electromagnetic shielding, photovoltaic devices and high-performance electrodes, as well as, a range of nanofiber-based sensors [84, 85].

In many of these applications the alignment, or controlled orientation, of the electrospun fibers is of great importance and large-scale commercialization of products will become viable only when sufficient control over fiber orientation can be obtained at high production rates. In the past few years research groups around the world have been focusing their attention on obtaining electrospun fibers in the form of yarns of continuous single nanofibers or uniaxial fiber bundles. Succeeding in this will allow the processing of nanofibers by traditional textile processing methods such as weaving, knitting and embroidery. This, in turn, not only will allow the significant commercialization of several of the applications cited above, but will also open the door to many other exciting new applications [86, 87].

Incorporating nanofibers into traditional textiles creates several opportunities. In the first instance, the replacement of only a small percentage of the fibers or yarns in a traditional textile fabric with yarns of similar diameter, but now made up of several thousands of nanofibers, can significantly increase the toughness and specific surface area of the fabric without increasing its overall mass [88]. Alternatively, the complete fabric can even be made from nanofiber yarns. This has important implications in protective clothing applications, where lightweight, breathable fabrics with protection against extreme temperatures, ballistics, and chemical or biological agents are often required. On an esthetic level, nanofiber textiles also exhibit extremely soft handling characteristics and have been proposed for use in the production of artificial leather and artificial cashmere [85]. In biomedical applications the similarity between certain electrospun polymeric nanofibers and the naturally occurring nanofibrous structures of connective tissues such as collagen and elastin gives rise to the opportunity of creating artificial biomimicking wound dressings and tissue engineering scaffolds. Several studies on nonwoven nanofiber webs of biocompatible polymers have already shown potential in this area. Simple three-dimensional constructs for vascular prostheses have also been manufactured by electrospinning onto preformed templates. Although these initial studies show that enhanced cell adhesion, cell proliferation and scaffold vascularization can be obtained on porous, nonwoven nanofiber webs, the simplicity of the constructs and the fragile nature of nonwoven webs still limit their applicability to small areas [89, 90].

Creating complex three-dimensional scaffold structures with fibers aligned in a controlled fashion along the directions of the forces that are usually present in dynamic tissue environments, as for instance in muscles and tendons, will lead to significant improvements in the performance of tissue engineering scaffolds. With continuous nanofiber yarns it will become possible to create such aligned fiber structures on a large scale, simply by weaving. In addition, the age-old techniques of knitting and embroidery can then be applied to create very complicated, three-dimensional scaffolds, with precisely controlled porosity, and yarns placed exactly along the lines of dynamic force [91].

Several other fields will also benefit from the availability of continuous yarns from electrospun fibers. Owing to the high fiber-aspect ratios and increased fiber–matrix adhesion caused by the high specific surface areas, aligned nanofiber yarns can lead to stronger and tougher, lightweight, fiber reinforced composite materials. The incorporation of nanofiber-based sensors into textiles can lead to new opportunities in the fields of smart and electronic textiles. Aligned nanofiber yarns of piezo-electric polymers and other microactuator materials may lead to better performance in advanced robotics applications [92].

Since the revival of electrospinning in the early 1990s, several research groups have worked on controlling the orientation of electrospun fibers. Those who have worked in the field of electrospinning over the past decade have come from various disciplinary backgrounds, including physics, chemistry and polymer science, chemical and mechanical engineering, and also from the traditional textiles field. The result of this has been that literature on the topic of electrospinning, and especially yarns from electrospun fibers, is plagued with terminology from different disciplines, which often leads to misunderstanding and even self-contradictory statements. So, for instance, in paper on the electrospinning of individual fibers of a novel polymer some authors might use the term yarn when they are actually referring to an individual fiber, or authors might refer to spinning a filament when they are actually spinning a yarn [93].

- **Fiber** – a single piece of a solid material, which is flexible and fine, and has a high aspect ratio (length/diameter ratio).
- **Filament** – a single fiber of indefinite length.
- **Tow** – an untwisted assembly of a large number of filaments; tows are cut up to produce staple fibers.
- **Sliver** – an assembly of fibers in continuous form without twist. The assembly of staple fibers, after carding but before twisting, is also known as a sliver.
- **Yarn** – a generic term for a continuous strand of textile fibers or filaments in a form suitable for knitting, weaving or otherwise intertwining to form a fabric.
- **Staple fiber** – short-length fibers, as distinct from continuous filaments, which are twisted together (spun) to form a coherent yarn. Most natural fibers are staple fibers, the main exception being silk which is a filament yarn. Most artificial staple fibers are produced in this form by slicing up a tow of continuous filaments.

- **Staple fiber yarn** – a yarn consisting of twisted together (spun) staple fibers.
- **Filament yarn** – a yarn normally consisting of a bundle of continuous filaments. The term also includes monofilaments.
- **Core-spun yarn** – a yarn consisting of an inner core yarn surrounded by staple fibers. A core-spun yarn combines the strength and/or elongation of the core thread and the characteristics of the staple fibers that form the surface.
- **Denier** – a measure of linear density: the weight in grams of 9000 meters of yarn.
- **Tex** – another measure of linear density: the weight in grams of 1000 meters of yarn.

1.6.5.2 Controlling Fiber Orientation

As stated in the previous section, achieving control over the orientation of electrospun fibers is an important step towards many of their potential applications. However, if one considers the fact that fiber formation occurs at very high rates (several hundreds of meters of fiber per second) and that the fiber formation process coincides with a very complicated three-dimensional whipping of the polymer jet (caused by electrostatic bending instability), it becomes clear that controlling the orientation of fibers formed by electrospinning is no simple task [94].

Various mechanical and electrostatic approaches have been taken in efforts to control fiber alignment. The two most successful methods are the following:

(a) Spinning Onto a Rapidly Rotating Surface

Several research groups have been routinely using this technique to obtain reasonably aligned fibers. The rapid rotation of a drum or disk and coinciding high linear velocity of the collector surface allows fast take-up of the electrospun fibers as they are formed. The 'point-to-plane' configuration of the electric field does, however, lead to fiber orientations that deviate from the preferred orientation.

A special instance of the rotating drum set-up involves spinning onto a rapidly rotating sharp-edged wheel, which uses an additional electrostatic effect, since the sharp edge of the wheel creates a stronger converging electrostatic field, or a 'point-to-point' configuration, which has a focusing effect on the collected fibers (Figure 1.1). This in turn leads to better alignment of the fibers [95, 96].

(b) The Gap Alignment Effect

Uniaxially aligned arrays of electrospun fibers can be obtained through the gap alignment effect, which occurs when charged electrospun fibers are deposited onto a collector that consists of two electrically conductive substrates, separated by an insulating gap. This electrostatic effect (Figure 1.2) has been observed by various groups. Recently this was investigated in more detail. Briefly, the lowest energy configuration for an array of highly charged fibers between two conductive substrates, separated by an insulating gap, is obtained when fibers align parallel to each other [97].

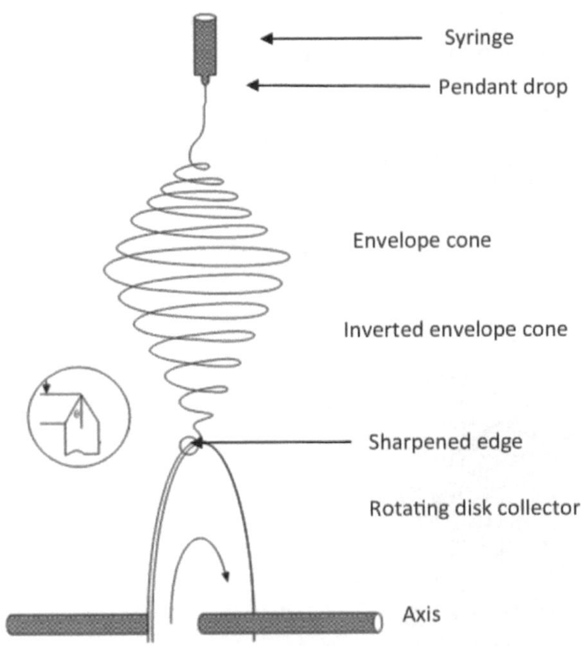

FIGURE 1.1 Converging electrostatic field on sharp-edged wheel electrode.

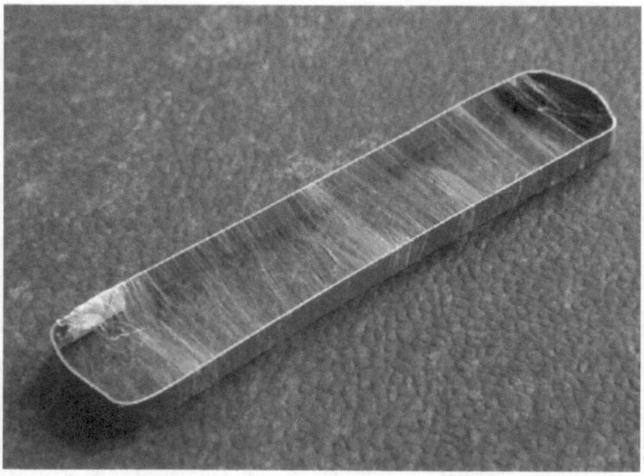

FIGURE 1.2 Aligned fibers obtained through the gap alignment effect.

1.6.5.3 Producing Noncontinuous or Short Yarns

Both spinning onto a rapidly rotating collector and the gap alignment effect have been used to obtain short yarns for experimental purposes [98].

- Rotating collector method

Fibers can be spun onto a rapidly rotating disk, where the shearing force of the rotating disk led to aligned fibrous assemblies with good orientation.

These oriented fibers could then be collected and manually twisted into a yarn [99].

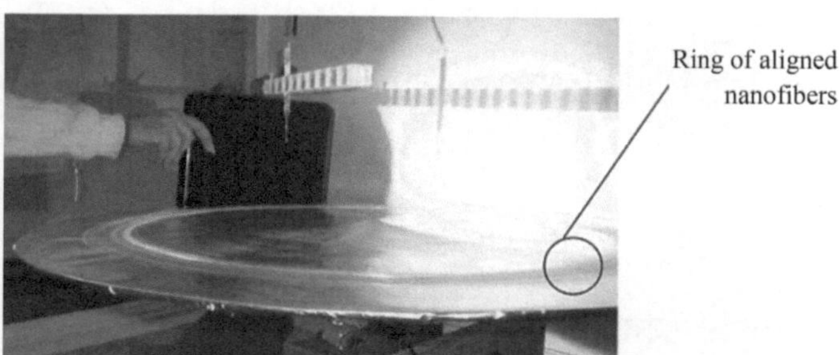

FIGURE 1.3 Aligned fiber tows on rotating disk collector.

- Twisted yarns obtained from tows spun on a rotating disk collector

The stress–strain behavior of the yarns can be examined and the modulus, ultimate strength and elongation at the ultimate strength can be measured as a function of twist angle.

- Gap alignment method

Short yarns can be made by quickly passing a wooden frame through the electrospinning jet several times (for up to an hour), in a process also known as 'combing,' resulting in a tow of reasonably aligned fibers, which were then 'gently twisted' to form a yarn [100].

1.6.5.4 Producing Continuous Yarns

A common misconception in recent electrospinning literature is that the first literature on electrospinning dates back to 1934 when Formhals patented a method for manufacturing yarns from electrospun fibers. Some of the first publications on electrospinning date back as far as 1902, when first Cooley and then Morton patented processes for dispersing fluids. In both patents, the authors describe processes for producing very fine artificial fibers by delivering a solution of a fiber-forming material, such as pyroxylin, a nitrated form of cellulose, dissolved in alcohol or ether, into a strong electric field [101, 102].

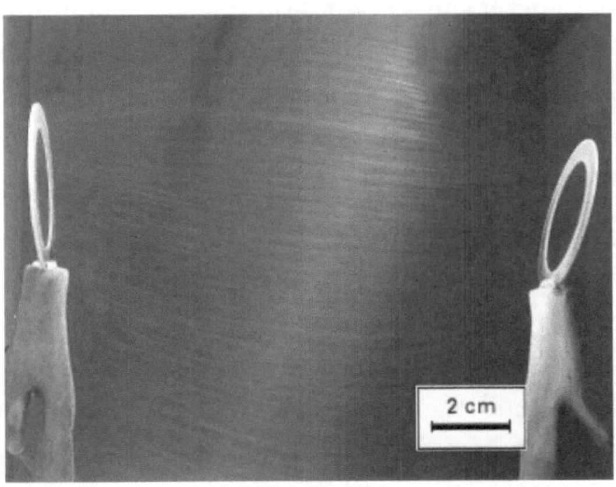

FIGURE 1.4 Fibers aligned between two parallel ring collectors.

The reason for this oversight is unclear, but it could possibly be blamed on differences in terminology since the term 'electrospinning' has only become popular with the revival of the process in the mid-1990s [100]. Another more puzzling oversight, which has recently led several authors to bemoan the lack of processes for making continuous yarns from electrospun fibers, is that Formhals actually registered a series of seven patents over a period of ten years between 1934 and 1944 and that all these patents describe processes and/or improvements to processes for the manufacture of continuous yarns from electrospun fibers. Since the youngest of these patents is more than 60 years old, one could speculate that these processes did not really work, which would explain the absence of commercially available electrospun fiber yarns. An alternative explanation could be that, since Formhals lived in Mainz, Germany, and since the last patent application was filed in 1940, the disruption of World War II and the ensuing years simply led to the processes being forgotten. Closer inspection of the patents, aided by more recent knowledge of the electrospinning process, also leads us to believe that at least some of the described processes are viable and that they deserve further consideration. The patents of Formhals show a gradual evolution of his yarn production process over time and in many instances he applied the same fundamental practical aspects of electrospinning that have re-occurred in the recent literature. These included obtaining aligned fibers by spinning onto conductive strips or rods that were separated in space from each other by an insulating material (gap alignment effect), increasing production rates by using multiple spinnerets, regulating the electric field between the source and the collector by adding additional electrostatic elements, using corona discharge to discharge the electrospun fibers, and post-treating the electrospun fibers by submerging them in a liquid bath [103].

- Rotating dual-collector yarn

Formhals's original patent relates to the manufacture of slivers of cellulose acetate fibers by electrospinning from a cogwheel source onto various collector set-ups. In these collector set-ups fibers are first spun onto a rotating collector and then removed in a continuous fashion, onto a second take-up roller. The first of these collector set-ups consisted of a solid conductive wheel or ring with string attached to the edge. In this set-up the wheel was rotated for a short period while fibers were spun onto its edge. The process was then stopped, the string was loosened and then drawn over rollers and/or through twist-imparting rings to a second take-up roller, and the spinning process restarted. The newly spun sliver or semi-twisted yarn was then drawn off continuously onto the second take-up roller. Another collector set-up consisted of a looped metal belt with fixtures to push or blow the fiber sliver off the belt before the fiber sliver was collected on a second take-up roller. The concept of using multiple spinnerets for increasing production rates was also introduced in this patent.

Formhals later identified several problems related to his first design and hence in subsequent patents he made various additions and/or alterations to the original design, which were intended to eliminate these problems. These problems and their solutions included the following [92, 104]:

- Problem: Fibers flying to-and-fro between the source and the collector Solution: In his second patent, Formhals claimed that one cause of the fibers flying to-and-fro between the jet and collector was that the collector was at too high a voltage of opposite polarity and that resulting corona discharge from the collector reversed the charge of the fibers while they were passing between the jet and the collector. This in turn caused them to change direction and fly back to the source. He proposed to eliminate this problem by adding a voltage regulator on the collector-side of the circuit in order to down-regulate the voltage of the collector.

- Problem: Fibers not drying sufficiently between source and collector. Solution: Formhals later designed various additions to his spinning system for regulating the shape and intensity of the electrical field in the vicinity of the spinning source. This was done in order to direct the formed fibers along a longer, predetermined and constant path towards the collecting electrode and was achieved by placing, in close proximity to the fiber stream, conductive strips, wires, plates and screens, which were connected to the same potential as the fiber source. These additions allowed a more thorough drying of fibers before they were deposited on the collector.

An additional problem, which is not specifically discussed in Formhals's patents, but which can be foreseen when examining his first collector design, is that fiber alignment would be less than ideal when spinning onto a solid wheel or belt. It appears, however, that Formhals did encounter this problem and that he overcame it by using the gap alignment effect in the design of subsequent collectors. The design consisted of a picket-fence-like belt, with individual, pointed electrodes, separated from each other by an air gap.

- Multi-collector yarn

In this patented process from the Korea Research Institute of Chemical Technology, continuous slivers or twisted yarns of different polymers, but especially of polyamide–polyimide copolymers, are claimed to be obtained by electrospinning first onto one stationary or rotating plate or conductive mesh collector, where the charges on the fibers are neutralized, and then continuously collecting the fibers from the first onto a second rotating collector.

A diagram depicting the process is given in Figure 1.5. The underlying principle of this process closely resembles the rotating dual-collector yarn process patented by Formhals in 1934 [105].

- Core-spun yarn

In his 1940 patent, Formhals described a method for making composite yarns by electrospinning onto existing cotton, wool or other pre-formed yarn. It was also proposed that a sliver of fibers, such as wool, could be coated with the electrospun fibers before twisting the product into an intimately blended yarn [103].

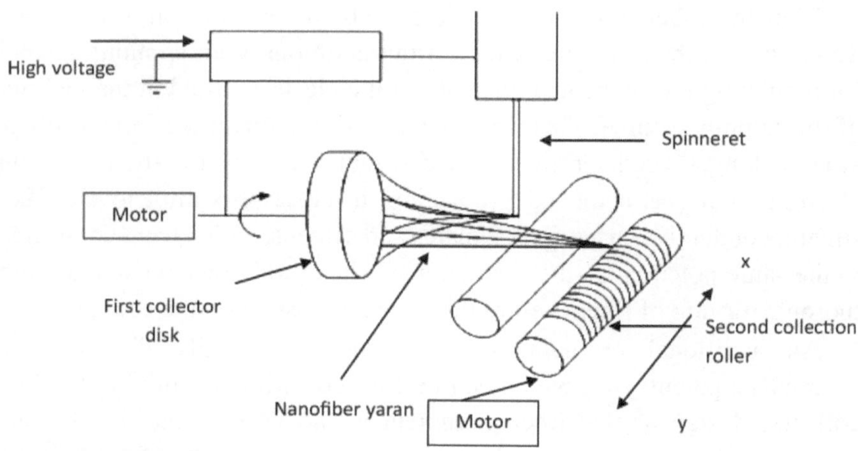

FIGURE 1.5 Multi-collector yarn process diagram.

- Staple fiber yarn

Formhals developed a method for controlling the length of the electrospun fibers with the main objective being to manufacture fibers with a controlled and comparatively short length. This goal was achieved by modulating the electric field using a spark gap. In this modulation, it was preferable to periodically switch the field strength to at least 35% and preferably 20% of its original voltage in order to interrupt the electrospinning process for a short period and thereby create a sliver of short fibers, which could then be spun into a staple fiber yarn [103].

- Continuous filament yarn

Instead of spinning directly onto the counter-electrode, Formhals altered the process so the polarity of the charge on the fibers was changed before reaching the counter-electrode. This was achieved by using high voltage of opposite polarity on a sharp-edged or thin wire counter-electrode. The high voltage led to corona discharge, which initially reduced and eventually inverted the charge on the fiber while it was traveling from the source to the collector [106].

This caused the fibers to turn away from the collector electrode and they could then be intercepted at a point below the counter-electrode and rolled up as a continuous filament yarn on a take-up roller. In a final improvement on the system, the entire spinning apparatus was encased in a box with earthed conductive siding. This avoided build-up of charges in the panels, which could lead to disturbance of the electric field inside the spinning chamber and disruption of the spinning process. In addition, variable voltage power on the source and collector electrodes allowed the tuning and moving of the position of the neutral zone in which the yarn formation process takes place, which in turn allowed better control over the continuity of the spinning process [107, 108].

- Self-assembled yarn

The self-assembled yarn process was developed by Ko [109] at Drexel University. When a solution of pure polymer, or a polymer-containing polymer blend, was electrospun onto a solid conductive collector under appropriate conditions, the fibers did not deposit on the collector in the form of a flat nonwoven web as is usually observed. Instead, initial fibers deposited on a relatively small area of the collector and then subsequent fibers started accumulating on top of them and then on top of each other, forming a self assembled yarn structure that rapidly grew upwards from the collector towards the spinneret. The formation of a self-assembled yarn is illustrated in Figure 1.6. The self-assembling yarn, suspended in the space between the spinneret and the collector, continued to grow in this fashion until it reached a critical point somewhere in the vicinity of the spinneret. At this critical point, a branched tree-like fiber structure formed and newly formed fibers deposited on the branches of the tree. The yarn could then be collected by slowly taking up the fibers collected on one of the tree branch structures, or by slowly moving the target electrode away from the spinneret. Post-processing of the yarn, including twisting, could be done in a second step [109].

It was proposed that the charge on the electrospun fibers, which is induced through the high voltage in the spinneret, is dissipated through the evaporation of the solvent during the electrospinning process, so that the fibers are essentially neutral when they reach the collector electrode. This could explain why the initial fibers deposit on such a small area on the collector. If the fibers on the collector are charged, they repel incoming fibers leading to an expanding random web. Neutral fibers would not have the same repelling effect on incoming fibers and so the fibers would collect on a smaller area. Neutral fibers, deposited on top of each other, and therefore closer to the spinneret than the target electrode

FIGURE 1.6 Self-assembled yarn formation.

surface, also form an attractive target for incoming fibers. This would explain why subsequent fibers selectively deposit on the tip of the self-assembling yarn [85].

- Conical collector yarn

A method for the production of hollow fibers by the electrospinning process was reported by scientists [45–47]. The conventional electrospinning device was modified to include a conical collector and an air-suction orifice to generate hollow and void-containing, uniaxially aligned electrospun fibers. Use of the conical collector allowed for the collection of aligned yarns with diameters of approximately 157 nm [110].

- Spin-bath collector yarn

In this recently published method, developed by our group at Stellenbosch University, continuous uniaxial fiber bundle yarns are obtained by electrospinning onto the surface of a liquid reservoir counter-electrode. The web of electrospun fibers, which forms on the surface of the spin-bath, is drawn at low linear velocity (ca. 0.05 m/s) over the liquid surface and onto a take-up roller. A diagrammatic representation of the electrospinning set-up is given in Figure 1.7. All the yarns obtained using this method exhibit very high degrees of fiber alignment and bent fiber loops are observed in all the yarns [47].

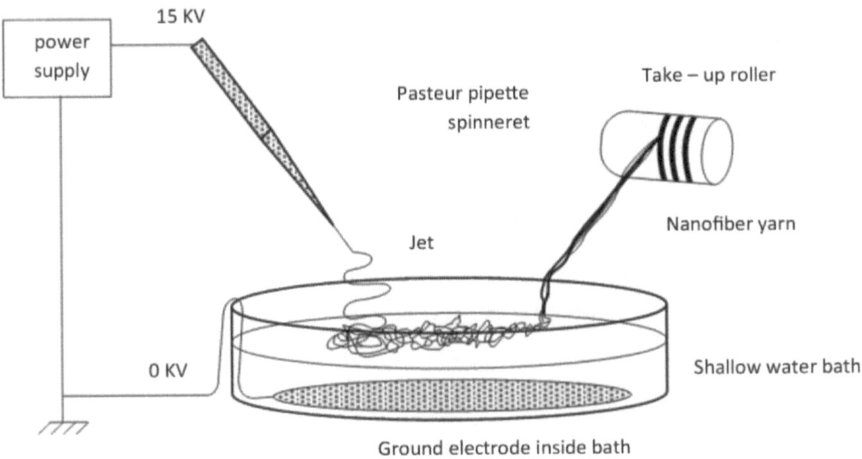

FIGURE 1.7 Yarn-spinning set-up with grounded spin-bath collector electrode.

The process of yarn formation is illustrated in Figure 1.8. It can be described in three phases. In the first phase, a flat web of randomly looped fibers forms on the surface of the liquid. In the second phase, when the fibers are drawn over or through the liquid, the web is elongated and alignment of the fibers takes place in the drawing direction. The third phase consists of drawing the web off the liquid and into air. The surface tension of the remaining liquid on the web pulls the fibers together into a three-dimensional, round yarn structure.

The average yarn obtained in a single-spinneret electrospinning set-up contains approximately 3720 fibers per cross-section and approximately 180 m of yarn can be spun per hour. The yarns obtained are very fine, with calculated linear densities in the order of 10.1 denier. Although higher linear densities can be obtained by reducing the yarn take-up rate, this is accompanied by a decrease in fiber alignment within the yarn. Currently investigations are focused on various options to overcome these challenges by, for instance, combining aligned yarns from multiple spinnerets into single yarns.

- Twisted nonwoven web yarn

This method, patented by Raisio Chemicals Korea Inc. involves electrospinning nanofibers through multiple nozzles to obtain a nonwoven nanofiber web, either directly in a ribbon form or in a larger form, which is then cut into ribbons, and subsequently passing the nanofiber web ribbons through an air twister to obtain a twisted nanofiber yarn. A diagram depicting the process is given in Figure 1.9 [92].

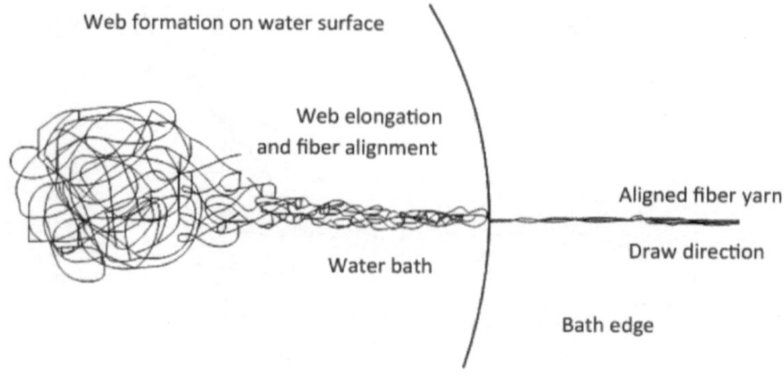

FIGURE 1.8 Top view of the yarn formation process.

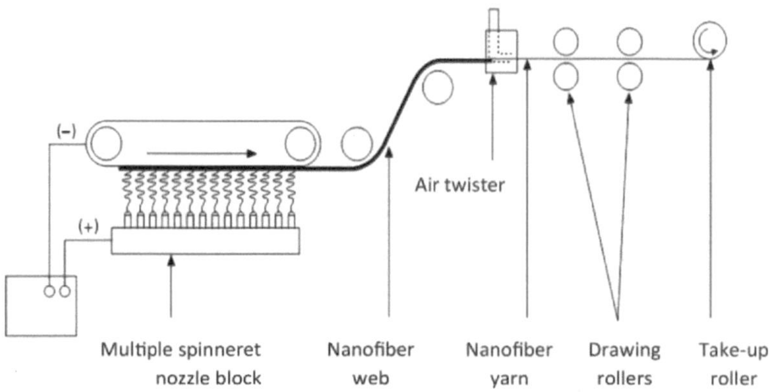

FIGURE 1.9 Spinning process used to prepare a twisted nonwoven web yarn.

- Grooved belt collector yarn

In a recent research works [92], a ribbon-shaped nanofiber web is prepared by electrospinning onto a collector consisting of an endless belt-type nonconductive plate with grooves formed at regular intervals along a lengthwise direction and a conductive plate inserted into the grooves of the nonconductive plate. The nanofiber webs are electrospun onto the conductive plates in the grooves and later separated from the collector, focused, drawn and wound into a yarn [92].

- Vortex bath collector yarn

In this patent pending process developed at the National University of Singapore, a basin with a hole at the bottom is used to allow water to flow out in such a manner that a vortex is created on the water surface.

Electrospinning is carried out over the top of the basin so that electrospun fibers are continuously deposited on the surface of the water. Owing to the presence of the vortex, the deposited fibers are drawn into a bundle as they flow through the water vortex. Generally, a higher feed rate or multiple spinnerets are required to deliver sufficient fibers on the surface of the water so that the resultant fiber yarn has sufficient strength to withstand the drawing and winding process. Figure 1.10 shows the set-up used for the yarn drawing process.

Yarn drawing speeds as high as 80 m/min have been achieved and yarns made of poly(vinylidene fluoride) (PVDF) and polycaprolactone have been fabricated using this process [103].

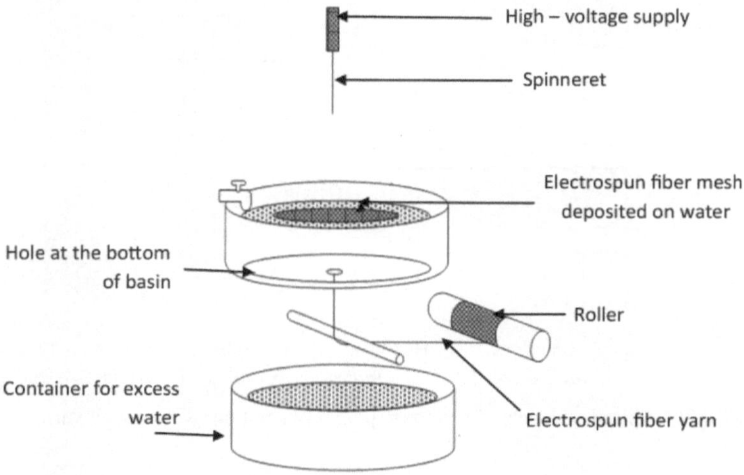

FIGURE 1.10 Vortex bath collector yarn process.

1.6.5.5 Gap-Separated Rotating Rod Yarn

This method developed by Doiphode and Reneker at the University of Akron uses the gap alignment effect in a very similar way to the work published by Dalton et al. [92] discussed before. The process is described with reference to Figure 1.11. Fibers are electrospun between a 2 mm

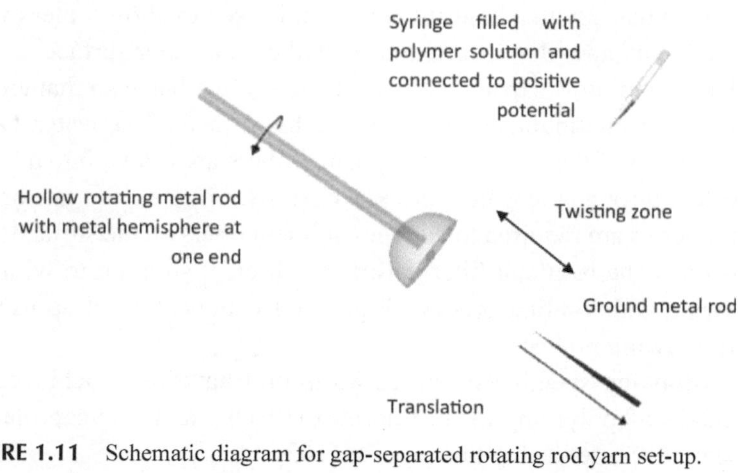

FIGURE 1.11 Schematic diagram for gap-separated rotating rod yarn set-up.

metal rod on the right and a hollow 25 mm metal rod with a hollow hemisphere attached to its end on the left. Both the geometries are grounded and placed at a distance of a few centimeters. Fibers are collected across the gap between these two collector surfaces and are given a twist by rotating the hemispherical collector. Yarn collected in this manner on the tip of the metal rod can be translated away from the rotating collector, thereby drawing the yarn and producing yarn continuously. Yarns with lengths up to 30 cm were produced by this method and the creators of the process believe that optimizing the winding mechanism can lead to production of continuous yarns [111].

- Conjugate electrospinning yarn

Methods for making continuous nanofiber yarns based on the principle of conjugate electrospinning were recently published [56–57] as well as Luming Li and co-workers [56] at Tsinghua University in Beijing. In conjugate electrospinning, two spinnerets or two groups of spinnerets are placed in an opposing configuration and connected to high voltage of positive and negative polarity, respectively. The process is presented diagrammatically in Figure 1.12. Oppositely charged fiber jets are ejected from the spinnerets and Coulombic attraction leads the oppositely charged fibers to collide with each other. The collision of the fibers leads to rapid neutralization of the charges on the fibers and rapid decrease in their flying speeds. In the processes described by both groups, the neutralized fibers are then collected onto take-up rollers to form yarns. Each continuous yarn contains a large quantity of nanofibers, which are well aligned along the longitudinal axis of the yarn. Conjugate electrospinning works for a variety of polymers, composites and ceramics [29].

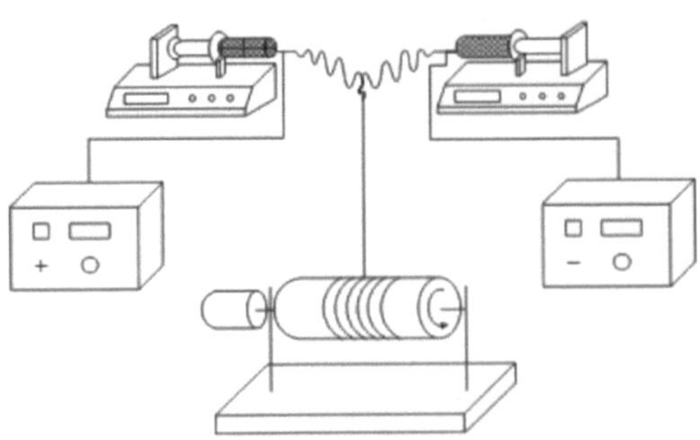

FIGURE 1.12 Conjugate electrospinning set-up.

1.7 NANOPARTICLES IN TEXTILES FINISHING

Fabric treated with nanoparticles of TiO_2 and MgO replaces fabrics with active carbon, previously used as chemical and biological protective materials. The photocatalytic activity of TiO_2 and MgO nanoparticles can break harmful and toxic chemicals and biological agents. These nanoparticles can be pre-engineered to adhere to textile substrates via spray coating or electrostatic methods. Textiles with nanoparticles finishing are used to convert fabrics into sensor-based materials, which has numerous applications. If nanocrystalline epiezoceramic particles are incorporated into fabrics, the finished fabric can convert exerted mechanical forces into electrical signals enabling the monitoring of bodily functions such as heart rhythm and pulse if they are worn next to skin [112–114].

1.8 CONCLUDING REMARK

Nanotechnology is growing by leaps and bounds and it has been introduced in many fields including the textile industries. Within the past decade, the textile industry has rediscovered and continues to develop technologies that enable production of extremely small fibers – nanofibers – using a process called electrospinning. The advantages of electrospun fibers make them very appealing for a broad array of potential applications in many industry segments. As it was debated in this chapter, there are various forms for each part of electrospinning instrument, which cause to make fibers with different qualities for various applications.

KEYWORDS

- **electrospinning**
- **nanotechnology**
- **textile history**

REFERENCES

1. Selin, C., *Expectations and the Emergence of Nanotechnology.* Science, Technology and Human Values, 2007, 32(2), 196–220.
2. Mansoori, G. A., *Principles of Nanotechnology: Molecular-Based Study of Condensed Matter in Small Systems.* 2005, World Scientific.
3. Peterson, C. L., *Nanotechnology: From Feynman to the Grand Challenge of Molecular Manufacturing.* Technology and Society Magazine, IEEE, 2004, 23(4), 9–15.
4. Gleiter, H., *Nanostructured Materials: Basic Concepts and Microstructure.* Acta Materialia, 2000, 48(1), 1–29.
5. Peterson, C. L., *Nanotechnology-Evolution of the Concept.* Journal of the British Interplanetary Society, 1992, 45, 395–400.
6. Yadugiri, V. T. and R. Malhotra, *'Plenty of Room'-Fifty Years after the Feynman Lecture.* Current Science, 2010, 99(7), 900–907.
7. Wilson, K., *History of Textiles.* 1979, Westview Press.
8. Harris, J., *Textiles, 5,000 Years: An International History and Illustrated Survey.* 1995, New York City, United States: Harry N. Abrams, Inc.
9. Sawhney, A. P. S., et al., *Modern Applications of Nanotechnology in Textiles.* Textile Research Journal, 2008, 78(8), 731–739.
10. Tao, X., *Smart Fibers, Fabrics and Clothing.* Vol. 20. 2001, Woodhead publishing.
11. Tyrer, R. B., *The Demographic and Economic History of the Audiencia of Quito: Indian Population and the Textile Industry, 1600–1800.* 1976, University of California, Berkeley.
12. Chapman, S. D. and S. Chassagne, *European Textile Printers in the Eighteenth Century: A Study of Peel and Oberkampf.* 1981, Heinemann Educational Books, Pasold Fund.
13. Jeremy, D. J., *Transatlantic Industrial Revolution: The Diffusion of Textile Technologies between Britain and America, 1790–1830s.* 1981, MIT Press.
14. Hekman, J. S., *The Product Cycle and New England Textiles.* The Quarterly Journal of Economics, 1980, 94(4), 697–717.
15. Mantoux, P., *The Industrial Revolution in the Eighteenth Century: An Outline of the Beginnings of the Modern Factory System in England.* 2013, Routledge.
16. French, G. J., *Life and Times of Samuel Crompton of Hall-in-the-Wood: Inventor of the Spinning Machine Called the Mule.* 1862, Charles Simms and Co.
17. O'brien, P., *The Micro Foundations of Macro Invention: The Case of the Reverend Edmund Cartwright.* Textile History, 1997, 28(2), 201–233.
18. Aspin, C., *The Cotton Industry.* 1981, Osprey Publishing.
19. Smout, T. C., *The Development and Enterprise of Glasgow 1556–1707.* Scottish Journal of Political Economy, 1959, 6(3), 194–212.
20. Sale, K., *Rebels against the Future: The Luddites and their War on the Industrial Revolution: Lessons for the Computer Age.* 1995, Basic Books.
21. Aspin, C., *The Woolen Industry.* 1982, Osprey Publishing.
22. Beckert, S., *Emancipation and Empire: Reconstructing the Worldwide Web of Cotton Production in the Age of the American Civil War.* The American Historical Review, 2004, 109(5), 1405–1438.
23. Heberlein, G., *Georges Heberlein*, 1935, Google Patents.

24. Townswnd, B. A., *Full Circle in Cellulose*. 1993, New York: Elsevier Applied Science.
25. Hicks, E. M., et al., *The Production of Synthetic-Polymer Fibers*. Textile Progress, 1971, 3(1), 1–108.
26. Cook, J. G., *Handbook of Textile Fibers: Man-Made Fibers*. Vol. 2. 1984, Elsevier.
27. Conrad, D. J., et al., *Ink-Printed, Low Basis Weight Nonwoven Fibrous Webs and Method*, 1995, Google Patents.
28. Sawhney, A. P. S., et al., *Modern Applications of Nanotechnology in Textiles*. Textile Research Journal, 2008, 78(8), 731–739.
29. Brown, P. and K. Stevens, *Nanofibers and Nanotechnology in Textiles*. 2007, Elsevier.
30. Valko, E. I., *Textile Material*, 1966, Google Patents.
31. Higgins, L. and M. E. Anand, *Textiles Materials and Products for Activewear and Sportswear*. Technical Textile Market, 2003, 52, 9–40.
32. Hunter, L., *Mohair, Cashmere and other Animal Hair Fibers*, in *Handbook of Natural Fibers*, R. M. Kozłowski, Editor. 2012, Woodhead Publishing. p. 196–290.
33. Kuffner, H. and C. Popescu, *Wool Fibers*, in *Handbook of Natural Fibers*, R. M. Kozłowski, Editor. 2012, Woodhead Publishing. p. 171–195.
34. Roff, W. J. and J. R. Scott, *Silk and Wool*, in *Fibers, Films, Plastics and Rubbers*, W. J. Roff and J. R. Scott, Editors. 1971, Butterworth-Heinemann. p. 188–196.
35. Gordon, S., *Identifying Plant Fibers in Textiles: The Case of Cotton*, in *Identification of Textile Fibers*, M. M. Houck, Editor. 2009, Woodhead Publishing. p. 239–258.
36. Shahid ul, I., M. Shahid, and F. Mohammad, *Perspectives for Natural Product Based Agents Derived from Industrial Plants in Textile Applications – A Review*. Journal of Cleaner Production, 2013, 57(0), 2–18.
37. Ansell, M. P. and L. Y. Mwaikambo, *The Structure of Cotton and other Plant Fibers*, in *Handbook of Textile Fiber Structure*, S. J. Eichhorn, et al., Editors. 2009, Woodhead Publishing. p. 62–94.
38. Brown, R. C., et al., *Pathogenetic Mechanisms of Asbestos and other Mineral Fibers*. Molecular Aspects of Medicine, 1990, 11(5), 325–349.
39. Porter, R. M., *Glass Fiber Compositions*, 1991, Google Patents.
40. Milašius, R. and V. Jonaitienė, *Synthetic Fibers for Interior Textiles*, in *Interior Textiles*, T. Rowe, Editor. 2009, Woodhead Publishing. p. 39–46.
41. Rugeley, E. W., T. A. Feild Jr, and J. L. Petrokubi, *Synthetic Textile Articles*, 1947, US Patent
42. Woolman, M. S. and E. B. McGowan, *Textiles: A Handbook for the Student and the Consumer*. 1915, Macmillan.
43. Todd, M. P., *Hand-loom Weaving: A Manual for School and Home*. 1902, Rand, McNally.
44. Horne, P. and S. Bowden, *Knitting and Crochet*. Design, 1974, 75(3), 38–38.
45. Burt, E. C., *Bark-Cloth in East Africa*. Textile History, 1995, 26(1), 75–88.
46. Halls, Z., *Machine-Made Lace 1780–1820 (Pillow and Bobbin to Bobbin and Carriage)*. Costume, 1970, 4(Supplement-1), 46–50.
47. Smith, C. W. and J. T. Cothren, *Cotton: Origin, History, Technology, and Production*. Vol. 4. 1999, John Wiley & Sons.

48. Kriger, C. E., *Guinea Cloth': Production and Consumption of Cotton Textiles in West Africa before and during the Atlantic Slave Trade.* The Spinning World. A Global History of Cotton Textiles, 2009, 1200–1850.
49. Betrabet, S. M., K. P. R. Pillay, and R. L. N. Iyengar, *Structural Properties of Cotton Fibers: Part II: Birefringence and Structural Reversals in Relation to Mechanical Properties.* Textile Research Journal, 1963, 33(9), 720–727.
50. Alston, J. M. and J. D. Mullen, *Economic Effects of Research into Traded Goods: The Case of Australian Wool.* Journal of Agricultural Economics, 1992, 43(2), 268–278.
51. Company, I. T. and I. C. Schools, *Wool, Wool Scouring, Wool Drying, Burr Picking, Carbonizing, Wool Mixing, Wool Oiling: Woolen Carding, Woolen Spinning, Woolen and Worsted Warp Preparation.* Vol. 79. 1905, International Textbook Company.
52. Parkes, C., *The Knitter's Book of Wool: The Ultimate Guide to Understanding, Using, and Loving this Most Fabulous Fiber.* 2011, Random House LLC.
53. D. Arden, L. D. and A. Pfister, *Woolen and Worsted Yarn for an Elizabethan Knitted Suite.* 2003.
54. Knapton, J. J. F., et al., *The Dimensional Properties of Knitted Wool Fabrics Part I: The Plain-Knitted Structure.* Textile Research Journal, 1968, 38(10), 999–1012.
55. Oikonomides, N., *Silk Trade and Production in Byzantium from the Sixth to the Ninth Century: The Seals of Kommerkiarioi.* Dumbarton Oaks Papers, 1986, 40, 33–53.
56. Lock, R. L., *Process for Making Silk Fibroin Fibers*, 1993, Google Patents.
57. Kaplan, D. L. *Silk Polymers.* in *Workshop on Silks: Biology, Structure, Properties, Genetics (1993, Charlottesville, Va.).* 1994, American Chemical Society.
58. Bolton, E. K., *Chemical Industry Medal. Development of Nylon.* Industrial & Engineering Chemistry, 1942, 34(1), 53–58.
59. Li, L., et al., *Formation and Properties of Nylon-6 and Nylon-6/Montmorillonite Composite Nanofibers.* Polymer, 2006, 47(17), 6208–6217.
60. Harazoe, H., M. Matsuno, and S. Noda, *Method of Manufacturing Polyesters*, 1996, Google Patents.
61. Mijatovic, D., J. C. T. Eijkel, and A. V. D. Berg, *Technologies for Nanofluidic Systems: Top–down vs. Bottom–up—A Review.* Lab on a Chip, 2005, 5(5), 492–500.
62. Patra, J. K. and S. Gouda, *Application of Nanotechnology in Textile Engineering: An Overview.* Journal of Engineering and Technology Research, 2013, 5(5), 104–111.
63. Mohanraj, V. J. and Y. Chen, *Nanoparticles-A Review.* Tropical Journal of Pharmaceutical Research, 2007, 5(1), 561–573.
64. Mehnert, W. and K. Mäder, *Solid Lipid Nanoparticles: Production, Characterization and Applications.* Advanced Drug Delivery Reviews, 2001, 47(2), 165–196.
65. Roney, C., et al., *Targeted Nanoparticles for Drug Delivery through the Blood–Brain Barrier for Alzheimer's Disease.* Journal of Controlled Release, 2005, 108(2), 193–214.
66. Joshi, M. and A. Bhattacharyya, *Nanotechnology–A New Route to High-Performance Functional Textiles.* Textile Progress, 2011, 43(3), 155–233.
67. Kaounides, L., H. Yu, and T. Harper, *Nanotechnology Innovation and Applications in Textiles Industry: Current Markets and Future Growth Trends.* Materials Science and Technology, 2007, 22(4), 209–237.
68. Kathiervelu, S. S., *Applications of Nanotechnology in Fiber Finishing.* Synthetic Fibers, 2003, 32(4), 20–22.

69. Wang, C. C. and C. C. Chen, *Physical Properties of the Cross-linked Cellulose Catalyzed with Nanotitanium Dioxide under UV Irradiation and Electronic Field.* Applied Catalysis A: General, 2005, 293, 171–179.
70. Price, D., et al., *Burning Behavior of Foam/Cotton Fabric Combinations in the Cone Calorimeter.* Polymer Degradation and Stability, 2002, 77(2), 213–220.
71. Lamba, N. M. K., K. A. Woodhouse, and S. L. Cooper, *Polyurethanes in Biomedical Applications*. 1997, CRC press.
72. Qian, L. and J. P. Hinestroza, *Application of Nanotechnology for High Performance Textiles.* Journal of Textile and Apparel, Technology and Management, 2004, 4(1), 1–7.
73. Singh, V. K., et al. *Applications and Future of Nanotechnology in Textiles*. in *National Cotton Council Beltwide Cotton Conference*. 2006.
74. Gibson, P., H. S. Gibson, and C. Pentheny, *Electrospinning Technology: Direct Application of Tailorable Ultrathin Membranes.* Journal of Industrial Textiles, 1998, 28(1), 63–72.
75. Snyder, R. G., *The Bionic Tailor: TC2's 3-D Body Scanner*, 2001, IEEE-INST Electrical Electronics Engineers INC 345 E 47TH ST NY 10017–2394 USA.
76. Okabe, H., et al. *Three Dimensional Apparel CAD System.* in *ACM SIGGRAPH Computer Graphics*. 1992, ACM.
77. Yan, H. and S. S. Fiorito, *Communication: CAD/CAM Adoption in US Textile and Apparel Industries.* International Journal of Clothing Science and Technology, 2002, 14(2), 132–140.
78. Xing, X., Y. Wang, and B. Li, *Nanofibers Drawing and Nanodevices Assembly in Poly (Trimethylene Terephthalate).* Optics Express, 2008, 16(14), 10815–10822.
79. Ikegame, M., K. Tajima, and T. Aida, *Template Synthesis of Polypyrrole Nanofibers Insulated within One-Dimensional Silicate Channels: Hexagonal versus Lamellar for Recombination of Polarons into Bipolarons.* Angewandte Chemie International Edition, 2003, 42(19), 2154–2157.
80. He, L., et al., *Fabrication and Characterization of Poly (L-Lactic Acid) 3D Nanofibrous Scaffolds with Controlled Architecture by Liquid–Liquid Phase Separation from a Ternary Polymer–Solvent System.* Polymer, 2009, 50(16), 4128–4138.
81. Hartgerink, J. D., E. Beniash, and S. I. Stupp, *Self-Assembly and Mineralization of Peptide-Amphiphile Nanofibers.* Science, 2001, 294(5547), 1684–1688.
82. Li, D. and Y. Xia, *Fabrication of Titania Nanofibers by Electrospinning.* Nano Letters, 2003, 3(4), 555–560.
83. Ramakrishna, S., et al., *Electrospun Nanofibers: Solving Global Issues.* Materials Today, 2006, 9(3), 40–50.
84. Reneker, D. H. and I. Chun, *Nanometer Diameter Fibers of Polymer, Produced by Electrospinning.* Nanotechnology, 1996, 7(3), 216–233.
85. Frenot, A. and I. S. Chronakis, *Polymer Nanofibers Assembled by Electrospinning.* Current Opinion in Colloid & Interface Science, 2003, 8(1), 64–75.
86. Dersch, R., et al., *Electrospun Nanofibers: Internal Structure and Intrinsic Orientation.* Journal of Polymer Science Part A: Polymer Chemistry, 2003, 41(4), 545–553.
87. Yang, F., et al., *Electrospinning of Nano/Micro Scale Poly (L-Lactic Acid) Aligned Fibers and their Potential in Neural Tissue Engineering.* Biomaterials, 2005, 26(15), 2603–2610.

88. Moroni, L., et al., *Fiber Diameter and Texture of Electrospun PEOT/PBT Scaffolds Influence Human Mesenchymal Stem Cell Proliferation and Morphology, and the Release of Incorporated Compounds.* Biomaterials, 2006, 27(28), 4911–4922.
89. Jayakumar, R., et al., *Novel Chitin and Chitosan Nanofibers in Biomedical Applications.* Biotechnology Advances, 2010, 28(1), 142–150.
90. Zhang, Y., et al., *Recent Development of Polymer Nanofibers for Biomedical and Biotechnological Applications.* Journal of Materials Science: Materials in Medicine, 2005, 16(10), 933–946.
91. Nazarov, R., H.-J. Jin, and D. L. Kaplan, *Porous 3-D scaffolds from regenerated silk fibroin.* Biomacromolecules, 2004, 5(3), 718–726.
92. Smit, E. A., *Studies towards High-throughput Production of Nanofiber Yarns*, 2008, Stellenbosch University: Stellenbosch.
93. Subbiah, T., et al., *Electrospinning of Nanofibers.* Journal of Applied Polymer Science, 2005, 96(2), 557–569.
94. Murugan, R. and S. Ramakrishna, *Design Strategies of Tissue Engineering Scaffolds with Controlled Fiber Orientation.* Tissue Engineering, 2007, 13(8), 1845–1866.
95. Badrossamay, M. R., et al., *Nanofiber Assembly by Rotary Jet-Spinning.* Nano letters, 2010, 10(6), 2257–2261.
96. Srary, J., *Method of and Apparatus for Ringless Spinning of Fibers*, 1970, Google Patents.
97. Katta, P., et al., *Continuous Electrospinning of Aligned Polymer Nanofibers onto a Wire Drum Collector.* Nano Letters, 2004, 4(11), 2215–2218.
98. Creight, M. D. J., et al., *Short Staple Yarn Manufacturing.* Vol. 700. 1997, NC, USA: Carolina Academic Press Durham, .
99. Pan, H., et al., *Continuous Aligned Polymer Fibers Produced by a Modified Electrospinning Method.* Polymer, 2006, 47(14), 4901–4904.
100. Dzenis, Y. A., *Spinning Continuous Fibers for Nanotechnology.* Science 2004, 304 p. 1917–1919.
101. Jacobs, V., R. D. Anandjiwala, and M. Maaza, *The Influence of Electrospinning Parameters on the Structural Morphology and Diameter of Electrospun Nanofibers.* Journal of Applied Polymer Science, 2010, 115(5), 3130–3136.
102. Sarkar, K., et al., *Electrospinning to Forcespinning.* Materials Today, 2010, 13(11), 12–14.
103. Smit, A. E., U. Buttner, and R. D. Sanderson, *Continuous Yarns from Electrospun Nanofibers.* Nanofibers and Nanotechnology in Textiles. 2007, 45.
104. Hong, Y. T., et al., *Filament Bundle Type Nano Fiber and Manufacturing Method Thereof*, 2010, Google Patents.
105. Zhou, F. L. and R. H. Gong, *Manufacturing Technologies of Polymeric Nanofibers and Nanofiber Yarns.* Polymer International, 2008, 57(6), 837–845.
106. Teo, W. E. and S. Ramakrishna, *A Review on Electrospinning Design and Nanofiber Assemblies.* Nanotechnology, 2006, 17(14), p. R89-R106.
107. Kleinmeyer, J., J. Deitzel, and J. Hirvonen, *Electro Spinning of Submicron Diameter Polymer Filaments*, 2003, Google Patents.
108. Childs, H. R., *Process of Electrostatic Spinning*, 1944, US Patent
109. Ko, F. K., *Nanofiber Technology: Bridging the Gap between Nano and Macro World*, in *Nanoengineered Nanofibrous Materials*. 2004, 1–18.

110. Murray, M. P., *Cone Collecting Techniques for Whitebark Pine.* Western Journal of Applied Forestry, 2007, 22(3), 153–155.
111. Hermes, J., *Apparatus for Winding Yarn*, 1986, Google Patents.
112. Yadav, A., et al., *Functional Finishing in Cotton Fabrics Using Zinc Oxide Nanoparticles.* Bulletin of Materials Science, 2006, 29(6), 641–645.
113. Perelshtein, I., et al., *A One-Step Process for the Antimicrobial Finishing of Textiles with Crystalline TiO_2 Nanoparticles.* Chemistry—A European Journal, 2012, 18(15), 4575–4582.
114. Jiu, J. T., et al., *The Preparation of MgO Nanoparticles Protected by Polymer.* Chinese Journal of Inorganic Chemistry, 2001, 17(3), 361–365.

CHAPTER 2

MODERN APPLICATIONS OF NANOTECHNOLOGY IN TEXTILES

A. AFZALI and SH. MAGHSOODLOU

University of Guilan, Rasht, Iran

CONTENTS

ABSTRACT

In this chapter nano-particles are represented for its outstanding performances in textile finishing. The bearing properties of chemical substances (nano-particles) can be encapsulated to the textile materials (yarn, fabric, etc.). Hence the textile materials will carry the properties of the particles.

2.1 FABRIC FINISHING BY USING NANOTECHNOLOGY

There are many ways in which the surface properties of a fabric can be manipulated and enhanced, by implementing appropriate surface finishing, coating, and/or altering techniques, using nanotechnology. A few representative applications of fabric finishing using nanotechnology are schematically displayed in the Figure 2.1.

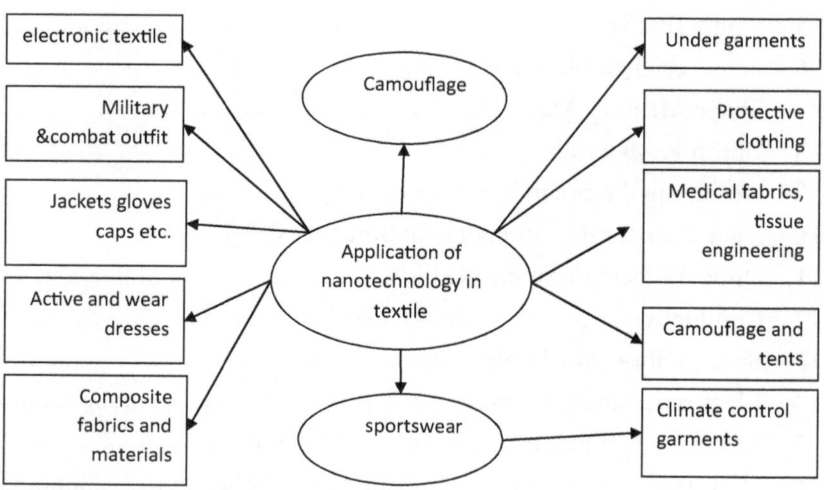

FIGURE 2.1 Some representative applications of nanotechnology in textiles.

In recent years, several attempts have been made by researchers and industries to use similar concepts of surface-engineered modifications through nanotechnology to develop high performance textile and smart textile. The concept of surface engineering and nano-textile develops hydrophobic fabric surfaces that are capable of repelling liquids and resisting stains, while complementing the other desirable fabric attributes, such as breathability, softness, and comfort [1–4].

2.2 TEXTILE MODIFICATION

Considering special advantages and high potentialities of the application of nanostructured materials in textile industry, especially for producing high performance textiles, here we reviewed the application of nano-structured materials for anti-bacterial modification of textile and polymeric materials. The modification of textile fibers is carried out by commonly used chemical or electro-chemical application methods. Many of the classical textile finishing techniques (e.g., hydrophobization, easy-care finishing) that are already used since decades are among these methods. Modification of textiles via producing polymeric nano-composites and also surface modification of textiles with metallic and inorganic nanostructured materials are developed due to their unique properties. Considering the fact that fiber and film processing are the most difficult procedures of molding polymeric materials, bulk modification of continuous multi-filament yarns is an extremely sensitive process. However, achieving optimum process conditions will present an economical technique [5, 6].

Different methods have been used for surface modifications of textiles by using poly carboxylic acids as spacers for attaching TiO_2 nano-particles to the fabrics [7] and argon plasma grafting nano-particles on wool surface [8]. Plasma pre-treatment has been used for the generation of active groups on the surface to be combined with TiO_2 nanoparticles [9]. The radical groups on the surface have also been generated using irradiation of the textile surfaces with UV light to bond the nano-particles [10]. Deposition of nanoparticles from their metallic salt solution on the surface pretreated with RF-plasma and vacuum-UV [11].

Nanotechnology holds great potential in the textile and clothing industry offering enhanced performance of textile manufacturing machines and processes so as to overcome the limitations of conventional methods. Nanofibers have good properties such as high surface area, a small fiber diameter, good filtration properties and high permeability. Nanofibers can be obtained via electro-spinning application or bicomponent extrusion (islands in the sea technique) [12]. One of the interesting areas for the application of nanotechnology in the textile industry are coating and finishing processes of textiles which is done by the techniques like sol-gel [13] and plasma [14]. Nanotech enhanced textiles include sporting industry, skincare, space technology and clothing as well as material technology exhibiting better healthcare systems, protective clothing and integrated electronics. By using nanotechnology, textiles with self-cleaning surfaces have attracted much attention, which is created by the lotus effect. In brief, nanoscaled structures similar to those of a lotus leaf create a surface that causes water and oil to be repelled, forming droplets, which will simply roll of the surface, taking any with them [15].

With the advent of nanoscience and technology, a new area has developed in the area of textile finishing called "Nanofinishing." Growing awareness of health and hygiene has increased the demand for bioactive or antimicrobial and UV-protecting textiles. Coating the active surfaces cause UV blocking, antimicrobial, flame retardant, water repellant and self-cleaning properties. The UV-blocking property of a fabric is enhanced when a dye, pigment, delustrant, or ultraviolet absorber finish is present that absorbs ultraviolet radiation and blocks its transmission through a fabric to the skin. Metal oxides like ZnO as UV-blocker are more stable when compared to organic UV-blocking agents. For antibacterial finishing, ZnO nanoparticles scores over nano-silver in cost-effectiveness, whiteness, and UV-blocking property [16, 17].

2.3 TYPES OF NANOMATERIALS

2.3.1 NANOCOMPOSITE FIBERS

A composite is a material that combines one or more separate components. Composites are designed to exhibit the best properties of each component. A large variety of systems combining one, two and three dimensional materials with amorphous materials mixed at the nanometer scale [18].

Nanostructure composite fibers are intensively used in automotive, aerospace and military applications. Nanocomposite fibers are produced by dispersing nanosize fillers into a fiber matrix. Due to their large surface area and high aspect ratio, nanofillers interact with polymer chain movement and thus reduce the chain mobility of the system. Being evenly distributed in polymer matrices, nanoparticles can carry load and increase the toughness and abrasion resistance. Most of the nanocomposite fibers use fillers such as nanosilicates, metal oxide nanoparticles, graphite nanofibers (GNF) as well as single-wall and multi-wall carbon nanotubes (CNT) [19, 20]. Some novel CNT reinforced polymer composite materials have been developed, which can be used for developing multifunctional textiles having superior strength, toughness, lightweight, and high electrical conductivity [21].

2.3.2 CARBON NANOFIBERS AND CARBON NANOPARTICLES

Carbon nanofibers and carbon black nanoparticles are among the most commonly used nanosize filling materials. Nanofibers can be defined as fibers with a diameter of less than 1 mm or 1000 nm and are characterized as having a high surface area to volume ratio and a small pore size in fabric form [22]. Carbon nanofibers can effectively increase the tensile strength of composite fibers due to its high aspect ratio, while carbon black nanoparticles can improve their abrasion resistance and toughness. Several fiber-forming polymers used as matrices have been investigated including polyester, nylon and polyethylene with the weight of the filler from 5 to 20% [23].

There are numerous applications in which nanofibers could be suited. The high surface area to volume ratio and small pore size allows viruses and spore-forming bacterium such as anthrax to be trapped. Filtration devices and wound dressings are just some of the applications in which nanofibers could be used. Researchers are investigating textile materials made from nanofibers, which can act as a filter for pathogens (bacteria, viruses), toxic gasses, or poisonous or harmful substances in the air. Medical staff, fire fighters, the emergency services or military personnel could all benefit from protective garments made from nanofibers materials [24].

2.3.3 CLAY NANOPARTICLES

Clay nanoparticles are resistant to heat, chemicals and electricity, and have the ability to block UV light. Incorporating clay nanoparticles into a textile can result in a fabric with improved tensile strength, tensile modulus, flexural strength and flexural modulus. Nanocomposite fibers, which use clay nanoparticles can be engineered to be flame, UV light resistant and anti-corrosive. Although there have been a number of flame retardant finishes available since the 1970s, the emission of toxic gasses when set ablaze make them somewhat hazardous. Clay nanoparticles have been incorporated into nylon to impart flame retardant characteristics to the textile without the emission of toxic gas. The addition of clay nanoparticles has made polypropylene dyeable. Metal oxide nanoparticles of TiO_2, Al_2O_3, ZnO and MgO exhibit photocatalytic ability, electrical conductivity, UV absorption and photo-oxidizing capacity against chemical and biological species. The main research efforts involving the use of nanoparticles of metal oxides have been focused on antimicrobial, self-decontaminating and UV blocking applications for both military protection gears and civilian health products [21, 22]. Nylon fibers filled with ZnO nanoparticles can provide UV shielding function and reduce static electricity on nylon fibers. A composite fiber with nanoparticle of TiO_2 or MgO can provide self-sterilizing function [23].

2.3.4 CARBON NANOTUBES

Carbon Nanotube is a tubular form of carbon with diameter as small as nanometer (nm). A carbon nanotube is configurationally equivalent to a two dimensional graphene sheet rolled into a tube. They can be metallic or semiconducting, depending on chirality. CNT are one of the most promising materials due to their high strength and high electrical conductivity. CNT consists of tiny shell(s) of graphite rolled up into a cylinder(s) [25, 26]. CNT exhibit 100 times the tensile strength of steel at one-sixth weight, thermal conductivity better than all but the purest diamond, and an electrical conductivity similar to copper, but with the ability to carry much higher currents. The potential applications of CNTs include conductive and high-strength composite fibers, energy storage and energy conversion devices, sensors, and field emission displays. Possible applications include screen displays, sensors, aircraft structures, explosion-proof blankets and electromagnetic shielding. The composite fibers have potential applications in safety harnesses, explosion-proof blankets, and electromagnetic shielding applications. Continuing research activities on CNT fibers involve study of different fiber polymer matrices, such as, polymethylmethacrylate (PMMA) and polyacrylonitrile (PNA) as well as CNT dispersion and orientation in polymers [27].

2.3.5 NANOCELLULAR FOAM STRUCTURE

Polymeric materials with nanosize porosity exhibit lightweight, good thermal insulation, as well as high cracking resistance at high temperature without sacrifices in mechanical strength. By choosing the pretreatment condition to the fiber, the transverse mechanical properties of the composite can be also enhanced through the molecular diffusion across the interface between the fiber and the matrix. The nanocomposites clearly surpass the mechanical properties of most comparable cellulosic materials, their greatest advantage being the fact that they are fully bio-based and biodegradable, but also of relatively high strength. A potential application of cellular structure is to encapsulate functional compounds such as pesticides and drugs inside of the nanosize cells. One of the approaches to fabricate nanocellular fibers is to make use of a thermodynamic instability during supercritical carbon dioxide extrusion and reduce the size of the cellular fibers that can be used as high-performance composite fibers as well as for sporting and aerospace materials [21, 26].

2.4 PROPERTIES OF NANO-TEXTILE FIBERS

2.4.1 WATER REPELLENCE

The water-repellent property of fabric created by nano-whiskers, which are hydrocarbons and 1/1000 of the size of a typical cotton fiber, when added to the fabric create a peach fuzz effect without lowering the strength of cotton. The spaces between the whiskers on the fabric are smaller than the typical drop of water, but still larger than water molecules; water thus, remains on the top of the whiskers and above the surface of the fabric. However, liquid can still pass through the fabric, if pressure is applied to it [3, 4].

Nanosphere impregnation involving a three-dimensional surface structure with gel forming additives which repel water and prevent dirt particles from attaching themselves are also used. Once water droplets fall onto them, water droplets bead up and, if the surface slopes slightly, will roll off. As a result, the surfaces stay dry even during a heavy shower. Furthermore, the droplets pick up small particles of dirt as they roll, and so the leaves of the lotus plant keep clean even during light rain. By altering the micro and nano-scale surface features on a fabric surface, a more robust control of wetting behavior can be attained. It has been demonstrated that by combining the nanoparticles of hydroxylapatite, TiO_2, ZnO and Fe_7O_3 with other organic and inorganic substances, the audio frequency plasma of fluorocarbon chemical was applied to deposit a nanoparticulate hydrophobic film onto a cotton fabric surface to improve its water repellent property. This sort of surface engineering, which is capable of replicating hydrophobic behavior, can be used in developing special chemical finishes for producing water-and/or stain- resistant fabrics while complementing the other desirable fabric attributes, such as breathability, softness and comfort. The surfaces of the textile fabrics can be appreciably modified to achieve considerably greater abrasion resistance, UV resistance, electromagnetic and infrared protection properties [28, 29].

2.4.2 UV-PROTECTION

Inorganic UV blockers are more preferable to organic UV blockers as they are non-toxic and chemically stable under exposure to both high temperatures and UV [30, 31]. Inorganic UV blockers are usually certain semiconductor oxides such as TiO_2, ZnO, SiO_2 and Al_2O_3. Among these semiconductor oxides, titanium dioxide (TiO_2) and zinc oxide (ZnO) are commonly used. It was determined that nano-sized titanium dioxide and zinc oxide are more efficient at absorbing and scattering UV radiation than the conventional size, and are thus better to provide protection against UV rays. This is due to the fact that nano-particles have a larger surface area per unit mass and volume than the conventional materials, leading to the increase of the effectiveness of blocking UV radiation [30, 32]. Various researchers have worked on the application of UV blocking treatment to fabric using nanotechnology. UV blocking treatment for cotton fabrics are developed using the sol-gel method. A thin layer of titanium dioxide is formed on the surface of the treated cotton fabric, which provides excellent UV protection; the effect can be maintained after 50 home launderings [31]. Apart from titanium dioxide, zinc oxide nano rods of 10–50 nm in length are also applied to cotton fabric to provide UV protection. According to the studies on the UV blocking effect, the fabric treated with zinc oxide nanorods were found to have demonstrated an excellent UV protective factor (UPF) rating. This effect can be further enhanced by using a different procedure for the application of nanoparticles on the fabric surface. When the process of padding is used for applying the nanoparticles on to the fabric, the nanoparticles get applied not only on the surface alone but also penetrates into the interstices of the yarns and the fabric, that is, some portion of the nanoparticles get penetrate into the fabric structure. Such Nanoparticles, which do not stay on the surface may not be very effective in shielding the UV rays. It is worthwhile that only the right (face) side of the fabric gets exposed to the rays and therefore, this surface alone needs to be covered with the nanoparticles for better UV protection. Spraying (using compressed air and spray gun) the fabric surface with the nanoparticles can be an alternate method of applying the nanoparticles [4].

2.4.3 ANTIMICROBIAL

Although many antimicrobial agents are already in used for textile, the major classes of antimicrobial for textile include organo-silicones, organometallics, phenols and quaternary ammonium salts. The bi- phenolic compounds exhibit a broad spectrum of antimicrobial activity. For imparting antibacterial properties, nano-sized silver, titanium dioxide, zinc oxide, triclosan and chitosan are used [4]. Nano-silver particles have an extremely large relative surface area, thus increasing their contact with bacteria or fungi and vastly improving their bactericidal and fungicidal effectiveness. Nano-silver is very reactive with protein and shows antimicrobial properties at concentrations as low as 0.0003–0.0005%. When contacting bacteria and fungi, it will adversely affect cellular metabolism and inhibits cell growth. It also suppresses respiration, the basal metabolism of the electron transfer system, and the transport of the substrate into the microbial cell membrane. Furthermore, it inhibits the multiplication and growth of those bacteria and fungi, which cause infection, odor, itchiness and sores [30]. Some synthetic antimicrobial nano particles, which are used in textiles are as follows. Triclosan, a chlorinated bi- phenol, is a synthetic, non-ionic and broad spectrum antimicrobial agent possessing mostly antibacterial alone with some antifungal and antiviral properties. Chitosan, a natural biopolymer, is effectively used as antibacterial, antifungal, antiviral, non-allergic and biocompatible. ZnO nanoparticles have been widely used for their antibacterial and UV-blocking properties.

2.4.5 ANTISTATIC

An antistatic agent is a compound used for treatment of materials or their surfaces in order to reduce or eliminate buildup of static electricity generally caused by the triboelectric effect. The molecules of an antistatic agent often have both hydrophilic and hydrophobic areas, similar to those of a surfactant; the hydrophobic side interacts with the surface of the material, while the hydrophilic side interacts with the air moisture and binds the water molecules [20]. As synthetic fibers provide poor anti-static properties, research work concerning the improvement of the anti-static properties of textiles by using nanotechnology has been at large. It was determined that nano-sized particles like titanium dioxide, zinc oxide whiskers, nanoantimony-doped tin oxide (ATO) and silanenanosol could impart anti-static properties to synthetic fibers. Such material helps to effectively dissipate the static charge, which is accumulated on the fabric [4]. On the other hand, silanenanosol improves anti-static properties, as the silica gel particles on fiber absorb water and moisture in the air by amino and hydroxyl groups and bound water. Electrically conductive nano-particles are durably anchored in the fibrils of the membrane of teflon, creating an electrically conductive network that prevents the formation of isolated chargeable areas and voltage peaks commonly found in conventional anti-static materials. This method can overcome the limitation of conventional methods, which is that the anti-static agent is easily washed off after a few laundry cycles [15].

2.4.6 WRINKLE RESISTANCE

To impart wrinkle resistance to fabric, resin is commonly used in conventional methods. However, there are limitations to applying resin, including a decrease in the tensile strength of fiber, abrasion resistance, water absorbency and dye-ability, as well as breathability. To overcome the limitations of using resin, some researchers employed nano-titanium dioxide and nano-silica to improve the wrinkle resistance of cotton and silk respectively [4]. Nano-titanium dioxide employed with carboxylic acid as a catalyst under UV irradiation to catalyzes the cross-linking reaction between the cellulose molecule and the acid. On the other hand, nano-silica when applied with maleic anhydride as a catalyst could successfully improve the wrinkle resistance of silk [2]. More over the wrinkle recovery of the fabrics can also be improved to a great extent by imparting techniques like padding and exhaustion beside the use of nano-materials to the fabrics. Studies also have suggest that treatment of fabrics with microwaves are more wrinkle resistant as comparable to oven curing, because it generates higher frequency and volumetric heating which minimizes the damage from over drying.

2.5 SMART TEXTILES

All applications textile materials have a vast number of clear advantages:

- they are omnipresent, everybody is familiar with them;
- they are easy to use and to maintain;
- clothes have a large contact with the body;
- they make us look nice;
- they are extremely versatile in terms of raw materials used, arrangement of the fibers, finishing treatments, shaping, etc.
- they can be made to fit typical applications where textile structures are to be preferred are:
  - long term or permanent contact without skin irritation,
  - home applications,
  - applications for children: in a discrete and careless way,
  - applications for the elderly: discretion, comfort and esthetics are important.

The multifunctional textiles such as fashion and environmental protection, ballistic and chemical protection, flame protection are all passive systems. The smart textiles are a new generation of fibers, yarns, fabrics and garments that are able to sense stimuli and changes in their environments, such as mechanical, thermal, chemical, electrical, magnetic and optical changes, and then respond to these changes in predetermined ways. They are multifunctional textile systems that can be classified into three categories of passive smart textiles, active smart textiles and very smart textiles [33].

The functionalities of smart textiles can be classified in five groups: sensoring, data processing, actuation, communication, energy.

At this moment, most of the progress has been achieved in the area of sensoring. Many type of parameters can be measured:

- Temperature
- Biopotentials: cardiogram, myographs, encephalographs
- Acoustic: heart, lungs, digestion, joints
- Ultrasound: blood flow
- Biological, chemical
- Motion: respiration, motion
- Pressure: blood
- Radiation: IR, spectroscopy
- Odor, sweat
- Mechanical skin parameters
- Electric (skin) parameters

Some of these parameters are well known, like cardiogram and temperature. Nevertheless, permanent monitoring also opens up new perspectives for these traditional parameters too. Indeed today evaluation is usually based on standards for global population groups. Permanent monitoring supported by self learning devices will allow the set up of personal profiles for each individual, so that conditions deviating from normal can be traced the soonest possible. Also diagnosis can be a lot more accurate.

Apart from the actual measuring devices data processing is a key feature in this respect. These type of data are new. They are numerous with multiple complex interrelationships and time dependent. New self learning techniques will be required. The introduction of such an approach will be slow, because no evidence of the benefits are available at this moment.

"We don't measure because we don't know the meaning, we don't know the meaning because we don't measure."

Actuation is another aspect. Identification of problems only makes sense when followed by an adequate reaction. This reaction can consist of reporting or calling for help, but also drug supply and physical treatment. A huge challenge in this respect is the development of high performance muscle like materials.

Smart textiles is a new aspect in textile that is a multidiscipline field of research in many sciences and technologies such as textile, physics, chemistry, medicine, electronics, polymers, biotechnology, telecommunications, information technology, microelectronics, wearable computers, nanotechnology and micro-electromechanical machines. Shape memory materials (SMMs), conductive materials, phase change materials (PCMs), chromic materials, photonic fibers, mechanical responsive materials, intelligent coating/membranes, micro and nanomaterials and piezoelectric materials are applied in smart textiles [34].

The objective of smart textile is to absorb a series of active components essentially without changing its characteristics of flexibility and comfort. In order to make a smart textile, firstly, conventional components such as sensors, devices and wires are being reshaped in order to fit in the textile, ultimately the research activities trend to manufacture active elements made of fibers, yarns and fabrics structures. Smart textiles are ideal vehicle for carrying active elements that permanently monitor our body and the environment, providing adequate reaction should something happen [35].

The smart textiles have some of the capabilities such as biological and chemical sensing and responding, power and data transmission from wearable computers and polymeric batteries, transmitting and receiving RF signals and automatic voice warning systems as to 'dangers ahead' that may be appropriate in military applications. Other than military applications of smart textiles, mountain climbers, sportsmen, businessmen, healthcare and medical personnel, police, and firemen will be benefitted from the smart textiles technologies.

A smart textile can be active in many other fields. Smart textiles as a carrier of sensor systems can measure heart rate, temperature, respiration, gesture and many other body parameters that can provide useful information on the health status of a person. The smart textiles can support the rehabilitation process and react adequately on hazardous conditions that may have been detected. The reaction can consist of warning, prevention or active protection. After an event has happened, the smart textile is able to analyze the situation and to provide first aid [36].

Wearable electronics and photonics, adaptive and responsive structures, biomimetics, bioprocessing, tissue engineering and chemical/drug releasing are some of the research areas in integrated processes and products of smart textiles. There are some areas that the research activities have reached the industrial application. Optical fibers, shape memory polymers, conductive polymers, textile fabrics and composites integrated with optical fiber sensors have been used to monitor the health of major bridges and buildings. The first generation of wearable motherboards has been developed, which has sensors integrated inside garments and is capable of detecting injury and health information of the wearer and transmitting such information remotely to a hospital. Shape memory polymers have been applied to textiles in fiber, film and foam forms, resulting in a range of high performance fabrics and garments, especially sea-going garments. Fiber sensors, which are capable of measuring temperature, strain/stress, gas, biological species and smell, are typical smart fibers that can be directly applied to textiles. Conductive polymer-based actuators have achieved very high levels of energy density. Clothing with its own senses and brain, like shoes and snow coats which are integrated with Global Positioning System (GPS) and mobile phone technology, can tell the position of the wearer and give him/her directions. Biological tissues and organs, like ears and noses, can be grown from textile scaffolds made from biodegradable fibers [37, 38].

2.6 PHASE CHANGE MATERIALS

Phase change materials are thermal storage materials that are used to regulate temperature fluctuations. The thermal energy transfer occurs when a material changes from a solid to a liquid or from a liquid to a solid. This is called a change in state, or phase.

Incorporating microcapsules of PCM into textile structures improves the thermal performance of the textiles. Phase change materials store energy when they change from solid to liquid and dissipate it when they change back from liquid to solid. It would be most ideal, if the excess heat a person produces could be stored intermediately somewhere in the clothing system and then, according to the requirement, activated again when it starts to get chilly.

The most widespread PCMs in textiles are paraffin-waxes with various phase change temperatures (melting and crystallization) depending on their carbon numbers. The characteristics of some of these PCMs are summarized in Table 2.1. These phase change materials are enclosed in microcapsules, which are 1–30 mm in diameter. Hydrated inorganic salts have also been used in clothes for cooling applications. PCM elements containing Glauber's salt (sodium sulfate) have been packed in the pockets of cooling vests.

PCM can be applied to fibers in a wet-spinning process, incorporated into foam or embedded into a binder and applied to fabric topically, or contained in a cell structure made of a textile reinforced synthetic material [39, 40].

In manufacturing the fiber, the selected PCM microcapsules are added to the liquid polymer or polymer solution, and the fiber is then expanded according to the conventional methods such as dry or wet spinning of polymer solutions and extrusion of polymer melts.

Fabrics can be formed from the fibers containing PCM by conventional weaving, knitting or nonwoven methods, and these fabrics can be applied to numerous clothing applications.

In this method, the PCMs are permanently locked within the fibers, the fiber is processed with no need for variations in yarn spinning, fabric knitting or dyeing and properties of fabrics (drape, softness, tenacity, etc.) are not altered in comparison with fabrics made from conventional fibers. The

TABLE 2.1 Phase Change Materials

Phase change material	Melting temperature, °C	Crystallization temperature, °C	Heat storage capacity in J/g
Eicosane	36.1	30.6	246
Nonadecane	32.1	26.4	222
Octadecane	28.2	25.4	244
Heptadecane	22.5	21.5	213
Hexadecane	18.5	16.2	237

microcapsules incorporated into the fibers in this method have an upper loading limit of 5–10% because the physical properties of the fibers begin to suffer above that limit, and the finest fiber is available. Due to the small content of microcapsules within the fibers, their thermal capacity is rather modest, about 8–12 J/g.

Usually PCM microcapsules are coated on the textile surface. Microcapsules are embedded in a coating compound such as acrylic, polyurethane and rubber latex, and applied to a fabric or foam. In lamination of foam containing PCMs onto a fabric, the selected PCMs microcapsules can be mixed into a polyurethane foam matrix, from which moisture is removed, and then the foam is laminated on a fabric. Typical concentrations of PCMs range from 20% to 60% by weight. Microcapsules should be added to the liquid polymer or elastomer prior to hardening. After foaming (fabricated from polyurethane) microcapsules will be embedded within the base material matrix. The application of the foam pad is particularly recommended because a greater amount of microcapsules can be introduced into the smart textile. In spite of this, different PCMs can be used, giving a broader range of regulation temperatures. Additionally, microcapsules may be anisotropically distributed in the layer of foam. The foam pad with PCMs may be used as a lining in a variety of clothing such as gloves, shoes, hats and outerwear. Before incorporation into clothing or footwear the foam pad is usually attached to the fabric, knitted or woven, by any conventional means such as glue, fusion or lamination [41, 42].

The PCM microcapsules are also applied to a fibrous substrate using a binder (e.g., acrylic resin). All common coating processes such as knife over roll, knife over air, screen-printing, gravure printing, dip coating may be adapted to apply the PCM microcapsules dispersed throughout a polymer binder to fabric. The conventional pad–mangle systems are also suitable for applying PCM microcapsules to fabrics. The formulation containing PCMs can be applied to the fabric by the direct nozzle spray technique.

There are many thermal benefits of treating textile structures with PCM microcapsules such as cooling, insulation and thermo regulating effect. Without phase change materials the thermal insulation capacity of clothing depends on the thickness and the density of the fabric (passive insulation). The application of PCM to a garment provides an active thermal insulation effect acting in addition to the passive thermal insulation effect of the garment system. The active thermal insulation of the PCM controls the heat flux through the garment layers and adjusts the heat flux to the thermal circumstances. The active thermal insulation effect of the PCM results in a substantial improvement of the garment's thermo-physiological wearing comfort [33]. Intensity and duration of the PCM's active thermal insulation effect depend mainly on the heat-storage capacity of the PCM microcapsules and their applied quantity. In order to ensure a suitable and durable effect of the PCM, it is necessary to apply proper PCM in sufficient quantity into the appropriate fibrous substrates of proper design.

The PCM quantity applied to the active wear garment should be matched with the level of activity and the duration of the garment use. Furthermore, the garment construction needs to be designed in a way, which assists the desired thermo-regulating effect. Thinner textiles with higher densities readily support the cooling process. In contrast, the use of thicker and less dense textile structures leads to a delayed and therefore more efficient heat release of the PCM. Further requirements on the textile substrate in a garment application include sufficient breathability, high flexibility, and mechanical stability [43].

In order to determine a sufficient PCM quantity, the heat generated by the human body has to be taken into account carrying out strenuous activities under which the active wear garments are worn. The heat generated by the body needs to be entirely released through the garment layers into the environment. The necessary PCM quantity is determined according to the amount of heat, which should be absorbed by the PCM to keep the heat balance equalized. It is mostly not necessary to put PCM in all parts of the garment. Applying PCM microcapsules to the areas that provide problems from a thermal standpoint and thermoregulating the heat flux through these areas is often enough. It is also advisable to use different PCM microcapsules in different quantities in distinct garment locations [42].

2.6.1 APPLICATIONS OF TEXTILES CONTAINING PCMS

Fabrics containing PCMs have been used in a variety of applications including apparel, home textiles and technical textiles (Table 2.2).

TABLE 2.2 Application of PMs in Textiles

Casual Clothing	Underwear, Jackets, Sport Garments
Professional clothing	Fire fighters protective clothing, Bullet proof fabrics, Space suits, Sailor suits
Medical uses	Surgical gauze, Bandage, Nappies, Bed lining, Gloves, Gowns, Caps, Blankets
Shoe linings	Ski boots, Golf shoes
Building materials	Second proof, In concrete
Life style apparel	Elegant fleece vests
Other uses	Automotive interiors, Battery warmers

Phase change materials are used both in winter and summer clothing. PCM is used not only in high-quality outerwear and footwear, but also in the underwear, socks, gloves, helmets and bedding of world-wide brand leaders. Seat covers in cars and chairs in offices can consist of phase change materials.

Currently, phase change materials are being used in a variety of outdoor apparel items such as smart jackets, vests, men's and women's hats and rainwear, outdoor active-wear jackets and jacket lining, golf shoes, trekking shoes, ski and snowboard gloves, boots, earmuffs and protective garments. In protective garments, the absorption of body heat surplus, insulation effect caused by heat emission of the PCM into the fibrous structure and thermo-regulating effect, which maintains the microclimate temperature nearly constant are the specified functions of PCM contained smart textile [44].

The addition of PCMs to fabric-backed foam significantly increases the weight, thickness, stiffness, flammability, insulation value, and evaporative resistance value of the material. It is more effective to have one layer of PCM on the outside of a tight-fitting, two layer ensemble than to have it as the inside layer. This may be because the PCMs closest to the body did not change phase [45].

PCM protective garments should improve the comfort of workers as they go through these environmental step changes (e.g., warm to cold to warm, etc.). For these applications, the PCM transition temperature should be set so that the PCMs are in the liquid phase when worn in the warm environment and in the solid phase in the cold environment. The effect of phase change materials in clothing on the physiological and subjective thermal responses of people would probably be maximized if the wearer was repeatedly going through temperature transients (i.e., going back and forth between a warm and cold environment) or intermittently touching hot or cold objects with PCM gloves [45].

One example of practical application of PCM smart textile is cooling vest. This is a comfort garment developed to prevent elevated body temperatures in people who work in hot environments or use extreme physical exertion. The cooling effect is obtained from the vest's 21 PCM elements containing Glauber's salt, which start absorbing heat at a particular temperature (28°C). Heat absorption from the body or from an external source continues until the elements have melted. After use the cooling vest has to be charged at room temperature (24°C) or lower. When all the PCMs are solidified the cooling vest is ready for further use [46].

A new generation of military fabrics feature PCMs which are able to absorb, store and release excess body heat when the body needs it resulting in less sweating and freezing, while the microclimate of the skin is influenced in a positive way and efficiency and performance are enhanced [47].

In the medical textiles field, a blanket with PCM can be useful for gently and controllably reheating hypothermia patients. Also, using PCMs in bed covers regulates the micro climate of the patient [48].

In domestic textiles, blinds and curtains with PCMs can be used for reduction of the heat flux through windows. In the summer month's large amounts of heat penetrate the buildings through windows during the day. At night in the winter months the windows are the main source of thermal loss. Results of the test carried out on curtains containing PCM have indicated a 30% reduction of the heat flux in comparison to curtains without PCM [49].

2.7 SHAPE MEMORY MATERIALS

Shape memory materials are able to 'remember' a shape, and return to it when stimulated, for example, with temperature, magnetic field, electric field, pH-value and UV light. An example of natural shape memory textile material is cotton, which expands when exposed to humidity and shrinks back when dried. Such behavior has not been used for esthetic effects because the changes, though physical, are in general not noticeable to the naked eye. The most common types of such SMMs materials are shape memory alloys and polymers, but ceramics and gels have also been developed. When sensing this material specific stimulus, SMMs can exhibit dramatic deformations in a stress free recovery. On the other hand, if the SMM is prevented from recovering this initial strain, a recovery stress (tensile stress) is induced, and the SMM actuator can perform work. This situation where SMM deforms under load is called restrained recovery [50].

Because of the wide variety of different activation stimuli and the ability to exhibit actuation or some other pre-determined response, SMMs can be used to control or tune many technical parameters in smart material systems in response to environmental changes –such as shape, position, strain, stiffness, natural frequency, damping, friction and water vapor penetration. Both the fundamental theories and engineering aspects of SMMs have been investigated extensively and a rather wide variety of different SMMs are presently commercial materials. Commercialized shape memory products have been based mainly on metallic shape memory alloys (SMAs), either taking advantage of the shape change due the shape memory effect or the super-elasticity of the material, the two main phenomena of SMAs. Shape memory polymers (SMPs) and shape memory gels are developed at a quick rate, and within the last few years also some products based on magnetic shape memory alloys have been commercialized. Shape memory ceramic (SMC) materials, which can be activated not only by temperature but also by elastic energy, electric or magnetic field, are mainly at the research stage [51].

2.7.1 APPLICATIONS OF TEXTILES CONTAINING SMMS

There are many potential applications of shape memory polymers in industrial components like automotive parts, building and construction products, intelligent packing, implantable medical devices, sensors and actuators, etc. SMPs are used in toys, handgrips of spoons, toothbrushes, razors and kitchen knives, also as an automatic choking device in small-size engines. One of the most well-known examples of SMP is a clothing application, a membrane called Diaplex. The membrane is based on polyurethane based shape memory polymers developed by Mitsubishi Heavy Industries [52].

Polyurethane is an example of shape memory polymers, which is based on the formation of a physical cross-linked network as a result of entanglements of the high molecular weight linear chains, and on the transition from the glassy state to the rubber-elastic state. Shape memory polyurethane (SMPU) is a class of polyurethane that is different from conventional polyurethane in that these have a segmented structure and a wide range of glass transition temperature (Tg). The long polymer chains entangle each other and a three-dimensional network is formed. The polymer network keeps the original shape even above Tg in the absence of stress. Under stress, the shape is deformed and the deformed shape is fixed when cooled below Tg. Above the glass transition temperature polymers show rubber-like behavior [53].

The material softens abruptly above the glass transition temperature Tg. If the chains are stretched quickly in this state and the material is rapidly cooled down again below the glass transition temperature the polynorbornene chains can neither slip over each other rapidly enough nor become disentangled. It is possible to freeze the induced elastic stress within the material by rapid cooling. The shape can be changed at will. In the glassy state the strain is frozen and the deformed shape is fixed. The decrease in the mobility of polymer chains in the glassy state maintains the transient shape in polynorbornene. The recovery of the material's original shape can be observed by heating again to a temperature above Tg. This occurs because of the thermally induced shape-memory effect. The disadvantage of this polymer is the difficulty of processing because of its high molecular weight [53, 54]. Some of the shape memory polymers are suitable for textiles applications (Table 2.3).

Shape memory polymers can be laminated, coated, foamed, and even straight converted to fibers. There are many possible end uses of these smart textiles. The smart fiber made from the shape memory polymer can be applied as stents, and screws for holding bones together.

Shape memory polymer coated or laminated materials can improve the thermophysiological comfort of surgical protective garments, bedding and incontinence products because of their temperature adaptive moisture management features.

TABLE 2.3 Some of the Shape Memory Polymers Are Suitable for Textiles Applications

Polymer	Physical interactions	
	Original shape	**Transient shape**
Polynorbomene	Chain entanglement	Glassy state
Polyurethane	Microcrystal	Glassy state
Polyethylene/nylon-6 graft copolymer	Crosslinking	Microcrystal
Styrene-1,4-butadiiene block copolymer	Microcrystal/glassy state of polystyrene	Microcrystal of poly(1,4-butadiene)
Ethylene oxide-ethylene terphethalate block copolymer	Microcrystal of PET	Microcrystal of PEO
Polymethylene-1,3-cyclopentane) polyethylene block copolymer	Microcrystal of PE	Microcrystal/glassy state of pmcp

Films of shape memory polymer can be incorporated in multilayer garments, such as those that are often used in the protective clothing or leisurewear industry. The shape memory polymer reverts within wide range temperatures. This offers great promise for making clothing with adaptable features. Using a composite film of shape memory polymer as an inter-liner in multilayer garments, outdoor clothing could have adaptable thermal insulation and be used as protective clothing. A shape memory polymer membrane and insulation materials keep the wearer warm. Molecular pores open and close in response to air or water temperature to increase or minimize heat loss. Apparel could be made with shape memory fiber. Forming the shape at a high temperature provides creases and pleats in such apparel as slacks and skirts. Other applications include fishing yarn, shirt neck bands, cap edges, casual clothing and sportswear. Also, using a composite film of shape memory polymers as an interlining provides apparel systems with variable tog values to protect against a variety of weather conditions [54].

2.8 CHROMIC MATERIALS

Chromic materials are the general term referring to materials, which their color changes by the external stimulus. Due to color changing properties, chromic materials are also called chameleon materials. This color changing phenomenon is caused by the external stimulus and chromic materials can be classified depending on the external stimulus of induction.

Photochromic, thermochromic, electrochromic, piezochromic, solvatechromic and carsolchromic are chromic materials that change their color by the external stimulus of heat, electricity, pressure, liquid and an electron beam, respectively. Photochromic materials are suitable for sun lens applications. Most photochromic materials are based on organic materials or silver particles. Thermochromic materials change color reversibly with changes in temperature. The liquid crystal type and the molecular rearrangement type are thermochromic systems in textiles. The thermochromic materials can be made as semiconductor compounds, from liquid crystals or metal compounds. The change in color occurs at a pre-determined temperature, which can be varied. Electrochromic materials are capable of changing their optical properties (transmittance and/or reflectance) under applied electric potentials. The variation of the optical properties is caused by insertion/extraction of cations in the electrochromic film. Piezochromism is the phenomenon where crystals undergo a major change of color due to mechanical grinding. The induced color reverts to the original color when the fractured crystals are kept in the dark or dissolved in an organic solvent [55].

Solvatechromism is the phenomenon, where color changes when it makes contact with a liquid, for example, water. Materials that respond to water by changing color are also called hydrochromic and this kind of textile material can be used, e.g., for swimsuits.

2.8.1 APPLICATIONS OF TEXTILES CONTAINING CHROMIC MATERIALS

The majority of applications for chromic materials in the textile sector today are in the fashion and design area, in leisure and sports garments. In workwear and the furnishing sector a variety of studies and investigations are in the process by industrial companies, universities and research centers. Chromic materials are one of the challenging material groups when thinking about future textiles. Color changing textiles are interesting, not only in fashion, where color changing phenomena will exploit for fun all the rainbow colors, but also in useful and significant applications in soldier and weapons camouflage, workwear and in technical and medical textiles. The combination of SMM and thermochromic coating is an interesting area, which produces shape, and color changes of the textile material at the same time [56].

2.9 DESIGNING THE SMART TEXTILE SYSTEMS

Comfort is very important in textiles because stresses lead to increased fatigue. The potential of smart textiles is to measure a number of body parameters such as skin temperature, humidity and conductivity and show the level of comfort through the textile sensors. To keep the comfort of textiles, adequate actuators are needed that can heat, cool, insulate, ventilate and regulate moisture. The use of the smart system should not require any additional effort. The weight of a smart textile system should not reduce operation time of the rescue worker.

Other key issues for the design of a smart textile system are [57]:

- Working conditions-relevant parameters: only relevant information should be provided in order to avoid additional workload; this includes indication of danger and need for help.
- Effective alarm generation: the rescue worker or a responsible person should be informed adequately on what needs to be done.
- System maintenance: it must be possible to treat the suit using usual maintenance procedures.
- Cost must be justified.
- Robustness.
- Energy constraints: energy requirements must be optimized.
- Long range transmission: transmission range must be adjusted to the situation of use.

Fighting a fire in a building is different from fighting one in an open field.

A wearable smart textile system basically comprises following components [36]:

- Sensors to detect body or environmental parameters;
- A data processing unit to collect and process the obtained data;
- An actuator that can give a signal to the wearer;
- An energy supply that enables working of the entire system;

- Interconnections that connect the different components;
- A communication device that establishes a wireless communication link with a nearby base station.

The main layers concerned with smart clothing are the skin layer and two clothing layers. Physically the closest clothing layer for a human user is an underwear layer, which transports perspiration away from the skin area. The function of this layer is to keep the interface between a user and the clothes comfortable and thus improve the overall wearing comfort. The second closest layer is an intermediate clothing layer, which consists of the clothes that are between the underclothes and outdoor clothing. The main purpose of this layer is considered to be an insulation layer for warming up the body. The outermost layer is an outerwear layer, which protects a human against hard weather conditions.

The skin layer is located in close proximity to the skin. In this layer we place components that need direct contact with skin or need to be very close to the skin. Therefore, the layer consists of different user interface devices and physiological measurement sensors. The number of the additional components in underwear is limited owing to the light structure of the clothing.

An inner clothing layer contains intermediate clothing equipped with electronic devices that do not need direct contact with skin and, on the other hand, do not need to be close to the surrounding environment. These components may also be larger in size and heavier in weight compared to components associated with underclothes. It is often beneficial to fasten components to the inner clothing layer, as they can be easily hidden. Surrounding clothes also protect electronic modules against cold, dirt and hard knocks.

Generally, the majority of electronic components can be placed on the inner clothing layer. These components include various sensors, a central processing unit (CPU) and communication equipment. Analogous to ordinary clothing, additional heating to warming up a person in cold weather conditions is also associated with this layer. Thus, the inner layer is the most suitable for batteries and power regulating equipment, which are also sources of heat.

The outer clothing layer contains sensors needed for environment measurements, positioning equipment that may need information from the surrounding environment and numerous other accessories. The physical surroundings of smart clothing components measure the environment and the virtual environment accessed by communication technologies. Soldier and weapons camouflage is possible by using chromic materials in outer layer of smart textiles [58].

2.10 DATA TRANSFER REQUIREMENTS IN SMART TEXTILES

The data transfer requirements can be divided into internal and external. The internal transfer services are divided into local health and security related measurements. Many of the services require or result in external communications between the smart clothing and its environment. Wired data transfer is in many cases a practical and straightforward solution. Thin wires routed through fabric are an inexpensive and high capacity medium for information and power transfer. The embedded wires inside clothing do not affect its appearance. However, wires form inflexible parts of clothing and the detaching and reconnecting of wires decrease user comfort and the usability of clothes. The cold winter environment especially stiffens the plastic shielding of wires. In hard usage and in cold weather conditions, cracking of wires also becomes a problem [59]. The connections between the electrical components placed on different pieces of clothing are another challenge when using wires. During dressing and undressing, connectors should be attached or detached, decreasing the usability of clothing. Connectors should be easily fastened, resulting in the need for new connector technologies.

A potential alternative to plastic shielded wires is to replace them with electrically conductive fibers. Conductive yarns twisted from fibers form a soft cable that naturally integrates in the clothing's structure keeping the system as clothing-like as possible. Fiber yarns provide durable, flexible and washable solutions. Also lightweight optical fibers are used in wearable applications, but their function has been closer to a sensor than a communication medium [60, 61]. The problem of conductive fibers is due to the reliable connections of them. Ordinary wires can be soldered directly to printed circuit boards, but the structure of the fiber yarn is more sensitive to breakage near the solder connections.

Protection materials that prevent the movement of the fiber yarn at the interface of the hard solder and the soft yarn must be used. Optical fibers are commonly used for health monitoring applications and also for lighting purposes.

Low-power wireless connections provide increased flexibility and also enable external data transfer within the personal space. Different existing and emerging WLAN and WPAN types of technologies are general purpose solutions for the external communications, providing both high speed transfer and low costs. For wider area communications and full mobility, cellular data networks are currently the only practical possibility [62].

2.11 OPTICAL FIBERS IN SMART TEXTILES

Optical fibers are currently being used in textile structures for several different applications. Optic sensors are attracting considerable interest for a number of sensing applications [63]. There is great interest in the multiplexed sensing of smart structures and materials, particularly for the real-time evaluation of physical measurements (e.g., temperature, strain) at critical monitoring points. One of the applications of the optical fibers in textile structures is to create flexible textile-based displays based on fabrics made of optical fibers and classic yarns [64]. The screen matrix is created during weaving, using the texture of the fabric. Integrated into the system is a small electronics interface that controls the LEDs that light groups of fibers. Each group provides light to one given area of the matrix. Specific control of the LEDs then enables various patterns to be displayed in a static or dynamic manner. This flexible textile-based displays are very thin size and ultra-light weight. This leads one to believe that such a device could quickly enable innovative solutions for numerous applications. Bending in optical fibers is a major concern since this causes signal attenuation at bending points. Integrating optical fibers into a woven perform requires bending because of the crimping that occurs as a result of weave interlacing. However, standard plastic optical fiber (POF) materials like poly methyl methacrylate, polycarbonate and polystyrene are rather stiff compared to standard textile fibers and therefore their integration into textiles usually leads to stiffen of the woven fabric and the textile touch is getting lost [65]. Alternative fibers with appropriate flexibility and transparency are not commercially available yet.

2.12 CONDUCTIVE MATERIALS IN SMART TEXTILES

Several conductive materials are in use in smart textiles. Conductive textiles include electrically conductive fibers fabrics and articles made from them. Flexible electrically conducting and semi-conducting materials, such as conductive polymers, conductive fibers, threads, yarns, coatings and ink are playing an important role in realizing light-weight, wireless and wearable interactive electronic textiles. Generally, conductive fibers can be divided into two categories such as naturally conductive fibers and treated conductive fibers.

Naturally, conductive fibers can be produced purely from inherently conductive materials, such as metals, metal alloys, carbon sources, and conjugated polymers (ICPs) [66].

However, it is important to point out that nanofibers produced from polymers do not in general show quantum effects usually associated with nanotechnology, with spatially confined matter. The reason simply is first of all that the nature of the electronic states of organic materials including polymer materials that control optical and electronic properties does not resemble the one known for semiconductors or conductors.

Electronic states that are not localized but rather extend throughout the bulk material are characteristics of such nonorganic materials, with the consequence that modifications first of all of the absolute size and secondly of the geometry of a body made from them have strong effects on properties particularly as the sizes approach the few tens to a few nanometers scale. Organic materials, on the other hand, display predominantly localized states for electronic excitations, electronic transport with the states being defined by molecular groups such as chromophore groups or complete molecules. The consequence is that the electronic states are not affected as the dimensions of the element such as a fiber element are reduced down into the nanometer scale [67].

Furthermore, both amorphous polymers and partially crystalline polymers have structures anyway even in the bulk, in macroscopic bodies that are restricted to the nanometer scale. The short-range order of amorphous polymers, as represented by the pair correlation function, does not extend beyond about 2 nm and the thickness of crystalline lamellae is typically in the range of a few nanometers or a few tens of nanometer, respectively.

So, the general conclusion is that the reduction of the diameter of fibers made from polymers or organic materials for that matter will affect neither optical nor electronic properties to a significant degree or the intrinsic structure. This will, of course, be different if we are concerned with fibers composed of metals, metal oxides, semiconductors as accessible via the precursor route. In such cases, one is well within the range of quantum effects and the properties of such fibers have to be discussed along the lines spelt out in textbooks on quantum effects [68].

Now, staying with the subject of polymer fibers the discussion given above should certainly not lead to the conclusion that nanofibers and nonwovens composed of nanofibers do not display unique properties and functions of interest both in the areas of basic science but also technical applications, just the contrary.

Taking the reduction of the fiber diameter into the nm scale as a first example it is readily obvious that the specific surface area increases dramatically as the fiber diameter approaches this range. In fact, it increases with the inverse of the fiber radius.

It is obvious that the specific surface increases from about 0.1 m^2/g for fibers with a thickness of about 50 micrometers (diameter of a human hair) to about 300 m^2/g for fibers with a thickness of 10 nm.

Secondly, the strength of fibers also scales inversely with the fiber diameter, thus increasing also very strongly with decreasing fiber diameter, following the Griffith criteria. The reason is that the strength tends to be controlled by surface flaws, the probability of which will decrease along a unit length of the fiber as the surface area decreases. So, a decrease of the fiber diameter from about 50 micrometer to about 10 nm is expected to increase the strength from about 300 N/mm^2 by a factor of about 1000 and more. Thirdly, the pores of nonwoven membranes composed of nanofibers reach the nm range as the fibers get smaller and smaller. A reduction of the fiber diameter from say about 50 micrometer, for which pores with diameter of about 500 micrometer are expected to about 100 nm will cause the pore diameter to decrease to about 1 micrometer. In a similar way, a further reduction to fiber diameters to about 10 nm will cause the average pores to display pore diameters around 100 nm. This will certainly show up in the selectivity of the filters with respect to solid particles, aerosols, etc. to be discussed further below [67].

The reduction in pore diameter is, furthermore, connected with strong modifications of the dynamics of gases and fluids within the nonwoven, located within or flowing through the nonwoven. Thermal insulations, for instance, in nonwovens containing a gas is controlled for larger pores – larger than the average free path length of the molecules – by just this free path length. However, as the pores get smaller the collision of the molecules with the pore walls – fiber surface for fiber-based nonwovens – takes over the control of thermal insulation with an increase of the thermal insulation that can amount to several orders of magnitude. This aspect will also be discussed below [69].

Finally, the flow of gases or fluids around a fiber changes very strongly in nature as the fiber diameter goes down to the nanometer range with a transition of the flow regime from the conventional one to the so-called Knudsen regime, to be discussed below in more detail.

All these effects are classical ones yet have major impact on nanofiber properties and applications. These examples show that nanofibers and nonwovens composed of them display unique properties of functions already based on classical phenomena. This suggests that such fibers/nonwovens can be used with great benefits in various types of applications. The spectrum of applications that can be envisioned for electrospun nanofibers is extremely broad due to their unique intrinsic structure, surface properties and functions [67].

Highly conductive flexible textiles can be prepared by weaving thin wires of various metals such as brass and aluminum. These textiles have been developed for higher degrees of conductivity.

Metal conductive fibers are very thin filaments with diameters ranging from 1 to 80 μm produced from conductive metals such as ferrous alloys, nickel, stainless steel, titanium, aluminum and copper. Since they are different from polymeric fibers, they may be hard to process and have problems of long-term stability. These highly conductive fibers are expensive, brittle, heavier and lower processability than most textile fibers.

Treated conductive fibers can be produced by the combination of two or more materials, such as non-conductive and conductive materials. This conductive textiles can be produced in various ways, such as by impregnating textile substrates with conductive carbon or metal powders, patterned printing, and so forth. Conducting polymers, such as polyacetylene (PA), polypyrrole (PPy), polythiophene (PTh) and polyaniline (PAn), offer an interesting alternative. Among them, polypyrrole has been widely investigated owing to its easy preparation, good electrical conductivity, good environmental stability in ambient conditions and because it poses few toxicological problems. PPy is formed by the oxidation of pyrrole or substituted pyrrole monomers. Electrical conductivity in PPy involves the movement of positively charged carriers or electrons along polymer chains and the hopping of these carriers between chains. The conductivity of PPy can reach the range 102 S cm^{-1}, which is next only to PA and PAn. With inherently versatile molecular structures, PPys are capable of undergoing many interactions [70, 71].

The conductive fibers obtained through special treatments such as mixing, blending, or coating are also known as conductive polymer composites (CPCs), can have a combination of the electrical and mechanical properties of the treated materials. Fibers containing metal, metal oxides and metal salts are a proper alternative for metal fibers. Polymer fibers may be coated with a conductive layer such as polypyrrole, copper or gold. The conductivity will be maintained as long as the layer is intact and adhering to the fiber. Chemical plating and dispersing metallic particles at a high concentration in a resin are two general methods of coating fibers with conductive metals [72].

The brittleness of PPy has limited the practical applications of it. The processability and mechanical properties of PPy can be improved by incorporating some polymers into PPy [73]. However, the incorporation of a sufficient amount of filler generally causes a significant deterioration in the mechanical properties of the conducting polymer, in order to exceed the percolation threshold of conductivity [74]. Another route to overcoming this deficiency is by coating the conducting polymer on flexible textile substrates to obtain a smooth and uniform electrically conductive coating that is relatively stable and can be easily handled [75].

Thus, PPy-based composites may overcome the deficiency in the mechanical properties of PPy, without adversely affecting the excellent physical properties of the substrate material, such as its mechanical strength and flexibility. The resulting products combine the usefulness of a textile substrate with electrical properties that are similar to metals or semi-conductors.

Due to electron-transport characteristics of Conjugated polymers or ICPs, they are regarded as semi conductors or even sometimes conductors. Due to their high conductivity, lower weight, and environmental stability, they have a very important place in the field of smart and interactive textiles [76].

The conductivity of materials is often affected by several parameters, which may be exploitable mechanisms for use as a sensor. Extension, heating, wetting and absorption of chemical compounds in general may increase or decrease conductivity. Swelling or shrinkage of composite fibers of carbon nanotubes alters the distance between the nanoparticles in the fibers, causing the conductivity to change.

Fibers containing conductive carbon are produced with several methods such as loading the whole fibers with a high concentration of carbon, incorporating the carbon into the core of a sheath–core bicomponent fiber, incorporating the carbon into one component of a side–side or modified side–side bicomponent fiber, suffusing the carbon into the surface of a fiber.

Nanoparticles such as carbon nanotubes can be added to the matrix for achieving conductivity. Semi-conducting metal oxides are often nearly colorless, so their use as conducting elements in fibers has been considered likely to lead to fewer problems with visibility than the use of conducting carbon. The oxide particles can be embedded in surfaces, or incorporated into sheath–core fibers, or react chemically with the material on the surface layer of fibers.

Conductive fibers can also be produced by coating fibers with metal salts such as copper sulfide and copper iodide. Metallic coatings produce highly conductive fibers; however, adhesion and corrosion resistance can present problems. It is also possible to coat and impregnate conventional fibers with conductive polymers, or to produce fibers from conductive polymers alone or in blends with other polymers.

Conductive fibers/yarns can be produced in filament or staple lengths and can be spun with traditional non-conductive fibers to create yarns that possess varying wearable electronics and photonics degrees of conductivity. Also, conductive yarns can be created by wrapping a nonconductive yarn with metallic copper, silver or gold foil and be used to produce electrically conductive textiles.

Conductive threads can be sewn to develop smart electronic textiles. Through processes such as electrodeless plating, evaporative deposition, sputtering, coating with a conductive polymer, filling or loading fibers and carbonizing, a conductive coating can be applied to the surface of fibers, yarns or fabrics. Electrodeless plating produces a uniform conductive coating, but is expensive. Evaporative deposition can produce a wide range of thicknesses of coating for varying levels of conductivity. Sputtering can achieve a uniform coating with good adhesion. Textiles coated with a conductive polymer, such as polyaniline and polypyrrole, are more conductive than metal and have good adhesion, but are difficult to process using conventional methods.

Adding metals to traditional printing inks creates conductive inks that can be printed onto various substrates to create electrically active patterns. The printed circuits on flexible textiles result in improvements in durability, reliability and circuit speeds and in a reduction in the size of the circuits. The printed conductive textiles exhibit good electrical properties after printing and abrading. The inks withstand bending without losing conductivity. However, after 20 washing cycles, the conductivity decreases considerably. Therefore, in order to improve washability, a protective polyurethane layer is put on top of the printed samples, which resulted in the good conductivity of the fabrics, even after washing. Currently, digital printing technologies promote the application of conductive inks on textiles [77].

2.12.1 APPLICATIONS OF CONDUCTIVE SMART TEXTILE

Electrically conductive textiles make it possible to produce interactive electronic textiles. They can be used for communication, entertainment, health care, safety, homeland security, computation, thermal purposes, protective clothing, wearable electronics and fashion. The application of conductive smart textile in combination with electronic advices is very widespread. In location and positioning, they can be used for child monitoring, geriatric monitoring, integrated GPS (global positioning system) monitoring, livestock monitoring, asset tracking, etc.

In infotainment, they can be used for integrated compact disc players, MP3 players, cell phones and pagers, electronic game panels, digital cameras, and video devices, etc.

In health and biophysical monitoring, they can be used for cardiovascular monitoring, monitoring the vital signs of infants, monitoring clinical trials, health and fitness, home healthcare, hospitals, medical centers, assisted-living units, etc.

They can be used for soldiers and personal support of them in the battlefield, space programs, protective textiles and public safety (fire-fighting, law enforcement), automotive, exposure-indicating textiles, etc.

They can be also used to show the environmental response such as color change, density change, heating change, etc. Fashion, gaming, residential interior design, commercial interior design and retail sites are other application of conductive smart textiles.

2.13 SMART FABRICS FOR HEALTH CARE

The continuous monitoring of vital signs of some patients and elderly people is an emerging concept of health care to provide assistance to patients as soon as possible either online or offline. A wearable smart textile can provide continuous remote monitoring of the health status of the patient. Wearable sensing systems will allow the user to perform everyday activities without discomfort. The simultaneous recording of vital signs would allow parameter extrapolation and inter-signal elaboration, contributing to the generation of alert messages and synoptic patient tables. In spite of this, a smart fabric is capable of recording body kinematic maps with no discomfort for several fields of application such as rehabilitation and sports [78].

2.14 ELECTRONIC SMART TEXTILES

The components of an electronic smart textile that provide several functions are sensors unit, network unit, processing unit, actuator unit and power unit. On the smart textile, several of these functions are combined to form services. Providing information, communication or assistance are possible services. Because mobility is now a fundamental aspect of many services and devices, these smart textiles can be used for health applications such as monitoring of vital signs of high-risk patients and elderly people, therapy and rehabilitee, knowledge applications such as instructions and navigation and entertainment applications such as audio and video devices. For communication between the different components of smart textile applications, both wired and wireless technologies are applicable. An applied solution for data transferring is often a compromise based on application requirements, operational environment, available and known technologies, and costs [37].

2.15 ELECTRICAL CIRCUITS IN SMART TEXTILE STRUCTURES

In order to form flexible circuit boards, printing of circuit patterns is carried out on polymeric substrates such as films. Fabric based circuits potentially offer additional benefits of higher flexibility in bending and shear, higher tear resistance, as well as better fatigue resistance in case of repeated deformation. Different processes that have been described in literature for the fabrication of fabric based circuits include embroidery of conductive threads on fabric substrates, weaving and knitting of conductive threads along with nonconductive threads, printing or deposition and chemical patterning of conductive elements on textile substrates.

The insulating fabric could be woven, non-woven, or knitted [79]. The conductive threads can be embroidered in any shape on the insulating fabric irrespective of the constituent yarn path in a fabric. One of the primary disadvantages of embroidery as a means of circuit formation is that it does not allow formation of multi-layered circuits involving conductive threads traversing through different layers as is possible in the case of woven circuits. Conductive threads can be either woven or knitted into a fabric structure along with nonconductive threads to form an electrical circuit. One of the limitations of using weaving for making electrical circuits is that the conductive threads have to be placed at predetermined locations in the warp direction while forming the warp beam or from a creel during set up of the machine. Different kinds of conductive threads can be supplied in the weft or filling direction and inserted using the weft selectors provided on a weaving machine. Some modifications to the yarn supply system of the machine may be needed in order to process the conductive threads that are more rigid [80].

In most conventional weft knitting machines, like a flatbed machine, the conductive threads can be knitted in the fabric only in one direction, that is, the course (or cross) direction. In order to keep the conductive element in a knit structure straight, one can insert a conductive thread in the course direction such that the conductive thread is embedded into the fabric between two courses formed from non-conductive threads.

Processes that have been employed to form a patterned conductive path on fabric surfaces include deposition of polymeric or nonpolymeric conducting materials and subsequent etching, reducing, or physical removal of the conductive materials from certain regions. Thus, the conductive material that is not removed forms a patterned electrical circuit or a region of higher conductivity. The biggest problem associated with patterning of circuits from thin conductive films (polymeric or metallic) deposited on fabric substrates is that use of an etching agent for forming a circuit pattern leads to non-uniform etching, as some of the etching liquid is absorbed by the threads of the underlying substrate fabric [81]. Another problem with deposition of conductive films on fabric substrates is that bending the fabric may lead to discontinuities in conductivity at certain points.

There are different device attachment methods like raised wire connectors, solders, snap connectors, and ribbon cable connectors in electronic smart textiles. Soldering produces reliable electrical connections to conductive threads of an electronic textile fabric but has the disadvantage of not being compatible with several conductive threads or materials like stainless steel. Moreover, soldering of electronic devices to threads that are insulated is a more complex process involving an initial step of removal of insulation from the conductive threads in the regions where the device attachments are desired and insulation of the soldered region after completion of the soldering process. The main advantage of employing snap connectors is the ease of attachment or removal of electronic devices from these connectors, whereas the main disadvantages are the large size of the device and the weak physical connection formed between the snap connectors and the devices. Ribbon cable connectors employ insulation displacement in order to form an interconnection with insulated conductor elements integrated into the textiles. A v-shaped contact cuts through the insulation to form a connection to the conductor. Firstly, the ribbon cable connector is attached to the conductive threads in an e-textile fabric and subsequent electronic devices and printed circuit boards are attached to the ribbon cable connector. One of the advantages of employing ribbon cable connectors for device attachment is the ease of attachment and removal of the electronic devices to form the electronic textiles [82, 83].

2.16 NEW TEXTILE MATERIALS FOR ENVIRONMENTAL PROTECTION

From prehistoric times till now, air pollution from hazardous chemical and biological particles is an essential threat to humans' health. Together with the development of civilization and escalation of the conflicts between nations, the risk of loss of health and even life due to polluted air increases considerably. Therefore the continuous development of the new materials used for protection of human respiratory tracts against hazardous particles is observed. The fibrous materials play a special role in this subject. Davies in his work 'Air Filtration' has presented an interesting review of the earliest literature considering problems connected with filtering polluted air [84].

For centuries, miners have used special clothes to protect nose and mouth against dust. Bernardino amazzini, who lived on the turn of the seventeenth century, in his work 'De morbis artificum' indicated the need for protection of the respiratory tracts against dusts of workers laboring in various professions listed by him. Brise Fradin developed in 1814 the first device, which provided durable protection of the respiratory tracts. It was composed of a container filled with cotton fibers, which was connected by a duct with the user's mouth. The first filtration respiratory mask was designed at the beginning of the nineteenth century with the aim of protecting the users against diseases transmitted by the breathing system. In these times, firemen began to use masks specially designed for them. The first construction of such a 'mask' was primitive: a leather helmet was connected with a hose, which supplied air from the ground level.

The construction was based on the observation that during fire, fewer amounts of toxic substances were at the ground level than at the level of the fireman's mouth. In addition, a layer of fibers protected the lower air inlet. John Tyndall, in 1868 designed a mask, which consisted of some layers of differentiated structure. A clay layer separated the first two layers of dry cotton fibers. Between the two next cotton fiber layers was inserted charcoal, and the last two cotton fiber layers were separated by a layer of wool fibers saturated with glycerin. The history of the development of filtration materials over the nineteenth century has been described in a work elaborated by Feldhaus [85].

The twentieth century left a lasting impression of the First World War, during which toxic gases were used for the first time. This was the reason that after 1914, the further history of the development of filtration materials was connected with absorbers of toxic substances manufactured with the use of charcoal and fibrous materials. The next discovery, which changed the approach to the designing of filtration materials, was done in 1930, Hansen, in his filter applied a mixture of fibers and resin as filtration materials. This caused an electrostatic field being created inside the material. The action of electrostatic forces on dust particles significantly increases the filtration efficiency of the materials manufactured.

The brief historical sketch presented above indicates that textile fibers were one of the material components, which protect the respiratory tracts, and have been applied from the dawn of history. From the beginning they had been used intuitively, without understanding the mechanism of filtration. The first attempts of scientific description of the filtration mechanism were presented by Albrecht [86], Kaufman [87], Langmuir [88], and recently by Brown [6] who characterized the four basic physical phenomena of mechanical deposition in the following way:

- direct interception occurs when a particle follows a streamline and is captured as a result of coming into contact with the fiber;
- inertial impaction is realized when the deposition is effected by the deviation of a particle from the streamline caused by its own inertia; in diffusive deposition the combined action of airflow and Brownian motion brings a particle into contact with the fiber; gravitational settling resulting from gravitation forces.

Illustration of the above mechanisms of filtration is presented in Figure 2.2.

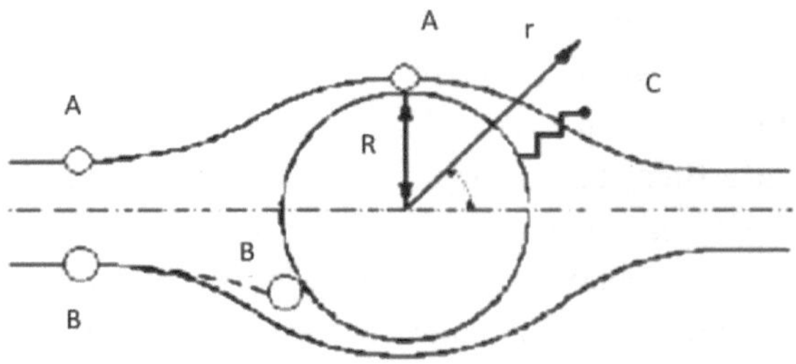

FIGURE 2.2 Particle capture mechanism: A – particle captured by interception; B – particle captured by inertial impaction; C – particle captured by diffusive deposition.

The analysis of equations determining the filter efficiency governed by the mechanisms specified above indicates that the most important parameters deciding about filtering efficiency are the thickness of filters, the diameter of fibers and the porosity of the filter. Identification of these phenomena was the basis for development of new technologies for filtering materials composed from ultra thin fibers. These technologies mainly are based on manufacturing the nonwovens directly from the dissolved or melted polymers using melt-blown technique, flash spinning and electrospinning.

Additionally theoretical consideration also indicates that the activity of fibers on particles significantly increases if an electrostatic field is formed inside the nonwoven. This is the reason that nonwovens are additional modified. Three following groups of fibrous electrostatic materials used can be distinguished, based upon their ability to generate an electrostatic charge:

- materials in which the charge is generated by corona discharge after fiber or web formation,
- materials in which the charge is generated by induction during spinning in an electrostatic field, and
- materials in which the charge is induced as the result of the triboelectric effect.

2.17 CONCLUDING REMARKS

The extremely fine structure and special properties of functional nanofibers make them exceedingly attractive, and possible new uses for the fibers in a wide range of advanced applications are being explored continuously. Smart textiles are considered to be a new niche for products with great potentials on the textile and apparel market. Technology and material solutions incorporate fiber, textile and electronics manufacturing technology with application of materials with electrical, chemical, mechanical, thermal and optical reaction. Generally, smart textiles are referred as textile products with additional value, they have the common properties of textiles, but insure additional functions, providing attractive solutions for a wide range of application fields. Functionalization of textiles may be processed at different levels, from fibers till fabric or even ready-made clothing.

Understanding of drivers, state-of-the-art and tendencies in smart textiles ensures further efficient development technology and its interaction with manufactures and consumers. The vision of Smart Textile is to create textile products that interact by combining smart materials and integrated computing power into textile applications. The introduction of smart materials and computing technology in textile structures offers an opportunity to develop textiles with a new type of behavior and functionality. Besides behavior like sense, react on and conducting electricity, the textile will be able to perform computational operations.

KEYWORDS

- **nanoparticles properties**
- **nanotechnology**
- **smart fabrics**
- **textile finishing**

REFERENCES

1. Soane, D. S., et al., *Nanoparticle-Based Permanent Treatments for Textiles*, 2003, Google Patents.
2. Song, X. Q., et al., *The Effect of Nano-Particle Concentration and Heating Time in the Anti-Crinkle Treatment of Silk.* Journal of Jilin Institute of Technology, 2001, 22, 24–27.
3. Russell, E., *Nanotechnologies and the Shrinking World of Textiles.* Textile Horizons, 2002, 9(10), 7–9.

4. Wong, Y. W. H., et al., *Selected Applications of Nanotechnology in Textiles.* AUTEX Research Journal, 2006, 6(1), 1–8.
5. Dastjerdi, R. and M. Montazer, *A Review on the Application of Inorganic Nano-Structured Materials in the Modification of Textiles: Focus on Anti-Microbial Properties.* Colloids and Surfaces B: Biointerfaces, 2010, 79(1), 5–18.
6. Dastjerdi, R., M. Montazer, and S. Shahsavan, *A New Method to Stabilize Nanoparticles on Textile Surfaces.* Colloids and Surfaces A: Physicochemical and Engineering Aspects, 2009, 345(1–3), 202–210.
7. Nazari, A., M. Montazer, and M. B. Moghadam, *Introducing Covalent and Ionic Cross-linking into Cotton through Polycarboxylic Acids and Nano TiO2.* Journal of The Textile Institute, 2012, 103(9), 985–996.
8. Gorjanc, M., et al., *The Influence of Water Vapor Plasma Treatment on Specific Properties of Bleached and Mercerized Cotton Fabric.* Textile Research Journal, 2010, 80(6), 557–567.
9. Mihailović, D., et al., *Functionalization of Cotton Fabrics with Corona/Air RF Plasma and Colloidal TiO2 Nanoparticles.* Cellulose, 2011, 18(3), 811–825.
10. Karimi, L., et al., *Effect of Nano TiO2 on Self - cleaning Property of Cross - linking Cotton Fabric with Succinic Acid Under UV Irradiation.* Photochemistry and photobiology, 2010, 86(5), 1030–1037.
11. Yuranova, T., et al., *Antibacterial Textiles Prepared by RF-Plasma and Vacuum-UV Mediated Deposition of Silver.* Journal of Photochemistry and Photobiology A: Chemistry, 2003, 161(1), 27–34.
12. M. Quaid, M. and P. Beesley, *Extreme Textiles: Designing for High Performance.* 2005, Princeton Architectural Press.
13. Mahltig, B., H. Haufe, and H. Böttcher, *Functionalization of Textiles by Inorganic Sol–Gel Coatings.* Journal of Materials Chemistry, 2005, 15(41), 4385–4398.
14. Kang, J. Y. and M. Sarmadi, *Textile Plasma Treatment Review–Natural Polymer-Based Textiles.* American Association of Textile Chemists and Colorists Review, 2004, 4(10), 28–32.
15. Patra, J. K. and S. Gouda, *Application of Nanotechnology in Textile Engineering: An Overview.* Journal of Engineering and Technology Research, 2013, 5(5), 104–111.
16. Rahman, M., et al., *Tool-Based Nanofinishing and Micromachining.* Journal of Materials Processing Technology, 2007, 185(1), 2–16.
17. Wang, R. H., J. H. Xin, and X. M. Tao, *UV-Blocking Property of Dumbbell-Shaped ZnO Crystallites on Cotton Fabrics.* Inorganic Chemistry, 2005, 44(11), 3926–3930.
18. Lee, H. J., S. Y. Yeo, and S. H. Jeong, *Antibacterial Effect of Nanosized Silver Colloidal Solution on Textile Fabrics.* Journal of Materials Science, 2003, 38(10), 2199–2204.
19. Sennett, M., et al., *Dispersion and Alignment of Carbon Nanotubes in Polycarbonate.* Applied Physics A, 2003, 76(1), 111–113.
20. Weiguo, D., *Research on Properties of Nano Polypropylene/TiO_2 Composite Fiber.* Journal of Textile Research, 2002, 23(1), 22–23.
21. Qian, L. and J. P. Hinestroza, *Application of Nanotechnology for High Performance Textiles.* Journal of Textile and Apparel, Technology and Management, 2004, 4(1), 1–7.
22. Joshi, M. and A. Bhattacharyya, *Nanotechnology–A New Route to High-Performance Functional Textiles.* Textile Progress, 2011, 43(3), 155–233.
23. Meier, U., *Carbon Fiber-Reinforced Polymers: Modern Materials in Bridge Engineering.* Structural Engineering International, 1992, 2(1), 7–12.

24. Huang, Z. M., et al., *A Review on Polymer Nanofibers by Electrospinning and their Applications in Nanocomposites.* Composites Science and Technology, 2003, 63(15), 2223–2253.
25. Wang, C. C. and C. C. Chen, *Physical Properties of the Cross-linked Cellulose Catalyzed with Nanotitanium Dioxide under UV Irradiation and Electronic Field.* Applied Catalysis A: General, 2005, 293, 171–179.
26. Daoud, W. A. and J. H. Xin, *Low Temperature Sol-Gel Processed Photocatalytic Titania Coating.* Journal of Sol-Gel Science and Technology, 2004, 29(1), 25–29.
27. Hartley, S. M., et al. *The next Generation of Chemical and Biological Protective Materials Utilizing Reactive Nanoparticles.* in *24th Army Science Conference.* 2004, Orlando, Florida, USA.
28. Wang, R. H., et al., *ZnO Nanorods Grown on Cotton Fabrics at Low Temperature.* Chemical Physics Letters, 2004, 398(1), 250–255.
29. Zhang, J., et al., *Hydrophobic Cotton Fabric Coated by a Thin Nanoparticulate Plasma Film.* Journal of Applied Polymer Science, 2003, 88(6), 1473–1481.
30. Yang, H., S. Zhu, and N. Pan, *Studying the Mechanisms of Titanium Dioxide as Ultraviolet-Blocking Additive for Films and Fabrics by an Improved Scheme.* Journal of Applied Polymer Science, 2004, 92(5), 3201–3210.
31. El-Molla, M. M., et al., *Nanotechnology to Improve Coloration and Antimicrobial Properties of Silk Fabrics.* Indian Journal of Fiber and Textile Research, 2011, 36(3), 266–271.
32. Kathiervelu, S. S., *Applications of Nanotechnology in Fiber Finishing.* Synthetic Fibers, 2003, 32(4), 20–22.
33. Mondal, S., *Phase Change Materials for Smart Textiles–An Overview.* Applied Thermal Engineering, 2008, 28(11), 1536–1550.
34. Hu, J., *Advances in Shape Memory Polymers.* 2013, Elsevier.
35. Spencer, B. F., M. E. R. Sandoval, and N. Kurata, *Smart Sensing Technology: Opportunities and Challenges.* Structural Control and Health Monitoring, 2004, 11(4), 349–368.
36. Diamond, D., et al., *Wireless Sensor Networks and Chemo-/Biosensing.* Chemical Reviews, 2008, 108(2), 652–679.
37. Tao, X. M., *Wearable Electronics and Photonics.* 2005, Elsevier.
38. Koncar, V., *Optical Fiber Fabric Displays.* Optics and Photonics News, 2005, 16(4), 40–44.
39. Sharma, A., et al., *Review on Thermal Energy Storage with Phase Change Materials and Applications.* Renewable and Sustainable Energy Reviews, 2009, 13(2), 318–345.
40. Zalba, B., et al., *Review on Thermal Energy Storage with Phase Change: Materials, Heat Transfer Analysis and Applications.* Applied Thermal Engineering, 2003, 23(3), 251–283.
41. Shin, Y., D. I. Yoo, and K. Son, *Development of Thermoregulating Textile Materials with Microencapsulated Phase Change Materials (PCM). IV. Performance Properties and Hand of Fabrics Treated with PCM Microcapsules.* Journal of Applied Polymer Science, 2005, 97(3), 910–915.
42. B. García, L., et al., *Phase Change Materials (PCM) Microcapsules with Different Shell Compositions: Preparation, Characterization and Thermal Stability.* Solar Energy Materials and Solar Cells, 2010, 94(7), 1235–1240.
43. Tyagi, V. V., et al., *Development of Phase Change Materials Based Microencapsulated Technology for Buildings: A Review.* Renewable and Sustainable Energy Reviews, 2011, 15(2), 1373–1391.

44. Nelson, G., *Application of Microencapsulation in Textiles.* International Journal of Pharmaceutics, 2002, 242(1), 55–62.
45. Shim, H., E. A. M. Cullough, and B. W. Jones, *Using Phase Change Materials in Clothing.* Textile Research Journal, 2001, 71(6), 495–502.
46. Gao, C., K. Kuklane, and I. Holmer, *Cooling Vests with Phase Change Material Packs: The Effects of Temperature Gradient, Mass and Covering Area.* Ergonomics, 2010, 53(5), 716–723.
47. Tang, S. L. P. and G. K. Stylios, *An Overview of Smart Technologies for Clothing Design and Engineering.* International Journal of Clothing Science and Technology, 2006, 18(2), 108–128.
48. Buckley, T. M., *Flexible Composite Material with Phase Change Thermal Storage*, 1999, Google Patents.
49. Soares, N., et al., *Review of Passive PCM Latent Heat Thermal Energy Storage Systems towards Buildings' Energy Efficiency.* Energy and Buildings, 2013, 59, 82–103.
50. Park, J. Y., et al., *Growth kinetics of nanograins in SnO_2 fibers and size dependent sensing properties.* Sensors and Actuators B: Chemical, 2011, 152(2), 254–260.
51. Cho, C. G., *Shape Memory Material*, in *Smart Clothing.* 2010, 189–221.
52. Wang, M., X. Luo, and D. Ma, *Dynamic Mechanical Behavior in the Ethylene Terephthalate-Ethylene Oxide Copolymer with Long Soft Segment as a Shape Memory Material.* European Polymer Journal, 1998, 34(1), 1–5.
53. Mother, P. T., H. G. Jeon, and T. S. Haddad, *Strain Recovery in POSS Hybrid Thermoplastics.* Polymer Preprints (USA), 2000, 41(1), 528–529.
54. Otsuka, K. and C. M. Wayman, *Shape Memory Materials.* 1999, Cambridge University Press.
55. B. Laurent, H. and H. Dürr, *Organic Photochromism (IUPAC Technical Report).* Pure and Applied Chemistry, 2001, 73(4), 639–665.
56. Mattila, H. R., *Intelligent Textiles and Clothing.* Vol. 3. 2006, CRC press England.
57. Kiekens, P. and S. Jayaraman, *Intelligent Textiles and Clothing for Ballistic and NBC Protection.* 2011, Springer.
58. Tao, X., *Smart Fibers, Fabrics and Clothing.* Vol. 20. 2001, Woodhead publishing.
59. Rantanen, J., et al., *Smart Clothing Prototype for the Arctic Environment.* Personal and Ubiquitous Computing, 2002, 6(1), 3–16.
60. Lind, E. J., et al. *A Sensate Liner for Personnel Monitoring Applications.* in *First International Symposium on Wearable Computers.* 1997, IEEE.
61. Lee, K. and D. S. Kwon. *Wearable Master Device Using Optical Fiber Curvature Sensors for the Disabled.* in *IEEE International Conference on Robotics and Automation.* 2001, IEEE.
62. Thomas, H. L., *Multi-Structure Ballistic Material*, 1998, Google Patents.
63. Rao, Y. J., *Fiber Bragg Grating Sensors: Principles and Applications*, in *Optical Fiber Sensor Technology.* 1998, Springer. p. 355–379.
64. Deflin, E. and V. Koncar. *For Communicating clothing: The Flexible Display of Glass fiber Fabrics is Reality.* in *2nd International Avantex Symposium.* 2002.
65. Rothmaier, M., M. Luong, and F. Clemens, *Textile Pressure Sensor Made of Flexible Plastic Optical Fibers.* Sensors, 2008, 8(7), 4318–4329.
66. Marchini, F., *Advanced Applications of Metallized Fibers for Electrostatic Discharge and Radiation Shielding.* Journal of Industrial Textiles, 1991, 20(3), 153–166.

67. Wendorff, J. H., S. Agarwal, and A. Greiner, *Electrospinning: Materials, Processing, and Applications*. 2012, John Wiley & Sons.
68. Brotin, T., et al., *[n]-Polyenovanillins (n= 1–6) as New Push-Pull Polyenes for Nonlinear Optics: Synthesis, Structural Studies, and Experimental and Theoretical Investigation of Their Spectroscopic Properties, Electronic Structures, and Quadratic Hyperpolarizabilities.* Chemistry of materials, 1996, 8(4), 890–906.
69. C. Jr, P. H., *Nonwoven Thermal Insulating Stretch Fabric and Method for Producing Same*, 1985, Google Patents.
70. Omastová, M., et al., *Synthesis, Electrical Properties and Stability of Polypyrrole - Containing Conducting Polymer Composites.* Polymer International, 1997, 43(2), 109–116.
71. Thieblemont, J. C., et al., *Thermal Analysis of Polypyrrole Oxidation in Air.* Polymer, 1995, 36(8), 1605–1610.
72. Bashir, T., *Conjugated Polymer-based Conductive Fibers for Smart Textile Applications*. 2013, Chalmers University of Technology.
73. Ruckenstein, E. and J. H. Chen, *Polypyrrole Conductive Composites Prepared by Coprecipitation.* Polymer, 1991, 32(7), 1230–1235.
74. Chen, Y., et al., *Morphological and Mechanical Behavior of an in Situ Polymerized Polypyrrole/Nylon 66 Composite Film.* Polymer Communications, 1991, 32(6), 189–192.
75. Gregory, R. V., W. C. Kimbrell, and H. H. Kuhn, *Electrically Conductive Non-Metallic Textile Coatings.* Journal of Industrial Textiles, 1991, 20(3), 167–175.
76. Batchelder, D. N., *Color and Chromism of Conjugated Polymers.* Contemporary Physics, 1988, 29(1), 3–31.
77. Kazani, I., et al., *Electrical Conductive Textiles Obtained by Screen Printing.* Fibers & Textiles in Eastern Europe, 2012, 20(1), 57–63.
78. Pacelli, M., et al., *Sensing Threads and Fabrics for Monitoring Body Kinematic and Vital Signs*, in *Fibers and Textiles for the Future*2001. p. 55–63.
79. Post, E. R., et al., *Electrically Active Textiles and Articles Made Therefrom*, 2001, Google Patents.
80. Jachimowicz, K. E. and M. S. Lebby, *Textile Fabric with Integrated Electrically Conductive Fibers and Clothing Fabricated Thereof*, 1999, Google Patents.
81. Marculescu, D., et al., *Electronic Textiles: A Platform for Pervasive Computing.* Proceedings of the IEEE, 2003, 91(12), 1995–2018.
82. Child, A. D. and A. R. D. Angelis, *Patterned Conductive Textiles*, 1999, Google Patents.
83. K. Jr, W. C. and H. H. Kuhn, *Electrically Conductive Textile Materials and Method for Making Same*, 1990, Google Patents.
84. Davies, C. N., *Air Filtration.* 1973, New York: Academic Press.
85. Kruiēska, I., E. Klata, and M. Chrzanowski, *New Textile Materials for Environmental Protection*, in *Intelligent Textiles for Personal Protection and Safety*. 2006, 41–53.
86. Albrecht, F., *Theoretische Untersuchungen über die Ablagerung von Staub aus strömender Luft und ihre Anwendung auf die Theorie der Staubfilter.* Physikalische Zeitschrift, 1931, 23, 48–56.
87. Walkenhorst, W., *Physikalische Eigenschaften von Stäuben sowie Grundlagen der Staubmessung und Staubbekämpfung*, in *Pneumokoniosen*. 1976, Springer, 11–70.
88. Langmuir, I., *Report on Smokes and Filters*, in *Section I,* 1942, US Office of Scientific Research and Development.

CHAPTER 3

NANOFIBER MEMBRANES: A PRACTICAL GUIDE

A. AFZALI and SH. MAGHSOODLOU

University of Guilan, Rasht, Iran

CONTENTS

ABSTRACT

Membrane filtration can be a very efficient and economical way of separating components that are suspended or dissolved in a liquid. The membrane is a physical barrier that allows certain compounds to pass through, depending on their physical and/or chemical properties. Polymeric membrane materials are intrinsically limited by a tradeoff between their permeability and their selectivity. One approach to increase the selectivity is to include dispersions of inorganic nanoparticles, such as zeolites, carbon molecular sieves, or carbon nanotubes, into the polymeric membranes – these membranes are classified as mixed-matrix membranes.

3.1 REVIEW OF TECHNIQUES FOR MANUFACTURING FIBROUS FILTERING MATERIALS

Current environmental problems are caused by human activities in the last 150 years. They are having serious negative impacts on us and this is likely to continue for a very long time. Effective solutions are urgently needed to protect our environment. Due to their high specific surface area, electrospun nanofibers are expected to be used to collect pollutants via physical blocking or chemical adsorption.

The first nonwovens using melted organic polymers were manufactured in the 1950s, using a method similar to air-blowing of the polymer melt. Application of this latter method enabled super-thin fibers to be obtained with a diameter smaller than 5 nm. The melt-blown technique of manufacturing nonwovens from super-fine fibers was developed by Wente at the Naval Research Laboratory in USA, [1]. Buntin, a worker at Exxon Research and Engineering introduced the melt-blown technique for processing PP into the industry [2]. Recently, the Nonwoven Technologies Inc., USA has announced the possibility of manufacturing melt-blown PP nonwovens composed from nano-size fibers of a diameter equal to 300 nm. To enhance the filtration efficiency, the melt -blown nonwovens are subjected to the process of activation, mainly using the corona discharge method [3].

An overview of flash-spinning technologies is presented by Wehman. Flashspun nonwovens made from fibers with very low linear density, which can be obtained using splittable fibers as a raw material for production of conventional webs [4].

Subsequently, webs can be subjected to the classical needle punching or spunlace process during which sacrificial polymer is removed and fibers of low linear density are obtained. The flash-spinning process can be also accomplished using such bicomponent melt-blown technology, which is based on spinning two incompatible polymers together and forming a web, which is then subjected to the splitting process.

Induction of electric charges is another mechanism used in filtering material technology. Induction consists of electric charge generation in a conductor placed in an electric field. Therefore, fine-fibers made from conductive solutions or melts, charged during electrostatic extrusion, belong to this group. Formation of nanofibers by the electrospinning method results from the reaction of a polymer solution drop subjected to an external electric field. This method enables manufacturing fibers with transversal dimensions of nanometers. Gilbert in 1600 made the first observation concerning the behavior of an electroconductive fluid under the action of an electrical field. He pointed out that a spherical drop of water on a dry surface is drowning up, taking the shape of a cone when a piece of rubbed amber is held above it. One of the first investigations into the phenomena of interaction of an electric field with a fluid drop was carried out by Zeleny [5]. He used the apparatus presented in Figure 3.1 in his experiment.

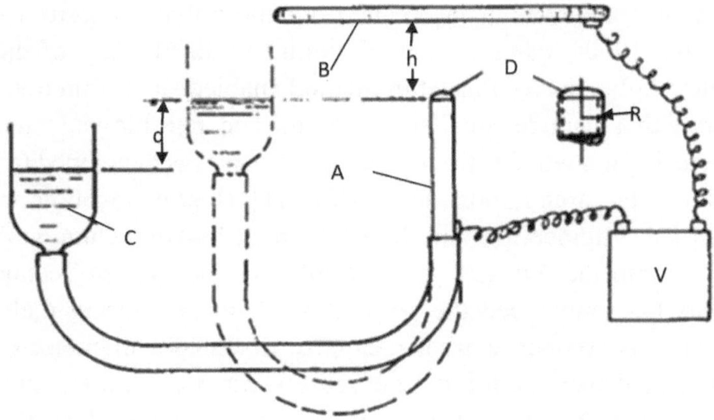

FIGURE 3.1 Scheme showing the idea of an one-plate apparatus for electrospinning.

The apparatus include an open-end capillary tube of metal or glass. The conductive fluid is delivered to this tube using the reservoir C. A plate B is mounted opposite to the capillary tube in a distance of h. The capillary tube and the plate are maintained at a given potential difference V using a high voltage generator. Formhals [6] used this kind of technology for spinning thin polymer filaments.

The electric charges, which diffuse in the liquid, forced by the electric field, cause a strong deformation of the liquid surface in order to minimize the system's total energy. The electric forces exceed the forces of surface tension in the regions of the maximum field strength and charge density, and the liquid forms a cone at the nozzle outlet. A thin stream of liquid particles is torn off from the end of the cone. Taylor [7] proved that for a given type of fluid, a critical value of the applied voltage exists, at which the drop of fluid, flowing from the capillary tube, is transformed into a cone under the influence of the electric field, and loss its stability. The critical value of this potential depends on the surface tension of the fluid and of the initial radius of drop. Zeleny's and Taylor's investigations have been an inspiration for many researchers who carried out observations of the behavior of different kinds of polymers in the electric field. These observations were the basis for the development of manufacturing technologies for a new generation of fibers with very small transversal dimensions. Schmidt demonstrated the possibility of application of electrospun polycarbonate fibers to enhance the dust filtration efficiency [8].

In the 1980s, the Carl Freudenberg Company used the electrospinning technology first commercially. Trouilhet and Weghmann presented a wide range of applications of electrospun webs especially in the filtration area. In that time the electrospinning method for manufacturing filtering materials did not find common application. The revival of this technology has been observed for the last 4 to 5 years. In 2000 Donaldson Inc., USA realized dust filters with a thin layer of nanofibers [9, 10].

A basic set for electrospinning consists form three major components, such as: a high voltage generator, a metal or glass capillary tube, and a collecting plate electrode, similar to the set designed by Zeleny. Such type of set is characterized by low productivity, usually less than 1 mLh^{-1}. To solve this problem, the array of multiply capillary tubes should be developed. Experiments carried out indicate that due to the interference between the electrical fields developed around such system an uniform electrical field strength cannot be ensured at the tip of each tube. For such a system, high probability of the tube clogging appears. To avoid such problems during the electrospinning process, some authors proposed to spin the fibers directly form the polymer solution surface. A new method with high productivity was developed by Jirsak at the Technical University of Liberec [11]. The proposed invention was commercialized by Elmarco company. The idea is very simple. The set is composed from two electrodes. The bottom electrode formed in the shape of a roll is immersed in the solution of a polymer, as shown in Figure 3.2.

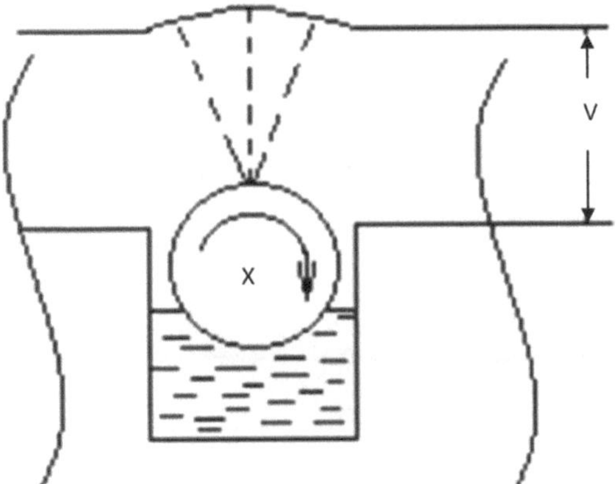

FIGURE 3.2 The idea of the electrospinning method developed by Jirsak.

A thin layer of polymer solution covers the rotating electrode, and multiple jets are formed due to the action of the electrical field. The Elmarco company offers a wide assortment of spun-bonded nonwovens covered by nanofiber membranes made of polyamide, polyurethane and polyvinyl alcohol. A further approach related to spinning directly from the solution surface was invented by Yarin and Zussman [12]. The proposed system is composed from two layers: a bottom layer in the form of ferro-magnetic suspension and an upper layer in the form of polymer solution. The two- layer system is subjected to the magnetic field provided by a permanent magnet. The scheme of this apparatus is presented in Figure 3.3.

Vertical spikes of magnetic suspension appear as the result of action of the magnetic field, what causes the perturbation of the free surface of the polymer solution. Under the action of the electrical field, perturbations of the free surface become the sites of jetting directed upward.

3.1.1 BASIC RESEARCH ON ELECTROSPUN NANOFIBERS IN FILTRATION

Usually, the particle filtration occurs via multiple collection mechanism such as sieving, direct interception, inertial impaction, diffusion, and electrostatic collection. In practice, sieving is not an important mechanism

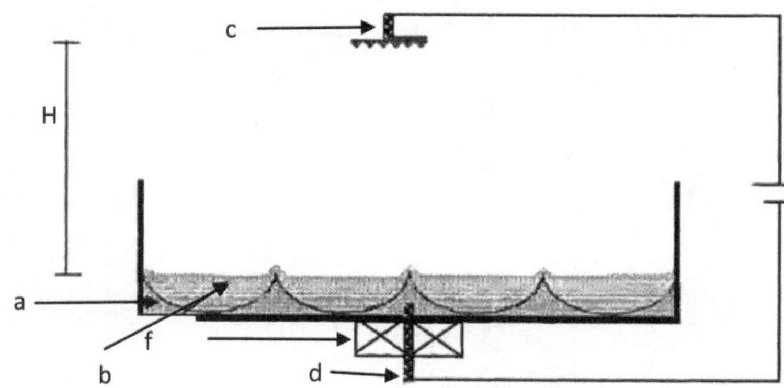

FIGURE 3.3 The idea of manufacturing electrospun nonwovens directly from the surface of a polymer solution: (a) ferromagnetic suspension, (b) polymer solution, (c) upper electrode, (d) lower electrode, (e) high voltage generator, and (f) permanent magnet.

in most air filtration application. Additionally, commercial nanofibers are electrically neutral. So, the remaining important mechanisms in mechanical filtration are direct interception, inertial impaction, and diffusion. The reasonable approximations of filtering media performance have been made using single-fiber filtration theory.

3.1.2 MEMBRANES FILTRATION

Membrane filtration is a mechanical filtration technique, which uses an absolute barrier to the passage of particulate material as any technology currently available in water treatment. The term "membrane" covers a wide range of processes, including those used for gas/gas, gas/liquid, liquid/liquid, gas/solid, and liquid/solid separations. Membrane production is a large-scale operation. There are two basic types of filters: depth filters and membrane filters.

Depth filters have a significant physical depth and the particles to be maintained are captured throughout the depth of the filter. Depth filters often have a flexuous three-dimensional structure, with multiple channels and heavy branching so that there is a large pathway through which the liquid must flow and by which the filter can retain particles. Depth filters have the advantages of low cost, high throughput, large particle retention capacity, and the ability to retain a variety of particle sizes. However, they can endure from entrainment of the filter medium, uncertainty regarding effective pore size, some ambiguity regarding the overall integrity of the filter, and the risk of particles being mobilized when the pressure differential across the filter is large.

The second type of filter is the membrane filter, in which depth is not considered momentous. The membrane filter uses a relatively thin material with a well-defined maximum pore size and the particle retaining effect takes place almost entirely at the surface. Membranes offer the advantage of having well-defined effective pore sizes, can be integrity tested more easily than depth filters, and can achieve more filtration of much smaller particles. They tend to be more expensive than depth filters and usually cannot achieve the throughput of a depth filter. Filtration technology has developed a well defined terminology that has been well addressed by commercial suppliers.

The term membrane has been defined in a number of ways. The most appealing definitions to us are the following:

"A selective separation barrier for one or several components in solution or suspension."

A thin layer of material that is capable of separating materials as a function of their physical and chemical properties when a driving force is applied across the membrane.

Membranes are important materials, which form part of our daily lives. Their long history and use in biological systems has been extensively studied throughout the scientific field. Membranes have proven themselves as promising separation candidates due to advantages offered by their high stability, efficiency, low energy requirement and ease of operation. Membranes with good thermal and mechanical stability combined with good solvent resistance are important for industrial processes [13].

The concept of membrane processes is relatively simple but nevertheless often unknown. Membranes might be described as conventional filters but with much finer mesh or much smaller pores to enable the separation of tiny particles, even molecules. In general, one can divide membranes into two groups: porous and nonporous. The former group is similar to classical filtration with pressure as the driving force; the separation of a mixture is achieved by the rejection of at least one component by the membrane and passing of the other components through the membrane (see Figure 3.4). However, it is important to note that nonporous membranes do not operate on a size exclusion mechanism.

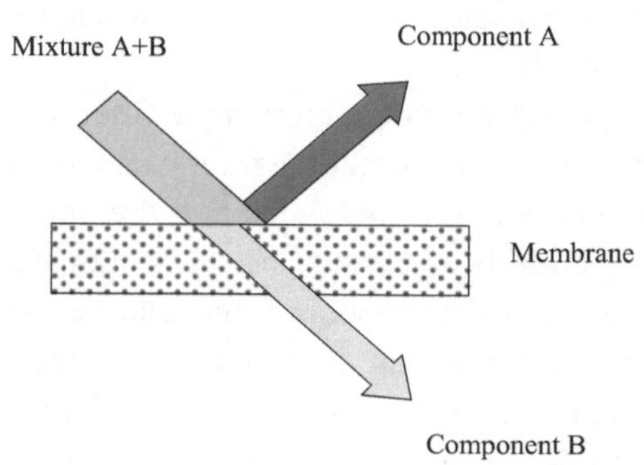

FIGURE 3.4 Basic principle of porous membrane processes.

Membrane separation processes can be used for a wide range of applications and can often offer significant advantages over conventional separation such as distillation and adsorption since the separation is based on a physical mechanism. Compared to conventional processes, therefore, no chemical, biological, or thermal change of the component is involved for most membrane processes. Hence membrane separation is particularly attractive to the processing of food, beverage, and bioproducts where the processed products can be sensitive to temperature (vs. distillation) and solvents (vs. extraction).

Synthetic membranes show a large variety in their structural forms. The material used in their production determines their function and their driving forces. Typically the driving force is pressure across the membrane barrier (see Table 3.1) [14–16]. Formation of a pressure gradient across the membrane allows separation in a bolter-like manner. Some other forms of separation that exist include charge effects and solution diffusion. In this separation, the smaller particles are allowed to pass through as permeates whereas the larger molecules (macromolecules) are retained. The retention or permeation of these species is ordained by the pore architecture as well as pore sizes of the membrane employed. Therefore based on the pore sizes, these pressure driven membranes can be divided into reverse osmosis (RO), nanofiltration (NF), ultrafiltration (UF), and microfiltration (MF), are already applied on an industrial scale to food and bioproduct processing [17–19].

A. Microfiltration (MF) Membranes

MF membranes have the largest pore sizes and thus use less pressure. They involve removing chemical and biological species with diameters

TABLE 3.1 Driving Forces and Their Membrane Processes

Driving force	Membrane process
Pressure difference	Microfiltration, Ultrafiltration, Nanofiltration, Reverse osmosis
Chemical potential difference	Pervaporation, Pertraction, Dialysis, Gas separation, Vapor permeation, Liquid membranes
Electrical potential difference	Electrodialysis, Membrane electrophoresis, Membrane electrolysis
Temperature difference	Membrane distillation

ranging between 100 and 10,000 nm and components smaller than this, pass through as permeates. MF is primarily used to separate particles and bacteria from other smaller solutes [16].

B. Ultrafiltration (UF) Membranes

UF membranes operate within the parameters of the micro- and nano-filtration membranes. Therefore UF membranes have smaller pores as compared to MF membranes. They involve retaining macromolecules and colloids from solution, which range between 2–100 nm and operating pressures between 1 and 10 bar. e.g. large organic molecules and proteins. UF is used to separate colloids such as proteins from small molecules such as sugars and salts[16].

C. Nanofiltration (NF) Membranes

NF membranes are distinguished by their pore sizes of between 0.5–2 nm and operating pressures between 5 and 40 bar. They are mainly used for the removal of small organic molecules and di- and multivalent ions. Additionally, NF membranes have surface charges that make them suitable for retaining ionic pollutants from solution. NF is used to achieve separation between sugars, other organic molecules, and multivalent salts on the one hand from monovalent salts and water on the other. Nanofiltration, however, does not remove dissolved compounds [16].

D. Reverse Osmosis (RO) Membranes

RO membranes are dense semi-permeable membranes mainly used for desalination of seawater [38]. Contrary to MF and UF membranes, RO membranes have no distinct pores. As a result, high pressures are applied to increase the permeability of the membranes [16]. The properties of the various types of membranes are summarized in Table 3.2.

The NF membrane is a type of pressure-driven membrane with properties in between RO and UF membranes. NF offers several advantages such as low operation pressure, high flux, high retention of multivalent anion salts and an organic molecular above 300, relatively low investment and low operation and maintenance costs. Because of these advantages, the applications of NF worldwide have increased [20]. In recent times, research in the application of nanofiltration techniques has been extended from separation of aqueous solutions to separation of organic solvents to homogeneous catalysis, separation of ionic liquids, food processing, etc. [21].

TABLE 3.2 Summary of Properties of Pressure Driven Membranes

	MF	UF	NF	RO
Permeability (L/h.m^2.bar)	1000	10–1000	1.5–30	0.05–1.5
Pressure (bar)	0.1–2	0.1–5	3–20	5–1120
Pore size (nm)	100–10,000	2–100	0.5–2	<0.5
Separation mechanism	Sieving	Sieving	Sieving, charge effects	Solution diffusion
Applications	Removal of bacteria	Removal of bacteria, fungi, viruses	Removal of multivalent ions	Desalination

Figure 3.5 presents a classification on the applicability of different membrane separation processes based on particle or molecular sizes. RO process is often used for desalination and pure water production, but it is the UF and MF that are widely used in food and bioprocessing.

While MF membranes target on the microorganism removal, and hence are given the absolute rating, namely, the diameter of the largest pore on the membrane surface, UF/NF membranes are characterized by the nominal rating due to their early applications of purifying biological solutions. The nominal rating is defined as the molecular weight cut-off (MWCO) that is the smallest molecular weight of species, of which the membrane has more than 90% rejection (see later for definitions). The separation mechanism in MF/UF/NF is mainly the size exclusion, which is indicated

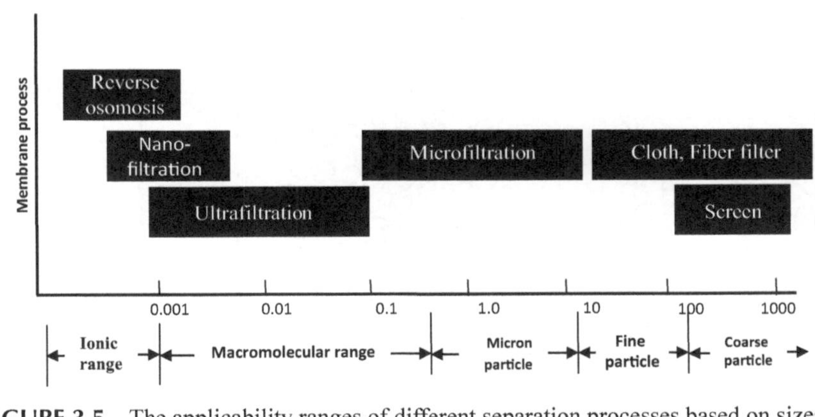

FIGURE 3.5 The applicability ranges of different separation processes based on sizes.

in the nominal ratings of the membranes. The other separation mechanism includes the electrostatic interactions between solutes and membranes, which depends on the surface and physiochemical properties of solutes and membranes [17]. Also, The principal types of membrane are shown schematically in Figure 3.6 and are described briefly in the following section. The characteristics of membrane process are shown in Figure 3.8.

3.2 THE RELATIONSHIP BETWEEN NANOTECHNOLOGY AND FILTRATION

Nowadays, nanomaterials have become the most interested topic of materials research and development due to their unique structural properties (unique chemical, biological, and physical properties as compared to

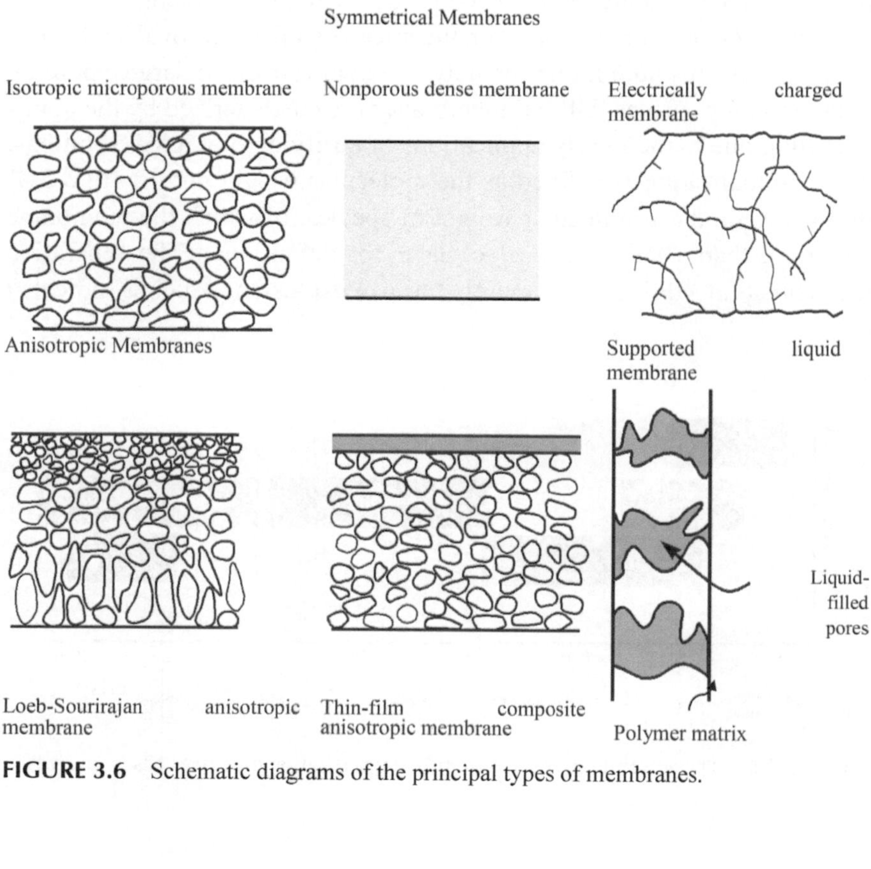

FIGURE 3.6 Schematic diagrams of the principal types of membranes.

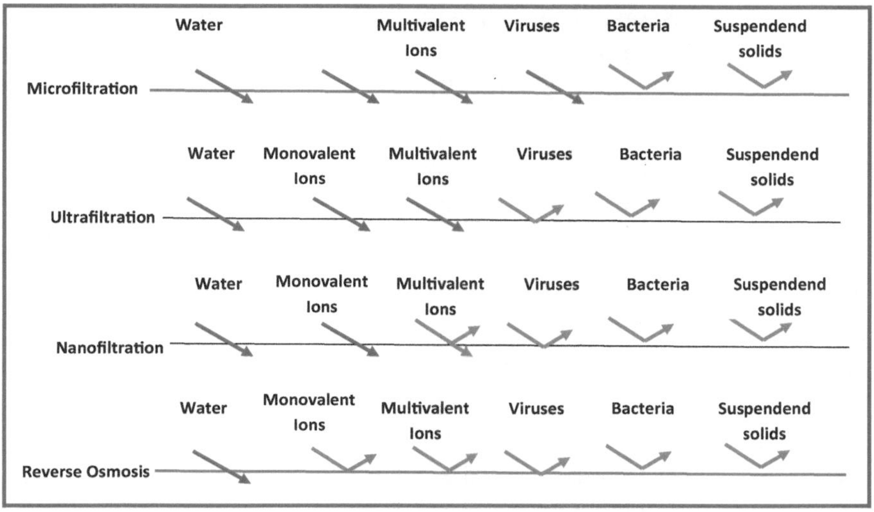

FIGURE 3.7 Membrane process characteristics.

larger particles of the same material) that cover their efficient uses in various fields, such as ion exchange and separation, catalysis, biomolecular isolation and purification as well as in chemical sensing [22]. However, the understanding of the potential risks (health and environmental effects) posed by nanomaterials hasn't increased as rapidly as research has regarding possible applications.

One of the ways to enhance their functional properties is to increase their specific surface area by the creation of a large number of nanostructured elements or by the synthesis of a highly porous material.

Classically, porous matter is seen as material containing three-dimensional voids, representing translational repetition, while no regularity is necessary for a material to be termed "porous." In general, the pores can be classified into two types: open pores, which connect to the surface of the material, and closed pores, which are isolated from the outside. If the material exhibits mainly open pores, which can be easily transpired, then one can consider its use in functional applications such as adsorption, catalysis and sensing. In turn, the closed pores can be used in sonic and thermal insulation, or lightweight structural applications. The use of porous materials offers also new opportunities in such areas as coverage chemistry, guest–host synthesis and molecular manipulations and reactions for manufacture of nanoparticles, nanowires and other quantum nanostructures. The International Union of Pure and Applied Chemistry (IUPAC) defines porosity scales as follows (Figure 3.8):

- Microporous materials 0–2-nm pores;
- Mesoporous materials 2–50-nm pores;
- Macroporous materials >50-nm pores.

This definition, it should be noted, is somewhat in conflict with the definition of nanoscale objects, which typically have large relative porosities (>0.4), and pore diameters between 1 and 100 nm. In order to classify porous materials according to the size of their pores the sorption analysis is one of the tools often used. This tool is based on the fact that pores of different sizes lead to totally different characteristics in sorption isotherms. The correlation between the vapor pressure and the pore size can be written as the Kelvin equation:

$$r_p\left(\frac{p}{p_0}\right)=\frac{2\gamma V_L}{RT\ln\left(3.{p}/{p_0}\right)}+t\left(3.\frac{p}{p_0}\right) \tag{3.1}$$

Therefore, the isotherms of microporous materials show a steep increase at very low pressures (relative pressures near zero) and reach a plateau

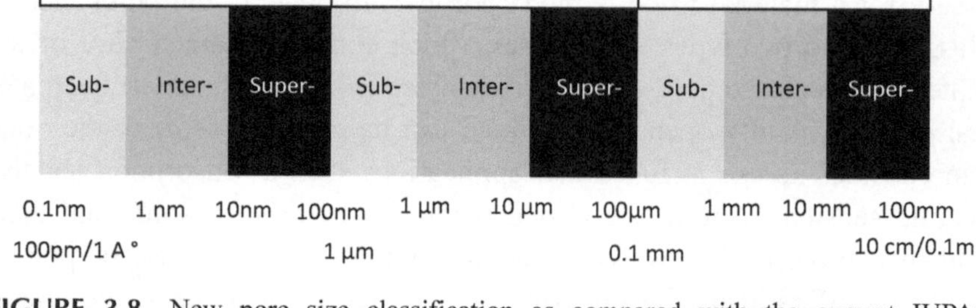

FIGURE 3.8 New pore size classification as compared with the current IUPAC nomenclature.

quickly. Mesoporous materials are characterized by a so-called capillary doping step and a hysteresis (a discrepancy between adsorption and desorption). Macroporous materials show a single or multiple adsorption steps near the pressure of the standard bulk condensed state (relative pressure approaches one) [22].

Nanoporous materials exuberate in nature, both in biological systems and in natural minerals. Some nanoporous materials have been used industrially for a long time. Recent progress in characterization and manipulation on the nanoscale has led to noticeable progression in understanding and making a variety of nanoporous materials: from the merely opportunistic to directed design. This is most strikingly the case in the creation of a wide variety of membranes where control over pore size is increasing dramatically, often to atomic levels of perfection, as is the ability to modify physical and chemical characteristics of the materials that make up the pores [23].

The available range of membrane materials includes polymeric, carbon, silica, zeolite and other ceramics, as well as composites. Each type of membrane can have a different porous structure, as illustrated in Figure 3.9. Membranes can be thought of as having a fixed (immovable) network of pores in which the molecule travels, with the exception of most polymeric membranes [24, 25]. Polymeric membranes are composed of an amorphous mix of polymer chains whose interactions involve mostly

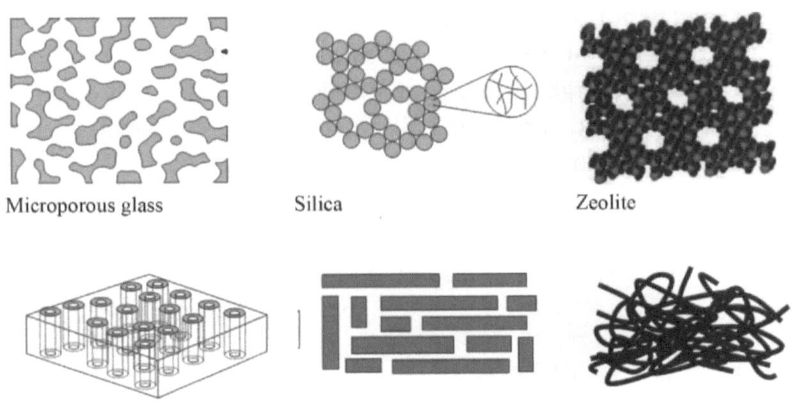

FIGURE 3.9 Porous structure within various types of membranes.

Van der Waals forces. However, some polymers manifest a behavior that is consistent with the idea of existence of opened pores within their matrix. This is especially true for high free volume, high permeability polymers, as has been proved by computer modeling, low activation energy of diffusion, negative activation energy of permeation, solubility controlled permeation [26, 27]. Although polymeric membranes have often been viewed as nonporous, in the modeling framework discussed here it is convenient to consider them nonetheless as porous. Glassy polymers have pores that can be considered as 'frozen' over short times scales, while rubbery polymers have dynamic fluctuating pores (or more correctly free volume elements) that move, shrink, expand and disappear [28].

Three nanotechnologies that are often used in the filtering processes and show great potential for applications in remediation are:

1. Nanofiltration (and its sibling technologies: reverse osmosis, ultrafiltration, and microfiltration), is a fully developed, commercially available membrane technology with a large number of vendors. Nanofiltration relies on the ability of membranes to discriminate between the physical size of particles or species in a mixture or solution and is primarily used for water pre-treatment, treatment, and purification). There are almost 600 companies in worldwide which offering membrane systems.
2. Electrospinning is a process used by the nanofiltration process, in which fibers are stretched and elongated down to a diameter of about 10 nm. The modified nanofibers that are produced are particularly useful in the filtration process as an ultra-concentrated filter with a very large surface area. Studies have found that electrospun nanofibers can capture metallic ions and are continually effective through re-filtration.
3. Surface modified membrane is a term used for membranes with altered makeup and configuration, though the basic properties of their underlying materials remain intact.

3.3 TYPES OF MEMBRANES

As it mentioned, membranes have achieved a momentous place in chemical technology and are used in a broad range of applications. The key property that is exploited is the ability of a membrane to control the permeation rate of a chemical species through the membrane. In essence, a membrane is nothing more than a discrete, thin interface that moderates the permeation of chemical species in contact with it. This interface may be molecularly homogeneous, that is completely uniform in composition and structure or it may be chemically or physically heterogeneous for example, containing holes or pores of finite dimensions or consisting of some form of layered structure. A normal filter meets this definition of a membrane, but, generally, the term filter is usually limited to structures that separate particulate suspensions larger than 1–10 μm [29].

The preparation of synthetic membranes is however a more recent invention which has received a great audience due to its applications [30]. Membrane technology like most other methods has undergone a developmental stage, which has validated the technique as a cost-effective treatment option for water. The level of performance of the membrane technologies is still developing and it is stimulated by the use of additives to improve the mechanical and thermal properties, as well as the permeability, selectivity, rejection and fouling of the membranes [31]. Membranes can be fabricated to possess different morphologies. However, most membranes that have found practical use are mainly of asymmetric structure. Separation in membrane processes takes place as a result of differences in the transport rates of different species through the membrane structure, which is usually polymeric or ceramic [32].

The versatility of membrane filtration has allowed their use in many processes where their properties are suitable in the feed stream. Although membrane separation does not provide the ultimate solution to water treatment, it can be economically connected to conventional treatment technologies by modifying and improving certain properties [33].

The performance of any polymeric membrane in a given process is highly dependent on both the chemical structure of the matrix and the physical arrangement of the membrane [34]. Moreover, the structural impeccability of a membrane is very important since it determines its permeation and selectivity efficiency. As such, polymer membranes should be seen as much more than just sieving filters, but as intrinsic complex structures which can either be homogenous (isotropic) or heterogeneous (anisotropic), porous or dense, liquid or solid, organic or inorganic [34, 35].

3.3.1 *ISOTROPIC MEMBRANES*

Isotropic membranes are typically homogeneous/uniform in composition and structure. They are divided into three subgroups, namely: microporous, dense and electrically charged membranes. Isotropic microporous membranes have evenly distributed pores (Figure 3.10a). Their pore diameters range between 0.01–10 μm and operate by the sieving mechanism. The microporous membranes are mainly prepared by the phase inversion method albeit other methods can be used. Conversely, isotropic dense membranes do not have pores and as a result they tend to be thicker than the microporous membranes (Figure 3.10b). Solutes are carried through the membrane by diffusion under a pressure, concentration or electrical potential gradient. Electrically charged membranes can either be porous or non-porous. However in most cases they are finely microporous with pore walls containing charged ions (Figure 3.10c) [20, 28].

3.3.2 *ANISOTROPIC MEMBRANES*

Anisotropic membranes are often referred to as Loeb-Sourirajan, based on the scientists who first synthesized them [36, 37]. They are the most widely used membranes in industries. The transport rate of a species through

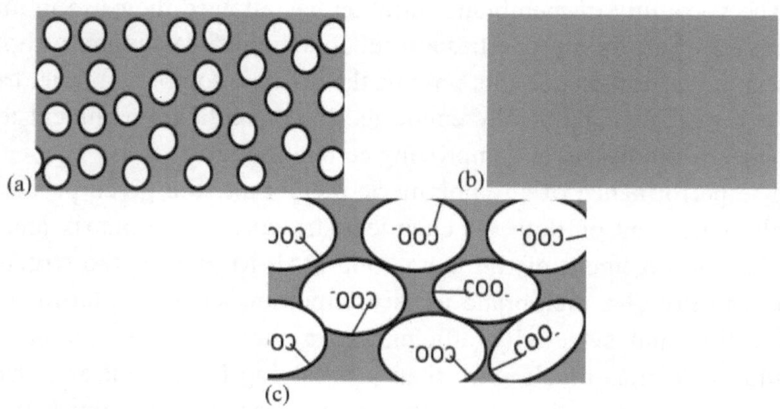

FIGURE 3.10 Schematic diagrams of isotropic membranes: (a) microporous; (b) dense; and (c) electrically charged membranes.

a membrane is inversely proportional to the membrane thickness. The membrane should be as thin as possible due to high transport rates are eligible in membrane separation processes for economic reasons. Contractual film fabrication technology limits manufacture of mechanically strong, defect-free films to thicknesses of about 20 μm. The development of novel membrane fabrication techniques to produce anisotropic membrane structures is one of the major breakthroughs of membrane technology. Anisotropic membranes consist of an extremely thin surface layer supported on a much thicker, porous substructure. The surface layer and its substructure may be formed in a single operation or separately [29]. They are represented by non-uniform structures, which consist of a thin active skin layer and a highly porous support layer. The active layer enjoins the efficiency of the membrane, whereas the porous support layer influences the mechanical stability of the membrane. Anisotropic membranes can be classified into two groups, namely: (i) integrally skinned membranes where the active layer is formed from the same substance as the supporting layer, (ii) composite membranes where the polymer of the active layer differs from that of the supporting sub-layer [37]. In composite membranes, the layers are usually made from different polymers. The separation properties and permeation rates of the membrane are determined particularly by the surface layer and the substructure functions as a mechanical support. The advantages of the higher fluxes provided by anisotropic membranes (Figure 3.11) are so great that almost all commercial processes use such membranes [29].

3.3.3 POROUS MEMBRANE

In Knudsen diffusion (Figure 3.12a), the pore size forces the penetrant molecules to collide more frequently with the pore wall than with other

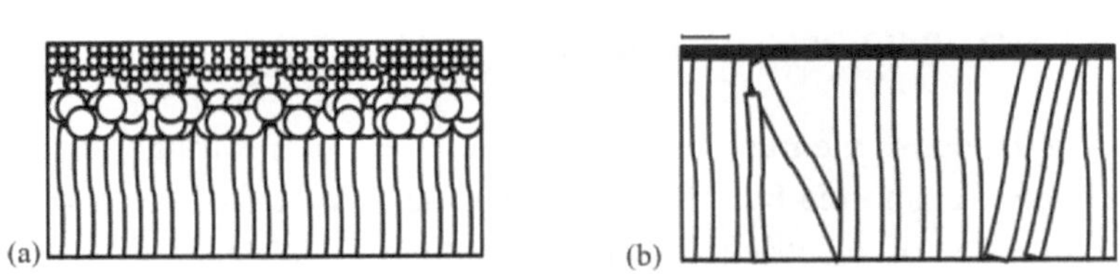

FIGURE 3.11 Schematic diagrams of anisotropic membranes: (a) Loeb-Sourirajan, and (b) thin film composite membranes.

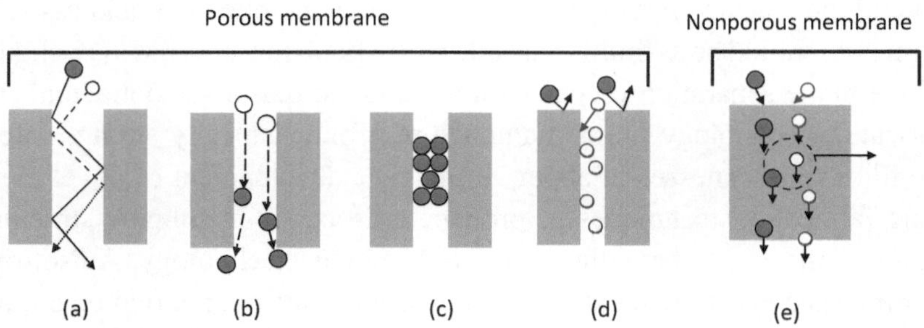

FIGURE 3.12 Schematic representation of membrane-based gas separations. (a) Knudsen-flow separation, (b) surface-diffusion, (c) capillary condensation, (d) molecular-sieving separation, and (e) solution-diffusion mechanism.

incisive species [38]. Except for some special applications as membrane reactors, Knudsen-selective membranes are not commercially attractive because of their low selectivity [39]. In surface diffusion mechanism (Figure 3.12b), the pervasive molecules adsorb on the surface of the pores so move from one site to another of lower concentration. Capillary condensation (Figure 3.12c) impresses the rate of diffusion across the membrane. It occurs when the pore size and the interactions of the penetrant with the pore walls induce penetrant condensation in the pore [40]. Molecular-sieve membranes in Figure 3.12d have gotten more attention because of their higher productivities and selectivity than solution-diffusion membranes. Molecular sieving membranes are means to polymeric membranes. They have ultra microporous (<7Å) with sufficiently small pores to barricade some molecules, while allowing others to pass through. Although they have several advantages such as permeation performance, chemical and thermal stability, they are still difficult to process because of some properties like fragile. Also they are expensive to fabricate.

3.3.4 NONPOROUS (DENSE) MEMBRANE

Nonporous, dense membranes consist of a dense film through which permeants are transported by diffusion under the driving force of a pressure, concentration, or electrical potential gradient. The separation of various components of a mixture is related directly to their relative transport rate within the membrane, which is determined by their diffusivity and solubility in the membrane material. Thus, nonporous, dense membranes can separate permeants of similar size if the permeant concentrations in the membrane material differ substantially. Reverse osmosis membranes use dense membranes to perform the separation. Usually these membranes have an anisotropic structure to improve the flux [29].

The mechanism of separation by non-porous membranes is different from that by porous membranes. The transport through nonporous polymeric membranes is usually described by a solution–diffusion mechanism (Figure 3.12a). The most current commercial polymeric membranes operate according to the solution–diffusion mechanism. The solution–diffusion mechanism has three steps: (1) the absorption or adsorption at the upstream boundary, (2) activated diffusion through the membrane, and (3) desorption or evaporation on the other side. This solution–diffusion mechanism is driven by a difference in the thermodynamic activities existing at the upstream and downstream faces of the membrane as well as the intermolecular forces acting between the permeating molecules and those making up the membrane material.

The concentration gradient causes the diffusion in the direction of decreasing activity. Differences in the permeability in dense membranes are caused not only by diffusivity differences of the various species but also by differences in the physicochemical interactions of the species within the polymer. The solution–diffusion model assumes that the pressure within a membrane is uniform and that the chemical potential gradient across the membrane is expressed only as a concentration gradient. This mechanism controls permeation in polymeric membranes for separations.

3.4 INTRODUCTION TO MORPHOLOGY AND POROSITY

Union of Pure and Applied Chemists (IUPAC) defines morphology as the "shape, optical appearance, or form of phase domains in substances, such as high polymers, polymer blends, composites, and crystals." Since this is a very broad and diffuse definition, two classes of morphology are set apart in this work. Shape and bulk morphology are distinguished, because both are very useful in the description of the porous networks. The former concerns the particle size, shape and pore structure, the latter classifies the polymers by the molecular architecture of the pore walls. Polymers have the advantage that they can be prepared in almost any micro- or macroscopic size and shape. This allows extensive tuning of the shape morphology to the desired application, while keeping the bulk morphological parameters unchanged and vice versa (Tables 3.3 and 3.4).

Classically, porous matter is seen as a material that has voids through and through. The voids show a translational repetition in 3-D space, while no regularity is necessary for a material to be termed " porous." A common and relatively simple porous system is one type of dispersion classically described in colloid science, namely foam or, better, solid foam. In correlation with this, the most typical way to think about a porous material is as a material with gas-solid interfaces as the most dominant characteristic. This already indicates that classical colloid and interface science as the creation of interfaces due to nucleation phenomena (in this

TABLE 3.3 Examples of Size-Dependent Shape Morphology

Size range	Shape morphology
Nanometer	Polymer brush
	Micelle
	Microgel
	Pores
Micrometer	Powders
	polyHIPE (high internal phase emulsion) Pores
Macroscopic	Beads
	Discs
	Membranes

TABLE 3.4 Examples of Cross-Link-Dependent Bulk Morphology

Content of cross-links	Bulk morphology
None	Soluble polymer
	Supported polymer brush
Low	Swellable polymer gels
High	Polymer networks
Extra-high	Hyper-cross-linked
	Polymers

case nucleation of wholes), decreasing interface energy, and stabilization of interfaces is of elemental importance in the formation process of nanoporous materials [23, 41–43].

These factors are often omitted because the final products are stable (they are metastable). This metastability is due to the rigid character of the void-surrounding network, which is covalently cross-linked in most cases. However, it must be noticed that most of the porous materials are not stable by thermodynamic means. As soon as kinetic energy boundaries are overcome, materials start to breakdown [23].

Porous materials have been extensively exploited for use in a broad range of applications: for example, as membranes for separation and purification [44], as high surface-area adsorbents, as solid supports for sensors [45] and catalysts, as materials with low dielectric constants in the fabrication of microelectronic devices [46], and as scaffolds to guide the growth of tissues in bioengineering [47].

Porous materials occur widely and have many important applications. They can, for example, offer a convenient method of imposing fine structure on adsorbed materials.

They can be used as substrates to support catalysts and can act as highly selective sieves or cages that only allow access to molecules up to a certain size.

Food is often finely structured. Many biologically active materials are porous, as are many construction and engineering materials. Porous geological materials are of great interest; high porosity rock may contain water, oil or gas; low porosity rock may act as a cap to porous rock, and is of importance for active waste sealing.

3.5 POROSITY

Porosity φ is the fraction of the total soil volume that is taken up by the pore space. Thus it is a single-value quantification of the amount of space available to fluid within a specific body of soil. Being simply a fraction of total volume, φ can range between 0 and 1, typically falling between 0.3 and 0.7 for soils. With the assumption that soil is a continuum, adopted here as in much of soil science literature, porosity can be considered a function of position.

3.6 POROSITY IN NATURAL SOILS

The porosity of a soil depends on several factors, including (i) packing density, (ii) the breadth of the particle size distribution (polydisperse vs. monodisperse), (iii) the shape of particles, and (iv) cementing. Mathematically considering an idealized soil of packed uniform spheres, φ must fall between 0.26 and 0.48, depending on the packing. Spheres randomly thrown together will have φ near the middle of this range, typically 0.30 to 0.35. Sand with grains nearly uniform in size (monodisperse) packs to about the same porosity as spheres. In polydisperse sand, the fitting of small grains within the pores between large ones can reduce φ, conceivably below the 0.26 uniform-sphere minimum. Figure 3.13 illustrates this concept. The particular sort of arrangement required to reduce φ to 0.26 or less is highly improbable, however, so φ also typically falls within the 0.30–0.35 for polydisperse sands. Particles more irregular in shape tend to have larger gaps between their nontouching surfaces, thus forming media of greater porosity. In porous rock such as sand-stone, cementation or welding of particles not only creates pores that are different in shape from those of particulate media, but also reduces the porosity as solid material takes up space that would otherwise be pore space. Porosity in such a case can easily be less than 0.3, even approaching 0. Cementing material can also have the opposite effect. In many soils, clay and organic substances cement particles together into aggregates. An individual aggregate might have 0.35 porosity within it but the medium as a whole has additional pore space in the form of gaps between aggregates, so that φ can be 0.5 or

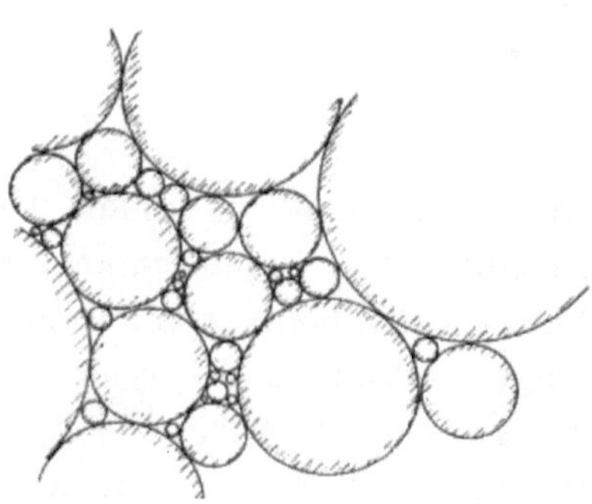

FIGURE 3.13 Dense packing of polydisperse spheres.

greater. Observed porosities can be as great as 0.8–0.9 in a peat (extremely high organic matter) soil.

Porosity is often conceptually partitioned into two components, most commonly called textural and structural porosity. The textural component is the value the porosity would have if the arrangement of the particles were random, as described above for granular material without cementing. That is, the textural porosity might be about 0.3 in a granular medium. The structural component represents nonrandom structural influences, including macropores and is arithmetically defined as the difference between the textural porosity and the total porosity.

The texture of the medium relates in a general way to the pore-size distribution, as large particles give rise to large pores between them, and therefore is a major influence on the soil water retention curve. Additionally, the structure of the medium, especially the pervasive-ness of aggregation, shrinkage cracks, worm-holes, etc. substantially influences water retention.

3.7 MEASUREMENT OF POROSITY

The technology of thin sections or of tomographic imaging can produce a visualization of pore space and solid material in a cross-sectional plane. The summed area of pore space divided by total area gives the areal porosity over that plane. An analogous procedure can be followed along a line through the sample, to yield a linear porosity. If the medium is isotropic, either of these would numerically equal the volumetric porosity as defined above.

The volume of water contained in a saturated sample of known volume can indicate porosity. The mass of saturated material less the oven-dry mass of the solids, divided by the density of water, gives the volume of water. This divided by the original sample volume gives porosity.

An analogous method is to determine the volume of gas in the pore space of a completely dry sample. Sampling and drying of the soil must be conducted so as not to com-press the soil or otherwise alter its porosity. A pycnometer can measure the air volume in the pore space. A gas-tight chamber encloses the sample so that the internal gas-occupied volume can be perturbed by a known amount while the gas pressure is measured. This is typically done with a small piston attached by a tube connection. Boyle's law indicates the total gas volume from the change in pressure resulting from the volume change. This total gas volume minus the volume within the piston, connectors, gaps at the chamber walls, and any other space not occupied by soil, yields the total pore volume to be divided by the sample volume.

To avoid having to saturate with water or air, one can calculate porosity from measurements of particle density and bulk density. From the definitions of bulk density as the solid mass per total volume of soil and particle density as the solid mass per solid volume, their ratio ρ_b / ρ_p is the complement of φ, so that:

$$\varnothing = 1 - \rho_b / \rho_p \tag{3.2}$$

Often the critical source of error is in the determination of total soil volume, which is harder to measure than the mass. This measurement can be based on the dimensions of a minimally disturbed sample in a regular geometric shape, usually a cylinder. Significant error can result from irregularities in the actual shape and from unavoidable compaction. Alternatively, the measured volume can be that of the excavation from which the soil sample originated. This can be done using measurements of a regular geometric shape, with the same problems as with measurements on an extracted sample. Additional methods, such as the balloon or sand-fill methods have other sources of error.

3.8 PORES AND PORE-SIZE DISTRIBUTION: THE NATURE OF A PORE

Because soil does not contain discrete objects with obvious boundaries that could be called individual pores, the precise delineation of a pore unavoidably requires artificial, subjectively established distinctions. This contrasts with soil particles, which are easily defined, being discrete material objects with obvious boundaries. The arbitrary criterion required to partition pore space into individual pores is often not explicitly stated when pores or their sizes are discussed. Because of this inherent arbitrariness, some scientists argue that the concepts of pore and pore size should be avoided. Much valuable theory of the behavior of the soil-water-air system, however, has been built on these concepts, defined using widely, if not universally, accepted criteria.

3.9 POROUS MATERIALS

Porous materials are solid forms of matter permeated by interconnected or non-interconnected pores (voids) of different kinds: channels, cavities or interstices; that can be divided into several classes.

According to the nomenclature suggested by the International Union of Pure and Applied Chemists (IUPAC), porous materials are usually classified into three different categories depending on the lateral dimensions of their pores: microporous (<2 nm), mesoporous (between 2 and 50 nm) and macroporous (>50 nm) [48].

Liquid and gaseous molecules have been known to exhibit characteristic transport behaviors in each type of porous material. For example, mass transport can be obtained via viscous flow and molecular diffusion in a macroporous material, through surface diffusion and capillary flow in a mesoporous material and by activated diffusion in a microporous material.

Pores from the nanoscopic to the macroscopic scale are generated depending on the method. A summary of selected methods that can be applied to styrene-codivinylbenzene polymers is given in Table 3.5 [49].

The internal structural architecture of the void space potentially controls the physical and chemical properties, such as reactivity, thermal

TABLE 3.5 Overview of Methods of Generating Porosity During Polymer Synthesis.

Method	Porogene	Porosity	Accessibility Typical size
Foaming	Gas, solvent, supercritical solvent	Open/closed	100 nm–1 mm
Phase separation	Solvent	Open	1 μm–1 mm
High internal phase emulsion polymer	Emulsion	Open	10 μm–100 μm
Soft templating by	Molecules (solvent)	Micelles	Bicontinuous microemulsion
Hard templating by	Colloidal crystals	Porous solids	Open

and electric conductivity, as well as the kinetics of numerous transport processes. The characterization of porous materials, therefore, has been of great practical interest in numerous areas including catalysis, adsorption, purification, separation, etc., where the essential aspects for such applications are pore accessibility, narrow pore size distribution (PSD), relatively high specific surface area and easily tunable pore sizes.

Ordered porous materials are judged to be much more interesting because of the control over pore sizes and pore shapes. Their disordered counterparts exhibit high polydispersity in pore sizes, and the shapes of the pores are irregular. Ordered porous materials seem to be much more homogeneous. In many cases a material possesses more than one porosity. This could be for microporous materials: an additional meso- or macroporosity caused by random grain packing for mesoporous materials: an additional macroporosity caused by random grain packing, or an additional microporosity in the continuous network. For macroporous materials: an additional meso- and microporosity, these factors should be taken into consideration when materials are classified according to their homogeneity. A material possessing just one type of pore, even when the pores are disordered, might be more homogenous than one having just a fraction of nicely ordered pores.

Ordered porous solids contain a regularly arranged pore system and it is desired to design materials with different cylindrical, window-like, spherical or slit-like pore shapes and sizes [50].

3.10 PROPERTIES OF POROUS MATERIALS

There are a number of important properties of porous materials such as:

- Porosity
- Specific surface area
- Permeability
- Breakthrough capillary pressure
- Diffusion properties of liquids in pores
- Pore size distribution
- Radial density function

3.11 MACROPOROUS MATERIALS AND THEIR USES

Macroporous materials are formed from the packing of monodisperse spheres (polystyrene or silica) into a three-dimensional ordered arrangement, to form face-centered cubic (FCC) or hexagonal close-packed (HCP) structures. The spaces between the packed spheres create a macroporous structure.

Glass or rubbery polymer that includes a large number of macropores (50 nm–1 μm in diameter) that persist when the polymer is immersed in solvents or in the dry state.

Macroporous polymers are often network polymers produced in bead form. However, linear polymers can also be prepared in the form of macroporous polymer beads. They swell only slightly in solvents.

Macroporous polymers can be used, for instance, as precursors for ion-exchange polymers, as adsorbents, as supports for catalysts or reagents, and as stationary phases in size-exclusion chromatography columns.

Macroporous materials have many applications in the field of engineering due to their large effective surface area, and can be used for purposes such as filters, catalysts, supports, heat exchangers, and fuel cells. Although microporous and mesoporous materials also possess large surface areas, their small pore diameters (less than 10 nm) do not allow larger molecules to pass through them. Hence, macroporous materials with larger pore diameters are preferred and are of more practical use. Through colloidal crystal templating techniques, three-dimensionally ordered macroporous materials with uniform pore size can be successfully synthesized, thus improving the efficiency of transport of materials through the pores. Furthermore, photonic crystals possessing optical band-gaps can be synthesized from these macroporous structures by infiltrating the macroporous material with a precursor fluid, followed by removal of the original spheres through calcination.

Photonic crystals are materials in which the dielectric constants vary periodically in space. Due to the alternating dielectric properties, photonic crystals are hence able to control the propagation of photons, by creating a frequency (band-gap) in which light is not able to propagate. Photonic crystals themselves have great potential use in the engineering field. Due to their ability to localize photons, photonic crystals can be used as wave guides in optical fibers, which would prove very valuable in optical communications for the transfer of information. With the advent of information technology and the need for faster, quicker and more efficient data transmission, the importance and potential of photonic crystals are ever more apparent.

3.12 MESOPOROUS MATERIALS

Mesoporous solids consists of inorganic or inorganic/organic hybrid units of long-range order with amorphous walls, tunable textural and structural properties with highly controllable pore geometry and narrow pore size distribution in the 2–50 nm range [51].

The pores can have different shapes such as spherical or cylindrical and be arranged in varying structures, see Figure 3.14. Some structures have pores that are larger than 50 nm in one dimension, see for example, the two first structures in Figure 3.14, but there the width of the pore is in the mesorange and the material is still considered to be mesoporous.

Mesoporous materials can have a wide range of compositions but mainly consists of oxides such as SiO_2, TiO_2, ZnO_2, Fe_2O or combinations of metal oxides, but also mesoporous carbon can be synthesized [52–55]. Most commonly is to use a micellar solution and grow oxide walls around the micelles. Both organic metal precursors such as alkoxides [56, 57] as well as inorganic salts such as metal chloride salts [53] can be used. Alternatively a mesoporous template can be used to grow another type of mesoporous material inside it. This is often used for synthesizing e.g. mesoporous carbon [55, 58, 59].

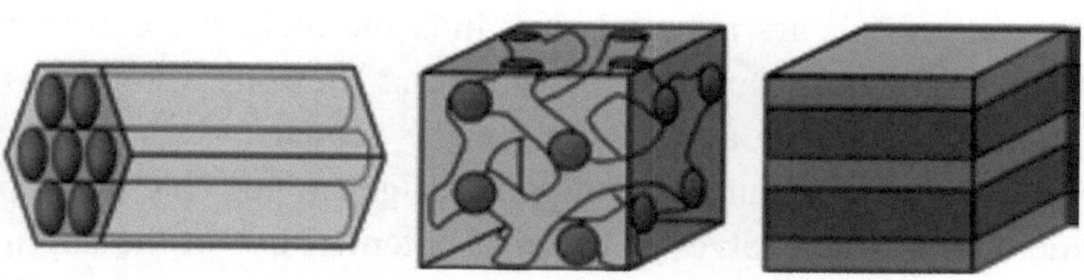

FIGURE 3.14 Different pore structures of mesoporous materials.

3.13 MICROPOROUS MATERIALS

Porous materials are networks of solid material, which contain void spaces. These materials can be further classified depending on the size of the pores present in the material. Microporous solids are materials that contain permanent cavities with diameters of less than 2 nm. Mesoporous materials contain pores ranging from 2 nm to 50 nm and macroporous materials contain pores of greater than 50 nm [60]. The field of microporous materials contains several classes, which are well known [61], including naturally occurring zeolites, activated carbons and silica. Synthetic microporous solids have recently emerged as a potentially important class of materials.

These include Metal Organic Frameworks (MOFs), Microporous Organic Polymers (MOPs) including Covalent Organic Frameworks (COFs) and Polymers of Intrinsic Microporosity (PIMs) [62]. It is the very large surface areas and very small pore sizes of these materials, which make them of specific interest. These two factors permit microporous materials to be useful in applications such as heterogeneous catalysis, separation chemistry, and potential uses in hydrogen or other gas storage [63, 64]. Most synthetic strategies to prepare microporous materials consist of linking together smaller units with di-topic or poly topic functionalities in order to form extended networks much like is displayed in the general diagram of Figure 3.15.

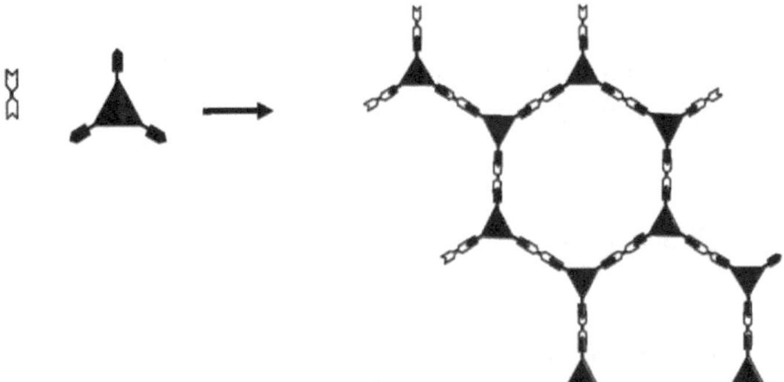

FIGURE 3.15 General schematic showing the linking of polytopic building blocks to form synthetic networks.

Whether a network formed is crystalline or amorphous is generally governed by whether the covalent bonds being formed involve reversible chemistry or not. Crystalline networks are typically formed under reversible reaction conditions that allow error corrections during network formation and produce thermodynamically stable networks. These types of reactions are commonly condensation reactions [65, 66]. On the other hand, when irreversible reactions are employed such as cross couplings, the networks tend to form in a disordered manner [67, 68], resulting in amorphous materials. This is because a carbon-carbon bond is formed irreversibly under conditions such as a Sonagashira or Yamamoto couplings resulting in amorphous networks. This is of course unless some other templating measure is taking into account while considering the reaction conditions [69].

Certain factors such as temperature, solvent and solvent-to-head-space ratio play an important role in the formation of a crystalline framework [66]. Certain solvents can be employed in order to form ordered networks by means of their ability to dissolve the monomer building blocks. If the concentration of a monomer in solution is controlled by a solvent in which it is slightly soluble, then a network is more likely to form under thermodynamic, instead of kinetic control [68]. Solvents could also be used based on their molecular size to act as templates for pores to form around [69]. While this idea of MOP/COF templating is generally understood in qualitative terms (which solvents produce crystalline networks) there has been no research into the quantitative effects (what solvent ratio is required to produce a well-structured network).

Finally, if a material is to exhibit microporosity, it must be composed of somewhat rigid building blocks in order to impart rigidity within the network and provide directionality for the formation of the network. This rigidity prevents collapse of the network upon itself and results in free volume, which becomes the pores within a framework.

3.14 NANOPOROUS POLYMERS

In the past few decades, nanomaterials have received substantial attention and efforts from academic and industrial world, due to the distinct properties at the nanoscale. Nanoporous materials as a subset of nanomaterials possess a set of unique properties: large specific surface volume ratio, high interior surface area, exclusive size sieving and shape selectivity, nanoscale space confinement, and specific gas/fluid permeability. Moreover, pore-filled nanoporous materials can offer synergistic properties that can never be reached by pure compounds. As a result, nanoporous materials are of scientific and technological importance and also considerable interest in a broad range of applications that include templating, sorting, sensing, isolating and releasing.

Nanoporous materials can be classified by pore geometry (size, shape, and order) or distinguished by type of bulk materials. Nanoporous materials are considered uniform if the pore size distribution is relatively narrow and the pore shape is relatively homogenous. The pores can be cylindrical, conical, slit-like, or irregular in shape. They can be well ordered with an alignment as opposed to a random network of tortuous pores. Nanoporous materials cover a wide variety of materials, which can be generally divided into inorganic, organic and composite materials. The majority of investigated nanoporous materials have been inorganic, including oxides, carbon, silicon, silicate, and metal. On the other side, polymers have been identified as materials that offer low cost, less toxicity, easy fabrication process, diverse chemical functionality, and extensive mechanical properties [70, 71].

Naturally, the success of inorganic materials to form nanoporous materials has promoted the development of analogous polymers. More importantly, advances in polymer synthesis and novel processing techniques have led to various nanoporous polymers.

3.15 NANOPOROUS MATERIALS CONNECTION

A whole variety of nanoporous materials in nature can be found in many different functions. The most common task for nanoporous materials in nature is to make inorganic material much lighter while preserving or improving the high structural stability of these compounds. Often, by filling the voids between inorganic matters the desired properties of the hybrid materials exceed the performances of the pure compounds by several orders of magnitude. Nanoporous materials in nature are organic-inorganic hybrids. Naturally occurring materials exhibit synergistic properties. Neither the organic material filling the void nor the inorganic network materials are able to achieve comparable performances by themselves [72–74].

It is seen that complex mechanisms are involved in the formation of these hierarchical materials. Similar to the structure motives on different length-scales cells, vesicles, supra-molecular structures, and biomolecules are involved in the structuring process of inorganic matter occurring in nature. This process is commonly known as biomineralization [73].

It is often not seen in this relation, but it will be shown later that ordered porous materials, and therefore artificial materials, are constructed according to very similar principles. A completely different area where nanoporous materials are highly important is in the lungs, where foam with a high surface area permits sufficient transfer of oxygen to the blood. Even the most recent developments in nanoporous materials, such as their application as photonic materials are already present in nature, the color of butterfly wings, for instance, originates from photonic effects [75, 76].

It can be concluded that nature applies the concept of nanoporous materials (either filled or unfilled) as a powerful tool for constructing all kinds of materials with advanced properties. So it is not surprising how much research has recently been devoted to porous materials in different areas such as chemistry, physics, and engineering. The current interests in nanoporous materials are now far behind their size-sieving properties.

3.16 CLASSIFICATION BY NETWORK MATERIAL

One of the most important goals in the field of nanoporous materials is to achieve any possible chemical composition in the network materials "hosting" the pores. It makes sense to divide the materials into two categories: (a) inorganic materials and (b) organic materials. Among the inorganic materials, the larger group, it could be found:

(i) Inorganic oxide-type materials. This is the field of the most commonly known porous silica, porous titania, and porous zirconia materials.

(ii) A category of its own is given for nanoporous carbon materials. In this category are the highly important active carbons and some examples for ordered mesoporous carbon materials.
(iii) Other binary compounds such as sulfides, nitrides. Into this category also fall the famous AIPO materials.
(iv) There are already some examples in addition to carbon where just one element (for instance, a metal) could be prepared in a nanoporous state. The most appropriate member of this class of materials is likely to be nanoporous silicon, with its luminescent properties [77, 78]. There are far fewer examples of nanoporous organic materials, such as polymers [79].

3.17 SUMMARY OF CLASSIFICATIONS

Three main criteria could be defined as:

a) size of pores;
b) type of network material; and
c) state of order: ordered or disordered materials.

3.18 ORIGIN AND CLASSIFICATION OF PORES IN SOLID MATERIALS

Solid materials have a cohesive structure, which depends on the interaction between the primary particles. The cohesive structure leads indispensably to a void space, which is not occupied by the composite particles such as atoms, ions, and line particles. Consequently, the state and population of such voids strongly depends on the inter-particle forces. The inter-particle forces are different from one system to another; chemical bonding, van der Waals force, electrostatic force, magnetic force, surface tension of adsorbed films on the primary particles, and so on. Even the single crystalline solid, which is composed of atoms or ions has intrinsic voids and defects. Therefore, pores in solids are classified into intra-particle pores and inter-particle pores (Table 3.6) [80].

TABLE 3.6 Classification of Pores from Origin, Pore Width *w*, and Accessibility

Origin and structure		
Intraparticle pore	Intrinsic intraparticle pore	Structurally intrinsic type
	Extrinsic intraparticle pore	Injected intrinsic type
		Pure type
		Pillared type
Interparticle pore	Rigid interparticle *pore* (Agglomerated)	
	Flexible interparticle pore (Aggregated)	
Pore width		
Macropore	$w < 50$ nm	
Mesopore	2 nm $< w <$ 50 nm	
Micropore	$w < 2$ nm	
	Supermicropore, $0.7 < w < 2$ nm	
	Ultramicropore, $W < 0.7$ nm	
	(Ultrapore, $w < 0.35$ nm in this review)	
Accessibility to surroundings		
Open pore	Communicating with external surface	
Closed pore	No communicating with surroundings	
Latent pore	Ultrapore and closed pore	

3.19 TRAPARTICLE PORES

3.19.1 INTRINSIC IN TRAPARTICLE PORE

Zeolites are the most representative porous solids whose pores arise from the intrinsic crystalline structure. Zeolites have a general composition of Al, Si, and O, where Al-O and Si-O tetrahedral units cannot occupy the space perfectly and therefore produce cavities. Zeolites have intrinsic pores of different connectivities according to their crystal structures [81]. These pores may be named intrinsic crystalize pores. The carbon nanotube has also an intrinsic crystalline pore [82].

Although all crystalline solids other than zeolites have more or less intrinsic crystalline pores, these are not so available for adsorption or diffusion due to their isolated state and extremely small size.

There are other types of pores in a single solid particle. α-FeOOH is a precursor material for magnetic tapes, a main component of surface deposits and atmospheric corrosion products of iron-based alloys, and a mineral. The α-FeOOH microcrystal is of thin elongated plate [83].

These new created intrinsic intra-particle pores should have their own name different from the intrinsic intra-particle pore. The latter is called a structurally intrinsic intra-particle pore, while the former is called an injected intrinsic intra-particle pore.

3.19.2 EXTRINSIC INTRA-PARTICLE PORE

When a foreign substance is impregnated in the parent material in advance this is removed by a modification procedure [84]. This type of pores should be called extrinsic intra-particle pores. Strictly speaking, as the material does not contain other components, extrinsic pure intra-particle pore is recommended. However, the extrinsic intra-pore can be regarded as the inter-particle pore in some cases.

It can be introduced a pore-forming agent into the structure of solids to produce voids or fissures which work as pores. This concept has been applied to layered compounds in which the interlayer bonding is very weak; some inserting substance swells the interlayer space. Graphite is a representative layered compound; the graphitic layers are weakly bound to each other by the van der Waals force [85]. If it heated in the presence of intercalants such as K atoms, the intercalants are inserted between the interlayer spaces to form a long periodic structure. K-intercalated graphite can adsorb a great amount of H_2 gas, while the original graphite cannot [86]. The interlayer space opened by intercalation is generally too narrow to be accessed by larger molecules. Intercalation produces not only pores, but also changes the electronic properties. Montmorillonite is a representative layered clay compound, which swells in solution to intricate hydrated ions or even surfactant molecules [87, 88]. Then some pillar materials such as metal hydroxides are intricate in the swollen interlayer space under wet conditions. As the pillar compound is not removed upon drying, the swollen structure can be preserved even under dry conditions.

The size of pillars can be more than several nm, being different from the above intercalants. As the graphite intercalation compounds and pillar ones need the help of foreign substances, they should be distinguished from the intrinsic intra-particle pore system. Their pores belong to extrinsic intra-particle pores. As the intercalation can be included in the pillar formation, it is better to say that both the pillared and intercalated compounds have pillared in traparticle pores [89].

3.20 INTER-PARTICLE PORES

Primary particles stick together to form a secondary particle, depending on their chemical composition, shape and size. In colloid chemistry, there are two gathering types of primary particles. One is aggregation and the other is agglomeration.

The aggregated particles are loosely bound to each other and the assemblage can be readily broken down. Heating or compressing the assemblage of primary particles brings about the more tightly bound agglomerate [90].

There are various interaction forces among primary particles, such as chemical bonding, van der Waals force, magnetic force, electrostatic force, and surface tension of the thick adsorbed layer on the particle surface. Sintering at the neck part of primary particles produces stable agglomerates having pores. The aggregate bound by the surface tension of adsorbed water film has flexible pores. Thus, inter-particle pores have wide varieties in stability, capacity, shape, and size, which depend on the packing of primary particles. They play an important role in nature and technology regardless of insufficient understanding. The inter-particle pores can be divided into rigid and flexible pores. The stability depends on the surroundings. Almost all inter-particle pores in agglomerates are rigid, whereas those in aggregates are flexible. Almost all sintered porous solids have rigid pores due to strong chemical bonding among the particles. The rigid inter-particle porous solids have been widely used and have been investigated as adsorbents or catalysts. Silica gel is a representative of inter-particle porous solids. Ultrafine spherical silica particles form the secondary particles, leading to porous solids [91, 92].

3.21 STRUCTURE OF PORES

The pore state and structure mainly depend on the origin. The pores communicating with the external surface are named open pores, which are accessible for molecules or ions in the surroundings. When the porous solids are insufficiently heated, some parts of pores near the outer shell are collapsed inducing closed pores without communication to the surroundings. Closed pores also remain by insufficient evolution of gaseous substance. The closed pore is not associated with adsorption and permeability of molecules, but it influences the mechanical properties of solid materials, the new concept of latent pores is necessary for the best description of the pore system. This is because the communication to the surroundings often depends on the probe size, in particular, in the case of molecular resolution porosimetry. The open pore with a pore width smaller than the probe molecular size must be regarded as a closed pore. Such effectively closed pores and chemically closed pores should be designated the latent pores [93]. The combined analysis of molecular resolution porosimetry and small angle X-ray scattering (SAXS) offers an effective method for separate determination of open and closed (or latent) pores, which will be described later. The porosity is defined as the ratio of the pore volume to the total solid volume [94].

The geometrical structure of pores is of great concern, but the three-dimensional description of pores is not established in less-crystalline porous solids. Only intrinsic crystalline intra-particle pores offer a good description of the structure. The hysteresis analysis of molecular adsorption isotherms and electron microscopic observation estimate the pore geometry such as cylinder (cylinder closed at one end or cylinder open at both ends), cone shape, slit shape, interstice between closed-packing spheres and ink bottle. However, these models concern with only the unit structures. The higher order structure of these unit pores such as the network structure should be taken into account. The simplest classification of the higher order structures is one-, two- and three-dimensional pores. Some zeolites and aluminophosphates have one-dimensional pores and activated carbons have basically two-dimensional slit-shaped pores with complicated network structures [95].

The IUPAC has tried to establish a classification of pores according to the pore width (the shortest pore diameter), because the geometry determination of pores is still very difficult and molecular adsorption can lead to the reliable parameter of the pore width. The pores are divided into three categories: macropores, mesopores, and micropores, as mentioned above. The fact that nanopores are often used instead of micropores should be noted.

These sizes can be determined from the aspect of N_2 adsorption at 77 K, and hence N_2 molecules are adsorbed by different mechanisms – multilayer adsorption, capillary condensation, and micropore filling for macropores, mesopores, and micropores, respectively (Figure 3.16). The critical widths of 50 and 2 nm are chosen from empirical and physical reasons. The pore width of 50 nm corresponds to the relative pressure of 0.96 for the N_2 adsorption isotherm. Adsorption experiments above that are considerably difficult and applicability of the capillary condensation theory is not sufficiently examined. The smaller critical width of 2 nm corresponds to the relative pressure of 0.39 through the Kelvin equation, where an unstable behavior of the N_2 adsorbed layer (tensile strength effect) is observed. The capillary condensation theory cannot be applied to pores having a smaller width than 2 nm. The micropores have two subgroups, namely ultra-micropores (0.7 nm) and super-micropores (0.7 nm < w < 2 nm). The statistical thickness of the adsorbed N2 layer on solid surfaces is 0.354 nm. The maximum size of ultra-micropores corresponds to the bilayer thickness of nitrogen molecules, and the adsorbed N_2 molecules near the entrance of the pores often block further adsorption. The ultra-micropore assessment by N_2

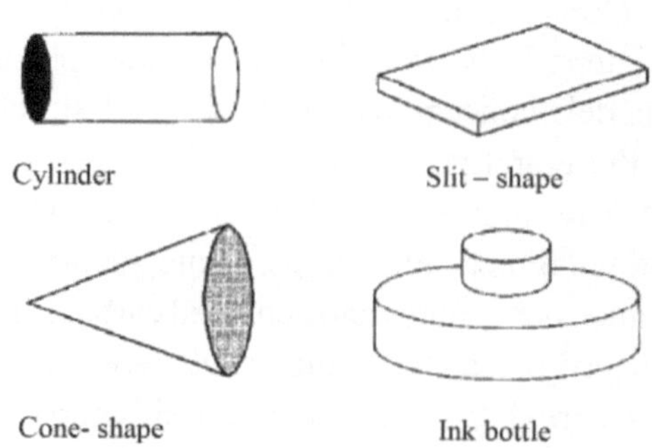

FIGURE 3.16 Pore shapes.

adsorption has an inevitable and serious problem. The micropores are divided into two groups.

Recently the molecular statistical theory has been used to examine the limitation of the Kelvin equation and predicts that the critical width between the micropore and the mesopore is 1.3–1.7 nm (corresponding to 4–5 layers of adsorbed N_2), which is smaller than 2 nm [96].

3.22 POROSITY AND PORE SIZE MEASUREMENT TECHNIQUES ON POROUS MEDIA

Crushing measure the volume of the porous material, crush it to remove the void space, and remeasure the volume.

- Optically this may involve filling the pores with a material such as black wax or Wood's metal, sectioning and inspecting with a microscope or scanning electron microscope.
- Imbibition weighing before and after filling the pores with a liquid.
- Gas Adsorption measure the change in pressure as a gas is adsorbed by the sample.
- Mercury Intrusion Measure the volume of mercury forced into the sample as a function of pressure.
- Thermoporosimetry fill the pores with a liquid, freeze it, then measure the heat evolved as the sample is warmed, until all the liquid is melted.
- NMR Cryoporometry fill the pores with a liquid, freeze it, then measure the amplitude of the NMR signal from the liquid component as the sample is warmed, until all the liquid is melted.
- Small Angle Neutron Scattering (SANS) scatter neutrons from the pores, then the smaller the dimensions of the variations in density distribution, the larger the angle through which the neutrons will be scattered.

Many of these methods give results that quite frequently differ from one another. This is often because they are in fact measuring different things some measurements are directly on the pores themselves. Others (such as mercury intrusion) are in effect measuring the necks that give access to the pores [97].

3.23 CARBON NANOTUBES-POLYMER MEMBRANE

Iijima discovered carbon nanotubes (CNTs) in 1991 and it was really a revolution in nanoscience because of their distinguished properties. CNTs have the unique electrical properties and extremely high thermal conductivity [98, 99] and high elastic modulus (>1 TPa), large elastic strain – up to 5%, and large breaking strain – up to 20%. Their excellent mechanical properties could lead to many applications [100]. For example, with their amazing strength and stiffness, plus the advantage of lightness, perspective future applications of CNTs are in aerospace engineering and virtual biodevices [101].

CNTs have been studied worldwide by scientists and engineers since their discovery, but a robust, theoretically precise and efficient prediction of the mechanical properties of CNTs has not yet been found. The problem is, when the size of an object is small to nanoscale, their many physical properties cannot be modeled and analyzed by using constitutive laws from traditional continuum theories, since the complex atomistic processes affect the results of their macroscopic behavior. Atomistic simulations can give more precise modeled results of the underlying physical properties. Due to atomistic simulations of a whole CNT are computationally infeasible at present, a new atomistic and continuum mixing modeling method is needed to solve the problem, which requires crossing the length and time scales. The research here is to develop a proper technique of spanning multi-scales from atomic to macroscopic space, in which the constitutive laws are derived from empirical atomistic potentials which deal with individual interactions between single atoms at the micro-level, whereas Cosserat continuum theories are adopted for a shell model through the application of the Cauchy-Born rule to give the properties which represent the averaged behavior of large volumes of atoms at the macro-level [102, 103]. Since experiments of CNTs are relatively expensive at present, and often unexpected manual errors could be involved, it will be very helpful to have a mature theoretical method for the study of mechanical properties of CNTs. Thus, if this research is successful, it could also be a reference for the research of all sorts of research at the nanoscale, and the results can be of interest to aerospace, biomedical engineering [104].

Subsequent investigations have shown that CNTs integrate amazing rigid and tough properties, such as exceptionally high elastic properties, large elastic strain, and fracture strain sustaining capability, which seem inconsistent and impossible in the previous materials. CNTs are the strongest fibers known. The Young's Modulus of SWNT is around 1TPa, which is 5 times greater than steel (200 GPa) while the density is only 1.2~1.4 g/cm^3. This means that materials made of nanotubes are lighter and more durable.

Beside their well-known extra-high mechanical properties, single-walled carbon nanotubes (SWNTs) offer either metallic or semiconductor characteristics based on the chiral structure of fullerene. They possess superior thermal and electrical properties so SWNTs are regarded as the most promising reinforcement material for the next generation of high performance structural and multifunctional composites, and evoke great interest in polymer based composites research. The SWNTs/polymer composites are theoretically predicted to have both exceptional mechanical and functional properties, which carbon fibers cannot offer [105].

3.23.1 CARBON NANOTUBES

Nanotubular materials are important "building blocks" of nanotechnology, in particular, the synthesis and applications of CNTs [82, 106, 107]. One application area has been the use of carbon nanotubes for molecular separations, owing to some of their unique properties. One such important property, extremely fast mass transport of molecules within carbon nanotubes associated with their low friction inner nanotube surfaces, has been demonstrated via computational and experimental studies [108, 109]. Furthermore, the behavior of adsorbate molecules in nano-confinement is fundamentally different than in the bulk phase, which could lead to the design of new sorbents [110].

Finally, their one-dimensional geometry could allow for alignment in desirable orientations for given separation devices to optimize the mass transport. Despite possessing such attractive properties, several intrinsic limitations of carbon nanotubes inhibit their application in large scale separation processes: the high cost of CNT synthesis and membrane formation (by microfabrication processes), as well as their lack of surface functionality, which significantly limits their molecular selectivity [111]. Although outer-surface modification of carbon nanotubes has been developed for nearly two decades, interior modification via covalent chemistry is still challenging due to the low reactivity of the inner-surface. Specifically, forming covalent bonds at inner walls of carbon nanotubes requires a transformation from sp^2 to sp^3 hybridization. The formation of sp^3 carbon is energetically unfavorable for concave surfaces [112].

Membrane is a potentially effective way to apply nanotubular materials in industrial-scale molecular transport and separation processes. Polymeric membranes are already prominent for separations applications due to their low fabrication and operation costs. However, the main challenge for using polymer membranes for future high-performance separations is to overcome the tradeoff between permeability and selectivity. A combination of the potentially high throughput and selectivity of nanotube materials with the process ability and mechanical strength of polymers may allow for the fabrication of scalable, high-performance membranes [113, 114].

3.23.2 STRUCTURE OF CARBON NANOTUBES

Two types of nanotubes exist in nature: multi-walled carbon nanotube (MWNTs), which were discovered by Iijima in 1991[82] and SWNTs, which were discovered by Bethune et al. in 1993 [115, 116].

Single-wall nanotube has only one single layer with diameters in the range of 0.6–1 nm and densities of 1.33–1.40 g/cm^3 [117] MWNTs are simply composed of concentric SWNTs with an inner diameter is from 1.5 to 15 nm and the outer diameter is from 2.5 nm to 30 nm [118]. SWNTs have better defined shapes of cylinder than MWNT, thus MWNTs have more possibilities of structure defects and their nanostructure is less stable. Their specific mechanical and electronic properties make them useful for future high strength/modulus materials and nanodevices. They exhibit low density, large elastic limit without breaking (of up to 20–30% strain before failure), exceptional elastic stiffness, greater than 1000GPa and their extreme strength which is more than 20 times higher than a high-strength steel alloy. Besides, they also posses superior thermal and elastic properties: thermal stability up to 2800°C in vacuum and up to 750°C in air, thermal conductivity about twice as high as diamond, electric current carrying capacity 1000 times higher than copper wire [119]. The properties of CNTs strongly depend on the size and the chirality and dramatically change when SWCNTs or MWCNTs are considered [120].

CNTs are formed from pure carbon bonds. Pure carbons only have two covalent bonds: sp^2 and sp^3. The former constitutes graphite and the latter constitutes diamond. The sp^2 hybridization, composed of one s orbital and two p orbitals, is a strong bond within a plane but weak between planes. When more bonds come together, they form six-fold structures, like honeycomb pattern, which is a plane structure, the same structure as graphite [121].

Graphite is stacked layer by layer so it is only stable for one single sheet. Wrapping these layers into cylinders and joining the edges, a tube of graphite is formed, called nanotube [122].

Atomic structure of nanotubes can be described in terms of tube chirality, or helicity, which is defined by the chiral vector, and the chiral angle, θ. Figure 3.17 shows visualized cutting a graphite sheet along the dotted lines and rolling the tube so that the tip of the chiral vector touches its tail. The chiral vector, often known as the roll-up vector, can be described by the following equation [123]:

$$C_h = na_1 + ma_2 \tag{3.3}$$

As shown in Figure 3.17, the integers (n, m) are the number of steps along the carbon bonds of the hexagonal lattice. Chiral angle determines the amount of "twist" in the tube. Two limiting cases exist where the chiral angle is at 0° and 30°. These limiting cases are referred to as zig-zag (0°)

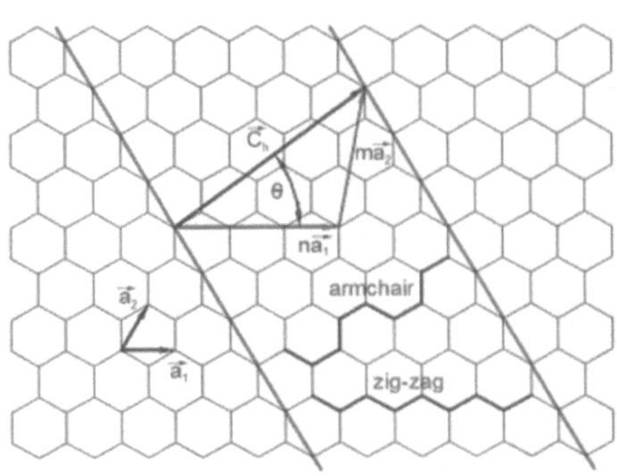

FIGURE 3.17 Schematic diagram showing how graphite sheet is 'rolled' to form CNT.

and armchair (30°), based on the geometry of the carbon bonds around the circumference of the nanotube. The difference in armchair and zig-zag nanotube structures is shown in Figure 3.18. In terms of the roll-up vector, the zig-zag nanotube is (n, 0) and the armchair nanotube is (n, n). The roll-up vector of the nanotube also defines the nanotube diameter since the inter-atomic spacing of the carbon atoms is known [105].

Chiral vector C_h is a vector that maps an atom of one end of the tube to the other. C_h can be an integer multiple a_1 of a_2, which are two basis vectors of the graphite cell. Then we have $C_h = a_1 + a_2$, with integer n and m, and the constructed CNT is called a (n, m) CNT, as shown in Figure 3.19. It can be proved that for armchair CNTs n=m, and for zigzag CNTs m=0. In Figure 3.19, the structure is designed to be a (4,0) zigzag SWCNT.

MWCNT can be considered as the structure of a bundle of concentric SWCNTs with different diameters. The length and diameter of MWCNTs are different from those of SWCNTs, which means, their properties differ significantly. MWCNTs can be modeled as a collection of SWCNTs, provided the interlayer interactions are modeled by Van der Waals forces in the simulation. A SWCNT can be modeled as a hollow cylinder by rolling a graphite sheet as presented in Figure 3.20.

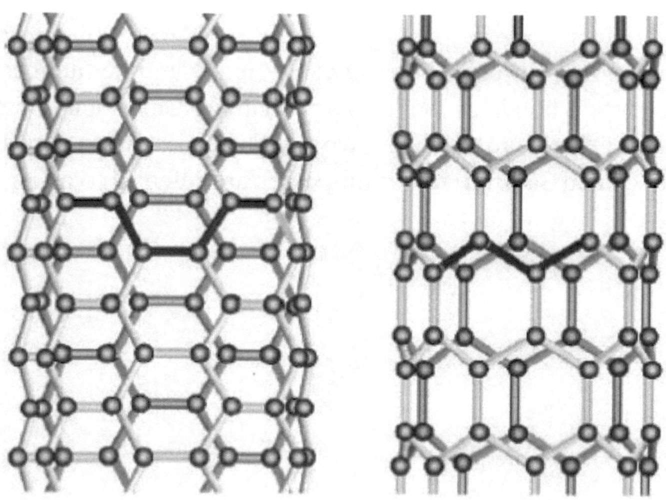

(a) (b)

FIGURE 3.18 Illustrations of the atomic structure (a) an armchair and (b) a zig-zag nanotube.

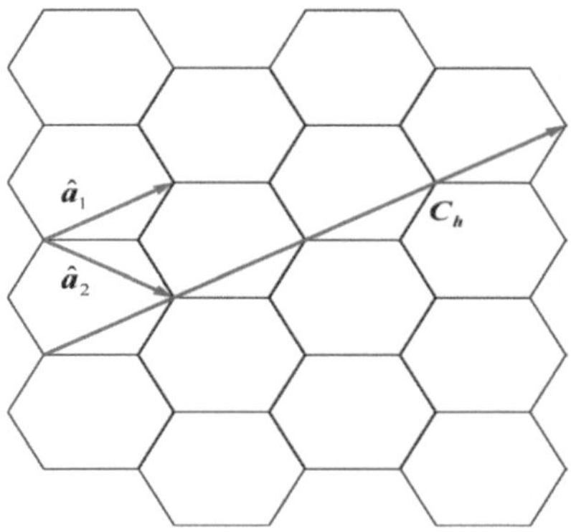

FIGURE 3.19 Basis vectors and chiral vector.

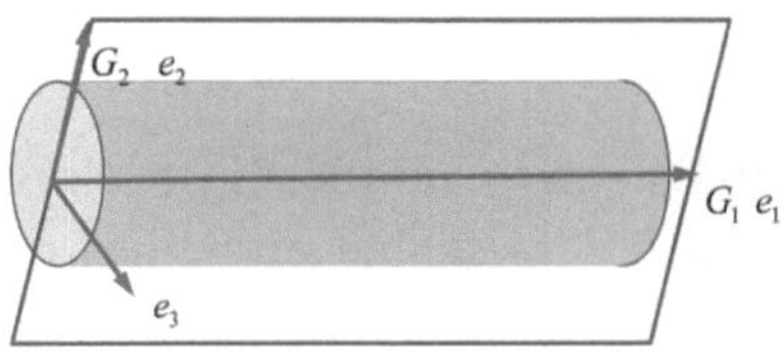

FIGURE 3.20 Illustration of a graphite sheet rolling to SWCNT.

If a planar graphite sheet is considered to be an undeformed configuration, and the SWCNT is defined as the current configuration, then the relationship between the SWCNT and the graphite sheet can be shown to be:

$$e_1 = G_1,\ \ e_2 = \mathrm{R}\sin\frac{\mathrm{G}_2}{\mathrm{R}},\quad e_3 = \mathrm{R}\cos\frac{\mathrm{G}_2}{\mathrm{R}} - \mathrm{R} \tag{3.4}$$

The relationship between the integer's n, m and the radius of SWCNT is given by:

$$R = a\sqrt{m^2 + mn + n^2}\ /\ 2\pi \tag{3.5}$$

where $a = \sqrt{3}a_0$, and a_0 is the length of a non-stretched C-C bond which is 0.142 nm [124].

As a graphite sheet can be 'rolled' into a SWCNT, we can 'unroll' the SWCNT to a plane graphite sheet. Since a SWCNT can be considered as a rectangular strip of hexagonal graphite monolayer rolling up to a cylindrical tube, the general idea is that it can be modeled as a cylindrical shell, a cylinder surface, or it can pull-back to be modeled as a plane sheet deforming into curved surface in three-dimensional space. A MWCNT can be modeled as a combination of a series of concentric SWCNTs with inter-layer inter-atomic reactions. Provided the continuum shell theory captures the deformation at the macro-level, the inner micro-structure can be described by finding the appropriate form of the potential function which is related to the position of the atoms at the atomistic level. Therefore, the SWCNT can be considered as a generalized continuum with microstructure [104].

3.23.4 CNT COMPOSITES

CNT composite materials cause significant development in nanoscience and nanotechnology. Their remarkable properties offer the potential for fabricating composites with substantially enhanced physical properties including conductivity, strength, elasticity, and toughness. Effective utilization of CNT in composite applications is dependent on the homogeneous distribution of CNTs throughout the matrix. Polymer-based nanocomposites are being developed for electronics applications such as thin-film capacitors in integrated circuits and solid polymer electrolytes for batteries. Research is being conducted throughout the world targeting the application of carbon nanotubes as materials for use in transistors, fuel cells, big TV screens, ultra-sensitive sensors, high-resolution Atomic Force Microscopy (AFM) probes, super-capacitor, transparent conducting film, drug carrier, catalysts, and composite material. Nowadays, there are more reports on the fluid transport through porous CNTs/polymer membrane.

3.23.5 STRUCTURAL DEVELOPMENT IN POLYMER/CNT FIBERS

The inherent properties of CNT assume that the structure is well preserved (large-aspect-ratio and without defects). The first step toward effective reinforcement of polymers using nano-fillers is to achieve a uniform dispersion of the fillers within the hosting matrix, and this is also related to the as-synthesized nano-carbon structure. Secondly, effective interfacial interaction and stress transfer between CNT and polymer is essential for improved mechanical properties of the fiber composite. Finally, similar to polymer molecules, the excellent intrinsic mechanical properties of CNT can be fully exploited only if an ideal uniaxial orientation is achieved. Therefore, during the fabrication of polymer/CNT fibers, four key areas need to be addressed and understood in order to successfully control the micro-structural development in these composites. These are: (i) CNT pristine structure, (ii) CNT dispersion, (iii) polymer–CNT interfacial interaction and (iv) orientation of the filler and matrix molecules (Figure 3.21). Figure 3.21 Four major factors affecting the micro-structural development in polymer/CNT composite fiber during processing [125].

Achieving homogenous dispersion of CNTs in the polymer matrix through strong interfacial interactions is crucial to the successful development of CNT/polymer nanocomposite [126]. As a result, various chemical or physical modifications can be applied to CNTs to improve its dispersion and compatibility with polymer matrix. Among these approaches acid treatment is considered most convenient, in which hydroxyl and carboxyl groups generated would concentrate on the ends of the CNT and at defect sites, making them more reactive and thus better dispersed [127, 128].

The incorporation of functionalized CNTs into composite membranes are mostly carried out on flat sheet membranes [129, 130]. For considering the potential influences of CNTs on the physicochemical properties of

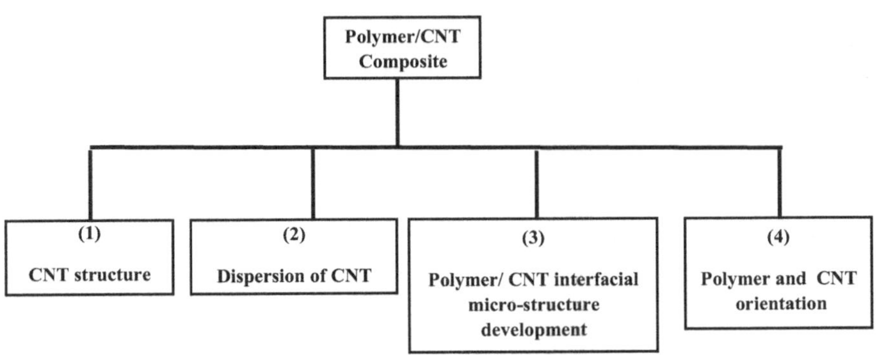

FIGURE 3.21 Four major factors affecting the micro-structural development in polymer/CNT composite fiber during processing.

dope solution [131] and change of membrane formation route originated from various additives [132], it is necessary to study the effects of CNTs on the morphology and performance.

3.23.6 GENERAL FABRICATION PROCEDURES FOR POLYMER/CNT FIBERS

In general, when discussing polymer/CNT composites, two major classes come to mind. First, the CNT nano-fillers are dispersed within a polymer at a specified concentration, and the entire mixture is fabricated into a composite. Secondly, as grown CNT are processed into fibers or films, and this macroscopic CNT material is then embedded into a polymer matrix [133]. The four major fiber-spinning methods (Figure 3.22) used for polymer/CNT composites from both the solution and melt include

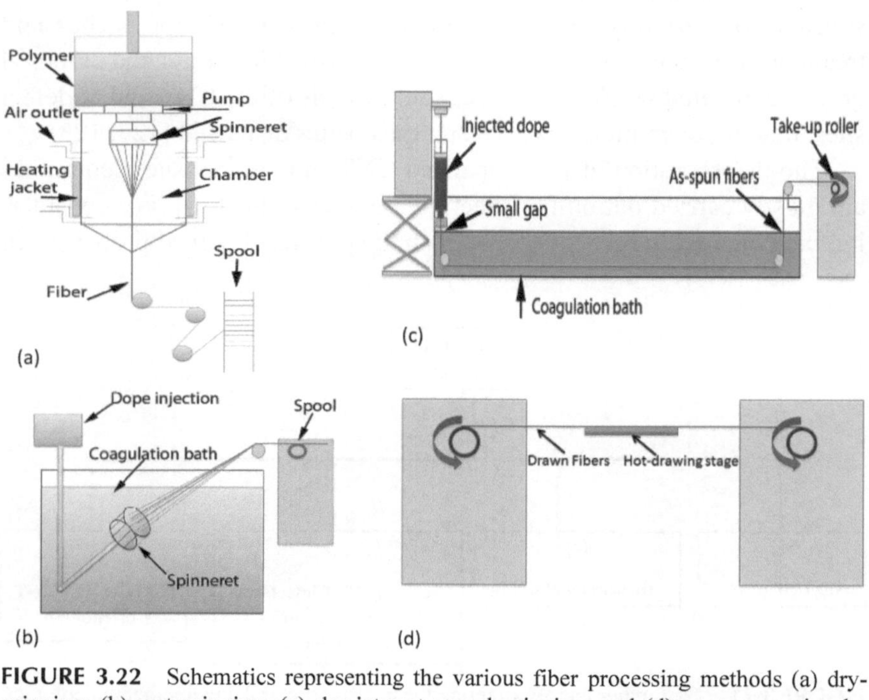

FIGURE 3.22 Schematics representing the various fiber processing methods (a) dry-spinning; (b) wet-spinning; (c) dry-jet wet or gel-spinning; and (d) post-processing by hot-stage drawing.

dry-spinning [134], wet-spinning [135], dry-jet wet spinning (gel-spinning), and electrospinning [136]. An ancient solid-state spinning approach has been used for fabricating 100% CNT fibers from both forests and aero gels. Irrespective of the processing technique, in order to develop high-quality fibers many parameters need to be well controlled.

All spinning procedures generally involve:

(i) fiber formation;
(ii) coagulation/gelation/solidification; and
(iii) drawing/alignment.

For all of these processes, the even dispersion of the CNT within the polymer solution or melt is very important. However, in terms of achieving excellent axial mechanical properties, alignment and orientation of the polymer chains and the CNT in the composite is necessary. Fiber alignment is accomplished in post-processing such as drawing/annealing and is key to increasing crystallinity, tensile strength, and stiffness [137].

3.24 FILTER APPLICATIONS

Nonwovens composed of fibers made from glass, paper or polymers are highly porous membranes – the total porosity typically being of the order of 80 to 95% – which can be used to remove solid particles, dust particles, aerosols, fine fluid droplets from a stream either composed of a gas or a fluid. Water filtration is a topic that is of enormous importance worldwide. Air filtration is highly important for a broad range of industrial applications including power plants, and the same holds for fuel filtration – a must in modern car engines – or coalescence filtration of gasoline for airplanes. Typical high-efficiency filter requirements are that the filters should capture all fluid or solid particles, respectively, surpassing a specified size and that the capture probability should be as high as possible, say in the range of 99 or 99.9% [138–141].

3.24.1 ANTIMICROBIAL AIR FILTER

It is well known that heating, ventilating, and air conditioning (HVAC) air filters usually operated in dark, damp, and ambient temperature conditions, which is susceptible for bacterial, mold, and fungal attacks, resulting in unpredictable deterioration and bad odor. To solve this problem, functionalization of the surface of filtering media with antimicrobial agents for long-lasting durable antimicrobial functionality is of current interest. In 2007, Jeong and Youk [142] explored the electrospun polyurethane cationomer (PUCs) nanofiber mats with different amounts of quaternary ammonium groups in antimicrobial air filter. They found that PUCs exhibited very strong antimicrobial activities against *Staphylococcus aureus* and *Escherichia coli*. Ramakrishna and co-workers [143] induced the silver nanoparticles based on different electrospun polymer [cellulose acetate (CA); polyacrylonitrile (PAN); and polyvinylchloride (PVC)] nanofiber for antimicrobial functionality owing to the remarkable antimicrobial ability of silver ions and silver compounds.

3.24.2 BASIC PROCESSES CONTROLLING FILTER EFFICIENCIES

It is helpful at this stage to recall some basic processes controlling filter efficiencies in general, that is, to a first approximation independent of the fiber diameters [144, 145]. What is known for conventional nonwovens composed of fibers with diameters well in the 10–100 micrometer range is that the filter efficiency is controlled by various types of capture processes happening within the nonwoven as the gas/fluids carrying particles pass through their pores. These basic processes are depicted in Figure 3.23.

The first process to be considered is the interception. Particles following the gas stream around the fiber as depicted in Figure 3.23a are intercepted by the fiber surfaces if the particles pass the fibers at a distance not larger than the particle diameter. It is obvious that larger particles tend to enhance the probability for such an interception.

The second process of importance is the impaction, as shown schematically in Figure 3.23b. Particles do not follow in this case the deflection of the gas stream due to the neighborhood of the fiber as a solid object but because of inertia effects follow the original path. This in turn causes the particle to impact on the fiber surface. Impaction tends to grow in importance as the flow velocity of the gas increases.

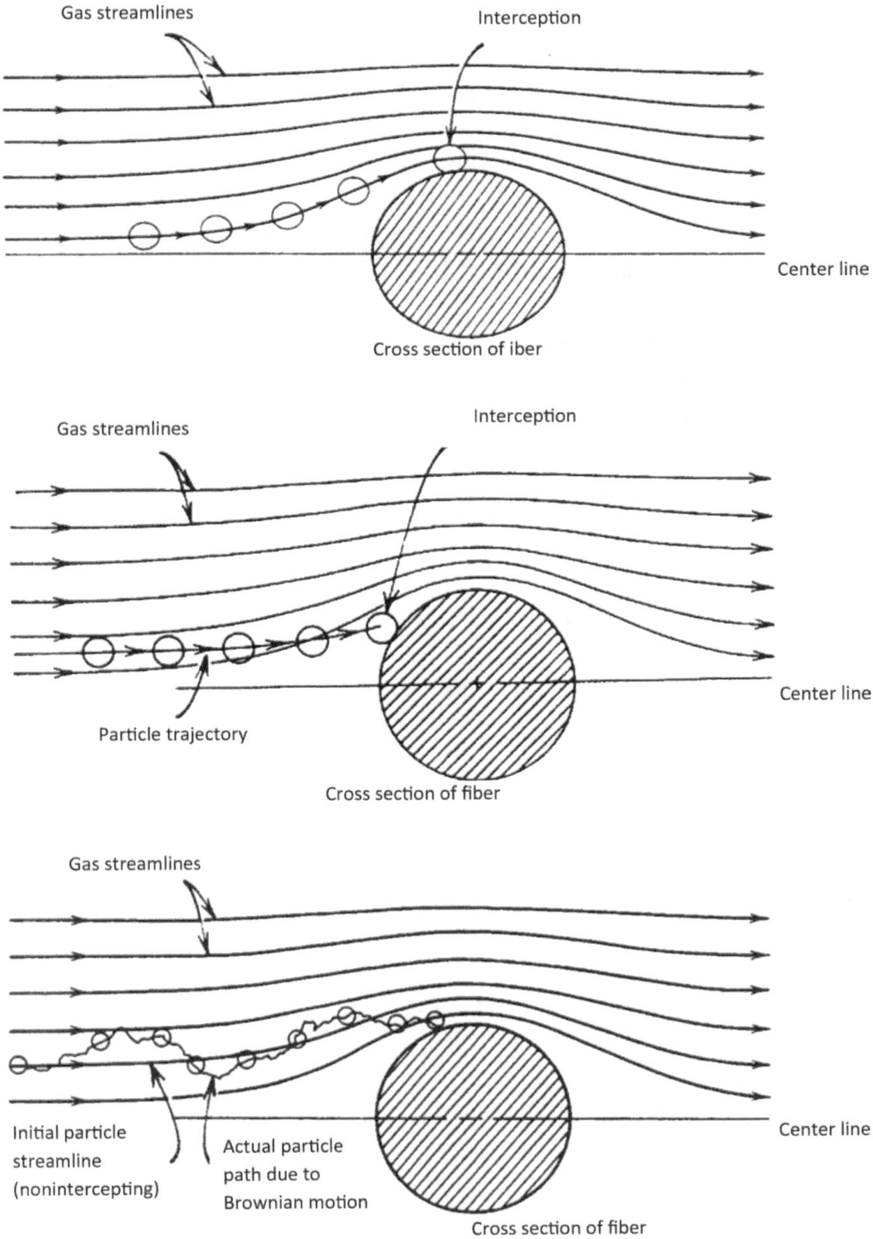

FIGURE 3.23 Molecular processes contributing to filtering effects: (a) interception, (b) impaction, (c) diffusion.

Finally, diffusion plays a role in controlling the capture efficiency. Here, particles carried by streamlines that pass the fiber at sufficient distance not to cause a direct interception nevertheless come into contact with the fiber surface because of diffusional motions, as depicted in Figure 3.23c. Diffusion tends to be of importance for smaller particles and low flow velocities.

Figure 3.24 gives a survey on the regimes in a flow-rate/particle-diameter diagram in which either the diffusion, the interception or finally the impaction dominate.

So, at low particle sizes diffusion more or less dominates the control of the filter efficiency, particularly for small flow velocities, whereas the impaction dominates for large flow velocities and particle sizes, with the three processes contributing in different ratios at intermediate particle sizes and flow velocities.

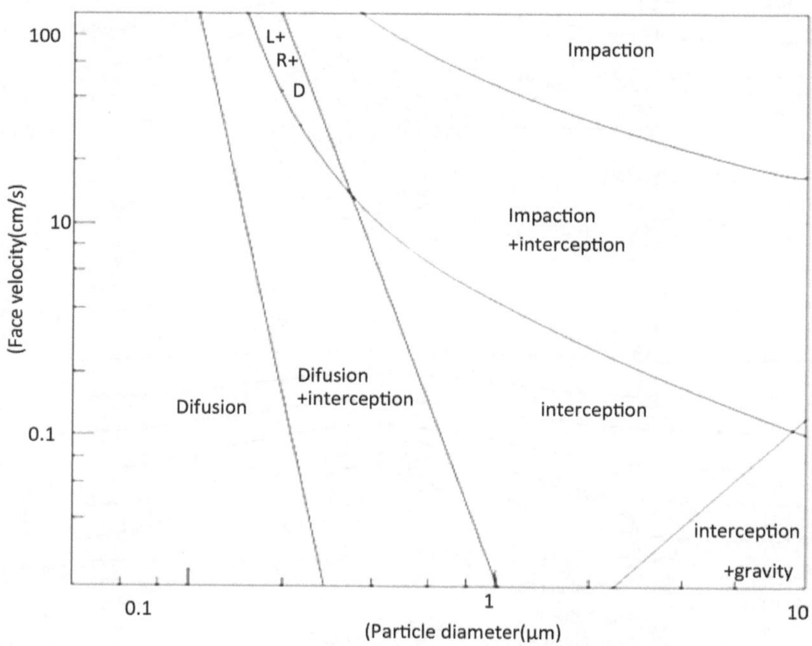

FIGURE 3.24 Survey on the regimes in a flow rate/particle diameter diagram in which either the diffusion, the interception or finally the impaction dominate diffusion.

3.25 CONCLUDING REMARK

In recent years, membrane separations have been applied in various industries, such as chemical, food, pharmaceutical, automobile-manufacturing, and metal-finishing industries. As the most popular membranes, polymeric membranes still have an inherent drawback—the tradeoff effect between permeability and selectivity, which means that more permeable membranes are generally less selective and vice versa. Hybrid membrane incorporating both organic and inorganic components is a convenient and efficient approach to avoid the tradeoff effect. Membranes commonly consist of a porous support layer with a thin dense layer on top that forms the actual membrane. Several researches are investigating the effects of incorporation of CNTs to develop mixed matrix membranes. In these membranes, the choice of both these components is a problem of materials selection, and also involves several fundamental issues, such as polymer-chain rigidity, free volume, and the altered interface – all of which influence transport through the membrane.

KEYWORDS

- **CNT membranes**
- **membrane**
- **membrane filtration**
- **membrane types**

REFERENCES

1. Wente, V. A., *Superfine Thermoplastic Fibers.* Industrial & Engineering Chemistry, 1956, 48(8), 1342–1346.
2. Buntin, R. R. and D. T. Lohkamp, *Melt Blowing-One-Step WEB Process for New Nonwoven Products.* Tappi, 1973, 56(4), 74–77.
3. Zhou, F. L. and R. H. Gong, *Manufacturing Technologies of Polymeric Nanofibers and Nanofiber Yarns.* Polymer International, 2008, 57(6), 837–845.

4. Angadjivand, S., R. Kinderman, and T. Wu, *High Efficiency Synthetic Filter Medium*, 2000, Google Patents.
5. Zeleny, J., *The electrical discharge from Liquid Points, and a Hydrostatic Method of measuring the electric intensity at their surface.* Physical Review, 1914, 3(2), 69–91.
6. Formhals, A., *Process and Apparatus Fob Pbepabing*, 1934, Google Patents.
7. Taylor, G., *Disintegration of Water Drops in an Electric Field.* Proceedings of the Royal Society of London. Series A. Mathematical and Physical Sciences, 1964, 280(1382), 383–397.
8. Gibson, P., H. S. Gibson, and D. Rivin, *Transport Properties of Porous Membranes Based on Electrospun Nanofibers.* Colloids and Surfaces A: Physicochemical and Engineering Aspects, 2001, 187, 469–481.
9. Patarin, J., B. Lebeau, and R. Zana, *Recent Advances in the Formation Mechanisms of Organized Mesoporous Materials.* Current Opinion in Colloid & Interface Science, 2002, 7(1), 107–115.
10. Weghmann, A. *Production of Electrostatic Spun Synthetic Microfiber Nonwovens and Applications in Filtration.* in *Proceedings of the 3rd World Filtration Congress, Filtration Society*. 1982, London.
11. Kruiēska, I., E. Klata, and M. Chrzanowski, *New Textile Materials for Environmental Protection*, in *Intelligent Textiles for Personal Protection and Safety*. 2006, 41–53.
12. Yarin, A. L. and E. Zussman, *Upward Needleless Electrospinning of Multiple Nanofibers.* Polymer, 2004, 45(9), 2977–2980.
13. Majeed, S., et al., *Multi-Walled Carbon Nanotubes (MWCNTs) Mixed Polyacrylonitrile (PAN) Ultrafiltration Membranes.* Journal of Membrane Science, 2012, 403, 101–109.
14. Macedonio, F. and E. Drioli, *Pressure-Driven Membrane Operations and Membrane Distillation Technology Integration for Water Purification.* Desalination, 2008, 223(1), 396–409.
15. Merdaw, A. A., A. O. Sharif, and G. A. W. Derwish, *Mass Transfer in Pressure-Driven Membrane Separation Processes, Part II.* Chemical Engineering Journal, 2011, 168(1), 229–240.
16. Van Der Bruggen, B., et al., *A Review of Pressure-Driven Membrane Processes in Wastewater Treatment and Drinking Water Production.* Environmental Progress, 2003, 22(1), 46–56.
17. Cui, Z. F. and H. S. Muralidhara, *Membrane Technology: A Practical Guide to Membrane Technology and Applications in Food and Bioprocessing*. 2010, Elsevier. 288.
18. Shirazi, S., C. J. Lin, and D. Chen, *Inorganic Fouling of Pressure-Driven Membrane Processes — A Critical Review.* Desalination, 2010, 250(1), 236–248.
19. Pendergast, M. M. and E. M. V. Hoek, *A Review of Water Treatment Membrane Nanotechnologies.* Energy & Environmental Science, 2011, 4(6), 1946–1971.
20. Hilal, N., et al., *A comprehensive review of nanofiltration membranes: Treatment, pretreatment, modeling, and atomic force microscopy.* Desalination, 2004, 170(3), 281–308.
21. Srivastava, A., S. Srivastava, and K. Kalaga, *Carbon Nanotube Membrane Filters*, in *Springer Handbook of Nanomaterials*. 2013, Springer. 1099–1116.
22. Colombo, L. and A. L. Fasolino, *Computer-Based Modeling of Novel Carbon Systems and Their Properties: Beyond Nanotubes*. Vol. 3. 2010, Springer. 258.

23. Polarz, S. and B. Smarsly, *Nanoporous Materials.* Journal of Nanoscience and Nanotechnology, 2002, 2(6), 581–612.
24. Gray-Weale, A. A., et al., *Transition-state theory model for the diffusion coefficients of small penetrants in glassy polymers.* Macromolecules, 1997, 30(23), 7296–7306.
25. Rigby, D. and R. Roe, *Molecular Dynamics Simulation of Polymer Liquid and Glass. I. Glass Transition.* The Journal of Chemical Physics, 1987, 87, 7285.
26. Freeman, B. D., Y. P. Yampolskii, and I. Pinnau, *Materials Science of Membranes for Gas and Vapor Separation.* 2006, Wiley.com. 466.
27. Hofmann, D., et al., *Molecular Modeling Investigation of Free Volume Distributions in Stiff Chain Polymers with Conventional and Ultrahigh Free Volume: Comparison Between Molecular Modeling and Positron Lifetime Studies.* Macromolecules, 2003, 36(22), 8528–8538.
28. Greenfield, M. L. and D. N. Theodorou, *Geometric Analysis of Diffusion Pathways in Glassy and Melt Atactic Polypropylene.* Macromolecules, 1993, 26(20), 5461–5472.
29. Baker, R. W., *Membrane Technology and Applications.* 2012, John Wiley & Sons. 592.
30. Strathmann, H., L. Giorno, and E. Drioli, *Introduction to Membrane Science and Technology.* 2011, Wiley-VCH Verlag & Company. 544.
31. Chen, J. P., et al., *Membrane Separation: Basics and Applications*, in *Membrane and Desalination Technologies*, L. K. Wang, et al., Editors. 2008, Humana Press. p. 271–332.
32. Mortazavi, S., *Application of Membrane Separation Technology to Mitigation of Mine Effluent and Acidic Drainage.* 2008, Natural Resources Canada. 194.
33. Porter, M. C., *Handbook of Industrial Membrane Technology.* 1990, Noyes Publications. 604.
34. Naylor, T. V., *Polymer Membranes: Materials, Structures and Separation Performance.* 1996, Rapra Technology Limited. 136.
35. Freeman, B. D., *Introduction to Membrane Science and Technology. By Heinrich Strathmann.* Angewandte Chemie International Edition, 2012, 51(38), 9485–9485.
36. Kim, I., H. Yoon, and K. M. Lee, *Formation of Integrally Skinned Asymmetric Polyetherimide Nanofiltration Membranes by Phase Inversion Process.* Journal of Applied Polymer Science, 2002, 84(6), 1300–1307.
37. Khulbe, K. C., C. Y. Feng, and T. Matsuura, *Synthetic Polymeric Membranes: Characterization by Atomic Force Microscopy.* 2007, Springer. 198.
38. Loeb, L. B., *The Kinetic Theory of Gases.* 2004, Courier Dover Publications. 678.
39. Koros, W. J. and G. K. Fleming, *Membrane-Based Gas Separation.* Journal of Membrane Science, 1993, 83(1), 1–80.
40. Perry, J. D., K. Nagai, and W. J. Koros, *Polymer Membranes for Hydrogen Separations.* MRS bulletin, 2006, 31(10), 745–749.
41. Hiemenz, P. C. and R. Rajagopalan, *Principles of Colloid and Surface Chemistry, revised and expanded.* Vol. 14. 1997, CRC Press.
42. McDowell-Boyer, L. M., J. R. Hunt, and N. Sitar, *Particle transport through porous media.* Water Resources Research, 1986, 22(13), 1901–1921.
43. Auset, M. and A. A. Keller, *Pore-scale processes that control dispersion of colloids in saturated porous media.* Water Resources Research, 2004, 40(3).

44. Bhave, R. R., *Inorganic membranes synthesis, characteristics, and applications*. Vol. 312. 1991, Springer.
45. Lin, V. S.-Y., et al., *A porous silicon-based optical interferometric biosensor.* Science, 1997, 278(5339), 840–843.
46. Hedrick, J., et al. *Templating nanoporosity in organosilicates using well-defined branched macromolecules.* in *Materials Research Society Symposium Proceedings.* 1998, Cambridge University Press.
47. Hubbell, J. A. and R. Langer, *Tissue engineering* Chem. Eng. News 1995, 13, 42–45.
48. Schaefer, D. W., *Engineered porous materials* MRS Bulletin 1994, 19, 14–17.
49. Hentze, H. P. and M. Antonietti, *Porous Polymers and Resins.* Handbook of Porous Solids: p. 1964–2013.
50. Endo, A., et al., *Synthesis of ordered microporous silica by the solvent evaporation method.* Journal of materials science, 2004, 39(3), 1117–1119.
51. Sing, K., et al., *Physical and biophysical chemistry division commission on colloid and surface chemistry including catalysis.* Pure and Applied Chemistry, 1985, 57(4), 603–619.
52. Kresge, C., et al., *Ordered mesoporous molecular sieves synthesized by a liquid-crystal template mechanism.* Nature, 1992, 359(6397), 710–712.
53. Yang, P., et al., *Generalized syntheses of large-pore mesoporous metal oxides with semicrystalline frameworks.* Nature, 1998, 396(6707), 152–155.
54. Jiao, F., K. M. Shaju, and P. G. Bruce, *Synthesis of Nanowire and Mesoporous Low-Temperature LiCoO2 by a Post-Templating Reaction.* Angewandte Chemie International Edition, 2005, 44(40), 6550–6553.
55. Ryoo, R., et al., *Ordered mesoporous carbons.* Advanced Materials, 2001, 13(9), 677–681.
56. Beck, J., et al., *Chu, DH Olson, EW Sheppard, SB McCullen, JB Higgins and JL Schlenker.* J. Am. Chem. Soc., 1992, 114(10), 834.
57. Zhao, D., et al., *Triblock copolymer syntheses of mesoporous silica with periodic 50 to 300 angstrom pores.* Science, 1998, 279(5350), 548–552.
58. Joo, S. H., et al., *Ordered nanoporous arrays of carbon supporting high dispersions of platinum nanoparticles. Nature,* 2001, 412(6843), 169–172.
59. Kruk, M., et al., *Synthesis and characterization of hexagonally ordered carbon nanopipes.* Chemistry of materials, 2003, 15(14), 2815–2823.
60. Rouquerol, J., et al., *Recommendations for the characterization of porous solids (Technical Report).* Pure and Applied Chemistry, 1994, 66(8), 1739–1758.
61. Schüth, F., K. S. W. Sing, and J. Weitkamp, *Handbook of Porous Solids*. 2002, Wiley-VCH.
62. Maly, K. E., *Assembly of nanoporous organic materials from molecular building blocks.* Journal of Materials Chemistry, 2009, 19(13), 1781–1787.
63. Davis, M. E., *Ordered porous materials for emerging applications. Nature,* 2002, 417(6891), 813–821.
64. Morris, R. E. and P. S. Wheatley, *Gas storage in nanoporous materials.* Angewandte Chemie International Edition, 2008, 47(27), 4966–4981.
65. Cote, A. P., et al., *Porous, crystalline, covalent organic frameworks.* Science, 2005, 310(5751), 1166–1170.

66. El-Kaderi, H. M., et al., *Designed synthesis of 3D covalent organic frameworks.* Science, 2007, 316(5822), 268–272.
67. Jiang, J. X., et al., *Conjugated microporous poly (aryleneethynylene) networks.* Angewandte Chemie International Edition, 2007, 46(45), 8574–8578.
68. Ben, T., et al., *Targeted synthesis of a porous aromatic framework with high stability and exceptionally high surface area.* Angewandte Chemie, 2009, 121(50), 9621–9624.
69. Eddaoudi, M., et al., *Modular chemistry: secondary building units as a basis for the design of highly porous and robust metal-organic carboxylate frameworks.* Accounts of Chemical Research, 2001, 34(4), 319–330.
70. Lu, G. Q., X. S. Zhao, and T. K. Wei, *Nanoporous Materials: Science and Engineering.* Vol. 4. 2004, Imperial College Press.
71. Holister, P., C. R. Vas, and T. Harper, *Nanocrystalline materials.* Technologie White Papers, 2003(4).
72. Smith, B. L., et al., *Molecular mechanistic origin of the toughness of natural adhesives, fibers and composites.* Nature, 1999, 399(6738), 761–763.
73. Mann, S. and G. A. Ozin, *Synthesis of inorganic materials with complex form. Nature,* 1996, 382(6589), 313–318.
74. Mann, S., *Molecular tectonics in biomineralization and biomimetic materials chemistry.* Nature, 1993, 365(6446), 499–505.
75. Busch, K. and S. John, *Photonic band gap formation in certain self-organizing systems.* Physical Review E, 1998, 58(3), 3896.
76. Argyros, A., et al., *Electron tomography and computer visualization of a three-dimensional 'photonic' crystal in a butterfly wing-scale.* Micron, 2002, 33(5), 483–487.
77. Sailor, M. J. and K. L. Kavanagh, *Porous silicon–what is responsible for the visible luminescence?* Advanced Materials, 1992, 4(6), 432–434.
78. Koshida, N. and B. Gelloz, *Wet and dry porous silicon.* Current opinion in colloid & interface science, 1999, 4(4), 309–313.
79. Hentze, H.-P. and M. Antonietti, *Template synthesis of porous organic polymers.* Current Opinion in Solid State and Materials Science, 2001, 5(4), 343–353.
80. Nakao, S.-i., *Determination of pore size and pore size distribution: 3. Filtration membranes.* Journal of Membrane Science, 1994, 96(1), 131–165.
81. Barrer, R. M., *Zeolites and their synthesis.* Zeolites, 1981, 1(3), 130–140.
82. Iijima, S., *Helical microtubules of graphitic carbon. Nature,* 1991, 354(6348), 56–58.
83. Kaneko, K. and K. Inouye, *Adsorption of water on FeOOH as studied by electrical conductivity measurements.* Bulletin of the Chemical Society of Japan, 1979, 52(2), 315–320.
84. Maeda, K., et al., *Control with polyethers of pore distribution of alumina by the sol-gel method.* Chem. & Ind., 1989(23), 807.
85. Matsuzaki, S., M. Taniguchi, and M. Sano, *Polymerization of benzene occluded in graphite-alkali metal intercalation compounds.* Synthetic metals, 1986, 16(3), 343–348.
86. Enoki, T., H. Inokuchi, and M. Sano, *ESR study of the hydrogen-potassium-graphite ternary intercalation compounds.* Physical Review B, 1988, 37(16), 9163.
87. Pinnavaia, T. J., *Intercalated clay catalysts.* Science, 1983, 220(4595), 365–371.
88. Yamanaka, S., et al., *High surface area solids obtained by intercalation of iron oxide pillars in montmorillonite.* Materials research bulletin, 1984, 19(2), 161–168.

89. Inagaki, S., Y. Fukushima, and K. Kuroda, *Synthesis of highly ordered mesoporous materials from a layered polysilicate.* J. Chem. Soc., Chem. Commun., 1993(8), 680–682.
90. Vallano, P. T. and V. T. Remcho, *Modeling interparticle and intraparticle (perfusive) electroosmotic flow in capillary electrochromatography.* Analytical Chemistry, 2000, 72(18), 4255–4265.
91. Levenspiel, O., *Chemical reaction engineering.* Vol. 2. 1972, Wiley New York etc.
92. Li, Q., et al., *Interparticle and intraparticle mass transfer in chromatographic separation.* Bioseparation, 1995, 5(4), 189–202.
93. Setoyama, N., et al., *Surface characterization of microporous solids with helium adsorption and small angle x-ray scattering.* Langmuir, 1993, 9(10), 2612–2617.
94. Marsh, H., *Introduction to carbon science.* 1989.
95. Kaneko, K., *Determination of pore size and pore size distribution: 1. Adsorbents and catalysts.* Journal of Membrane Science, 1994, 96(1), 59–89.
96. Seaton, N. and J. Walton, *A new analysis method for the determination of the pore size distribution of porous carbons from nitrogen adsorption measurements.* Carbon, 1989, 27(6), 853–861.
97. Dullien, F. A., *Porous media: fluid transport and pore structure.* 1991, Academic press.
98. Yang, W., et al., *Carbon Nanotubes for Biological and Biomedical Applications.* Nanotechnology, 2007, 18(41), 412001.
99. Bianco, A., et al., *Biomedical Applications of Functionalized Carbon Nanotubes.* Chemical Communications, 2005(5), 571–577.
100. Salvetat, J., et al., *Mechanical Properties of Carbon Nanotubes.* Applied Physics A, 1999, 69(3), 255–260.
101. Zhang, X., et al., *Ultrastrong, Stiff, and Lightweight Carbon-Nanotube Fibers.* Advanced Materials, 2007, 19(23), 4198–4201.
102. Arroyo, M. and T. Belytschko, *Finite Crystal Elasticity of Carbon Nanotubes Based on the Exponential Cauchy-Born Rule.* Physical Review B, 2004, 69(11), 115415.
103. Wang, J., et al., *Energy and Mechanical Properties of Single-Walled Carbon Nanotubes Predicted Using the Higher Order Cauchy-Born rule.* Physical Review B, 2006, 73(11), 115428.
104. Zhang, Y., *Single-walled carbon nanotube modeling based on one-and two-dimensional Cosserat continua,* 2011, University of Nottingham.
105. Wang, S., *Functionalization of Carbon Nanotubes: Characterization, Modeling and Composite Applications.* 2006, Florida State University. 193.
106. Lau, K.-t., C. Gu, and D. Hui, *A critical review on nanotube and nanotube/nanoclay related polymer composite materials.* Composites Part B: Engineering, 2006, 37(6), 425–436.
107. Choi, W., et al., *Carbon Nanotube-Guided Thermopower Waves.* Materials Today, 2010, 13(10), 22–33.
108. Sholl, D. S. and J. Johnson, *Making High-Flux Membranes with Carbon Nanotubes.* Science, 2006, 312(5776), 1003–1004.
109. Zang, J., et al., *Self-Diffusion of Water and Simple Alcohols in Single-Walled Aluminosilicate Nanotubes.* ACS nano, 2009, 3(6), 1548–1556.
110. Talapatra, S., V. Krungleviciute, and A. D. Migone, *Higher Coverage Gas Adsorption on the Surface of Carbon Nanotubes: Evidence for a Possible New Phase in the Second Layer.* Physical Review Letters, 2002, 89(24), 246106.

111. Pujari, S., et al., *Orientation Dynamics in Multiwalled Carbon Nanotube Dispersions Under Shear Flow.* The Journal of Chemical Physics, 2009, 130, 214903.
112. Singh, S. and P. Kruse, *Carbon Nanotube Surface Science.* International Journal of Nanotechnology, 2008, 5(9), 900–929.
113. Baker, R. W., *Future Directions of Membrane Gas Separation Technology.* Industrial & Engineering Chemistry Research, 2002, 41(6), 1393–1411.
114. Erucar, I. and S. Keskin, *Screening Metal–Organic Framework-Based Mixed-Matrix Membranes for CO_2/CH4 Separations.* Industrial & Engineering Chemistry Research, 2011, 50(22), 12606–12616.
115. Bethune, D. S., et al., *Cobalt-Catalyzed Growth of Carbon Nanotubes with Single-Atomic-Layer Walls.* Nature 1993, 363, 605–607.
116. Iijima, S. and T. Ichihashi, *Single-Shell Carbon Nanotubes of 1-nm Diameter.* Nature, 1993, 363, 603–605.
117. Treacy, M., T. Ebbesen, and J. Gibson, *Exceptionally high Young's modulus observed for individual carbon nanotubes.* 1996.
118. Wong, E. W., P. E. Sheehan, and C. Lieber, *Nanobeam Mechanics: Elasticity, Strength, and Toughness of Nanorods and Nanotubes.* Science, 1997, 277(5334), 1971–1975.
119. Thostenson, E. T., C. Li, and T. W. Chou, *Nanocomposites in Context.* Composites Science and Technology, 2005, 65(3), 491–516.
120. Barski, M., P. Kędziora, and M. Chwał, *Carbon Nanotube/Polymer Nanocomposites: A Brief Modeling Overview.* Key Engineering Materials, 2013, 542, 29–42.
121. Dresselhaus, M. S., G. Dresselhaus, and P. C. Eklund, *Science of Fullerenes and Carbon nanotubes: Their Properties and Applications.* 1996, Academic Press. 965.
122. Yakobson, B. and R. E. Smalley, *Some Unusual New Molecules—Long, Hollow Fibers with Tantalizing Electronic and Mechanical Properties—have Joined Diamonds and Graphite in the Carbon Family.* Am Scientist, 1997, 85, 324–337.
123. Guo, Y. and W. Guo, *Mechanical and Electrostatic Properties of Carbon Nanotubes under Tensile Loading and Electric Field.* Journal of Physics D: Applied Physics, 2003, 36(7), 805.
124. Berger, C., et al., *Electronic Confinement and Coherence in Patterned Epitaxial Graphene.* Science, 2006, 312(5777), 1191–1196.
125. Song, K., et al., *Structural Polymer-Based Carbon Nanotube Composite Fibers: Understanding the Processing–Structure–Performance Relationship.* Materials, 2013, 6(6), 2543–2577.
126. Park, O. K., et al., *Effect of Surface Treatment with Potassium Persulfate on Dispersion Stability of Multi-Walled Carbon Nanotubes.* Materials Letters, 2010, 64(6), 718–721.
127. Banerjee, S., T. Hemraj-Benny, and S. S. Wong, *Covalent Surface Chemistry of Single-Walled Carbon Nanotubes.* Advanced Materials, 2005, 17(1), 17–29.
128. Balasubramanian, K. and M. Burghard, *Chemically Functionalized Carbon Nanotubes.* Small, 2005, 1(2), 180–192.
129. Xu, Z. L. and F. Alsalhy Qusay, *Polyethersulfone (PES) Hollow Fiber Ultrafiltration Membranes Prepared by PES/non-Solvent/NMP Solution.* Journal of Membrane Science, 2004, 233(1–2), 101–111.
130. Chung, T. S., J. J. Qin, and J. Gu, *Effect of Shear Rate Within the Spinneret on Morphology, Separation Performance and Mechanical Properties of Ultrafiltration*

Polyethersulfone Hollow Fiber Membranes. Chemical Engineering Science, 2000, 55(6), 1077–1091.
131. Choi, J. H., J. Jegal, and W. N. Kim, *Modification of Performances of Various Membranes Using MWNTs as a Modifier.* Macromolecular Symposia, 2007, 249–250(1), 610–617.
132. Wang, Z. and J. Ma, *The Role of Nonsolvent in-Diffusion Velocity in Determining Polymeric Membrane Morphology.* Desalination, 2012, 286(0), 69–79.
133. Vilatela, J. J., R. Khare, and A. H. Windle, *The Hierarchical Structure and Properties of Multifunctional Carbon Nanotube Fiber Composites.* Carbon, 2012, 50(3), 1227–1234.
134. Benavides, R. E., S. C. Jana, and D. H. Reneker, *Nanofibers from Scalable Gas Jet Process.* ACS Macro Letters, 2012, 1(8), 1032–1036.
135. Gupta, V. B. and V. K. Kothari, *Manufactured Fiber Technology.* 1997, Springer. 661.
136. Wang, T. and S. Kumar, *Electrospinning of Polyacrylonitrile Nanofibers.* Journal of Applied Polymer Science, 2006, 102(2), 1023–1029.
137. Song, K., et al., *Lubrication of Poly (vinyl alcohol) Chain Orientation by Carbon nano-chips in Composite Tapes.* Journal of Applied Polymer Science, 2013, 127(4), 2977–2982.
138. Filatov, Y., A. Budyka, and V. Kirichenko, *Electrospinning of micro-and nanofibers: fundamentals in separation and filtration processes.* 2007, Begell House Inc., Redding, CT.
139. Brown, R. C., *Air filtration: an integrated approach to the theory and applications of fibrous filters.* Vol. 650. 1993, Pergamon Press Oxford.
140. Hinds, W. C., *Aerosol technology: properties, behavior, and measurement of airborne particles.* 1982.
141. Greiner, A. and J. Wendorff, *Functional self-assembled nanofibers by electrospinning,* in *Self-Assembled Nanomaterials I.* 2008, Springer. p. 107–171.
142. Jeong, E. H., J. Yang, and J. H. Youk, *Preparation of polyurethane cationomer nanofiber mats for use in antimicrobial nanofilter applications.* Materials Letters, 2007, 61(18), 3991–3994.
143. Lala, N. L., et al., *Fabrication of nanofibers with antimicrobial functionality used as filters: protection against bacterial contaminants.* Biotechnology and Bioengineering, 2007, 97(6), 1357–1365.
144. Maze, B., et al., *A simulation of unsteady-state filtration via nanofiber media at reduced operating pressures.* Journal of Aerosol Science, 2007, 38(5), 550–571.
145. Payet, S., et al., *Penetration and pressure drop of a HEPA filter during loading with submicron liquid particles.* Journal of Aerosol Science, 1992, 23(7), 723–735.

CHAPTER 4

NANOTEXTILE AND TISSUE ENGINEERING FROM A BIOLOGICAL PERSPECTIVE

A. AFZALI and SH. MAGHSOODLOU

University of Guilan, Rasht, Iran

CONTENTS

ABSTRACT

Nanofibers have yield potential applications in areas such as filtration, recovery of metal ions, drug release, dentistry, tissue engineering, catalysts and enzyme carriers, wound healing, protective clothing, cosmetics, biosensors, medical implants and energy storage. Improvement of catalytic efficiency of immobilized enzymes via materials engineering is demonstrated through the preparation of bioactive nanofibers. The nanofibers are produced by electrospinning, can be followed by the chemical attachment of a model enzyme. On the other hand, from a biological perspective, almost all human tissues and organs are deposited in some kind of nanofibrous form or structure.

4.1 CATALYST AND ENZYME CARRIERS

A carrier for catalyst in chemistry and biology is used to preserve high catalysis activity, increase the stability, and simplify the reaction process. An inert porous material with a large surface area and high permeability to reactants could be a promising candidate for efficient catalyst carriers. Using an electrospun nanofiber mat as catalyst carrier, the extremely large surface could provide a huge number of active sites, thus enhancing the catalytic capability. The well-interconnected small pores in the nanofiber mat warrant effective interactions between the reactant and catalyst, which is valuable for continuous flow chemical reactions or biological processes. The catalyst can also be grafted onto the electrospun nanofiber surface via coating or surface modification [1].

4.1.1 *CATALYSIS*

It is well known that nanostructured materials have opened new possibilities for creating and mastering nanoobjects for novel advanced catalytic materials. In general, catalysis is a molecular phenomenon and the reaction occurs on an active site. A crucial step in catalysis is how to remove and recycle the catalyst after the reaction. The immobilization of catalysts in materials with large surface area advances an interesting solution to this problem. Taking the large surface area and high porosities, electrospun nanofibers, as a novel catalysts or supports for catalysts, have been widely investigated in catalytic field.

4.1.2 ELECTROCHEMICAL CATALYSTS

In 2009, Su and Lei [2] investigated the electrochemical catalytical properties based on Pd/polyamide (Pd/PA6) electrospun nanofiber mats for the oxidation of ethanol in alkaline medium in which Pd/PA6 was directly used as electrocatalytic electrodes.

Simultaneously, Kim and co-workers [3] also explored the electrochemical catalytic properties based on electrospun Pt and PtRh nanowires for dehydrogenative oxidation of cyclohexane to benzene. In contrast to the conventional Pt nanoparticle catalysts (e.g., carbon/Pt or Pt black), Pt and PtRh electrospun nanowires electrocatalysts exhibited higher catalytic activities with the same metal loading amount. Furthermore, PtRh nanowires performed the best catalytic activities with a maximum power density of ca. 23 $mWcm^{-2}$. Such higher electrocatalytic performances than Pt nanoparticles are attributed to the inherent physicochemical and electrical properties of 1D nanostructures (Figure 4.1).

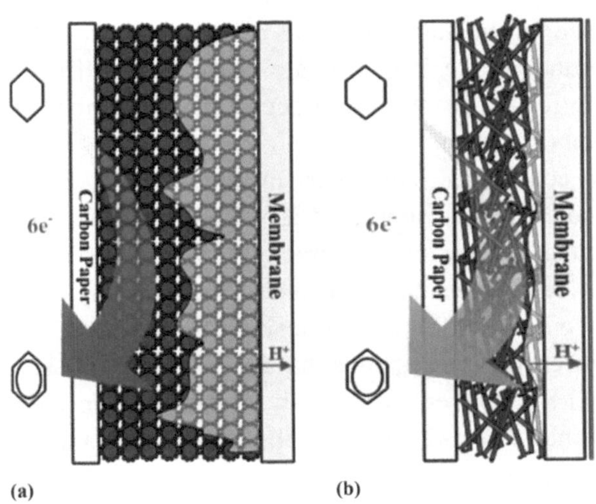

FIGURE 4.1 Schematic illustration for cyclohexane electrooxidation over the nanoparticles (a) and nanowire (b) catalysts.

1. Nanowire catalysts could provide facile pathways for the electron transfer by reducing the number of interfaces between the electrocatalysts, whereas the nanoparticles are likely to impose more impedance for electrons to transfer particle to particle.
2. Adding Rh to Pt to form the PtRh alloy can facilitate the adsorption/desorption properties of benzene and cyclohexane along with the modification of the C–H bond breaking ability.
3. The rougher morphology of PtRh nanowires comprise of small nanoparticles, which can provide high catalytic area.

4.1.3 CATALYSIS

The combination of nanofibers and catalysis seems, at first, to be rather odd. However, considering homogeneous catalysis as a first example it is well known that a huge problem is the removal and recycling of the catalytic agent after the reaction, either from the reaction solution or from the product [4]. Complex separation methods involving in some cases several processing steps have been used for this purpose. The implantation of homogeneous or also heterogeneous catalysts into nanofibers poses an interesting solution for these problems. Now of course, the very nature of homogeneous reactions requires that the catalyst is molecularly dispersed in the same phase as the reaction compounds, so that these compounds come into intimate contact via diffusional motions, thus allowing for the reaction to proceed. The common phase is generally a solution or a melt.

However, considering nanofibers made from polymers in which the catalyst is molecularly dispersed it is well known that smaller but also larger molecules can perform surprisingly rapid diffusional processes in polymer matrices in the amorphous phase, in a partially crystalline phase, above and even below the glass-transition temperature. So, it is highly probable that reaction compounds that are dispersed in a solution or melt surrounding the nanofibers with catalysts dispersed in them can diffuse into the fiber matrix, make contact with the catalyst via diffusion.

Finally, the product molecules diffuse out of the fiber again. In fact, experiments to be discussed in the following have shown that this actually is the case.

One problem that has to be solved is to keep the catalyst within the fiber, despite allowing it to diffuse in the polymer matrix. By choosing the nature of the polymer carrier appropriately to induce specific interactions between carrier and catalysts, by attaching the catalyst via flexible spacers to the polymer backbone, one is able to achieve this goal.

The reaction mixture can circulate around the fibers, as is the case, for example, in the continuously working microreaction technique, or the fibers fixed on a carrier can be immersed repeatedly into a reaction vessel to catalyze the content of the vessel. In addition to the short diffusion distances within nanofibers, the specific pore structures and high surface areas of nanofiber nonwovens allow a rapid access of the reaction components to the catalysts and of the products back into the reaction mixture.

For homogeneous catalysis, systems consisting of core-shell nanofibers combined with proline and $Sc(OTf)_3$ ($TfO = CF_3SO_3$) catalysts were fabricated, for instance, by template methods described above (TUFT—tubes by fiber template process method). In contrast to conventional catalysis in homogeneous solution or in microemulsions, for which the conversion is 80%, the fiber systems achieved complete conversion in the same or shorter reaction times. The fibers can be used several times without loss of activity. Furthermore, nanofibers were used as carriers for enzymes, where the enzymes were either chemically attached to the electrospun fibers or directly dispersed in the nanofibers during the electrospinning process. High catalyst activities were reported in this case as well. Current activity will certainly lead to a broad range of catalytic systems.

The use of polymer nanofibers in heterogeneous catalysis was analyzed for nanofibers loaded with monometallic or bimetallic nanoparticles (such as Rh, Pt, Pd, Rh/Pd, and Pd/Pt) has been reported in the literature. These catalyst systems can be applied in hydrogenation reactions, for example. To fabricate such fibrous catalyst systems, polymer nanofibers are typically electrospun from solutions containing metal salts (such as $Pd(OAc)_2$) as precursors. In the next step, the salts incorporated in the fibers are reduced, either purely thermally in air or in the presence of a reducing agent such as H_2 or hydrazine). The Pd nanoparticles formed have diameters in the range of 5–15 nm, depending on the fabrication method. The catalytic properties of these mono- or bimetallic nanofiber catalysts were investigated in several model hydrogenations, which demonstrated that the catalyst systems are highly effective.

4.1.4 ENZYMES

Chemical reactions using enzymes as catalysts have high selectivity and require mild reaction conditions. For easy separation from the reaction solution, enzymes are normally immobilized with a carrier. The immobilization efficiency mainly depends on the porous structure and enzyme-matrix interaction. To immobilize enzyme on electrospun nanofibers, many approaches have been used, including grafting enzyme on fiber surface, physical adsorption, and incorporating enzyme into nanofiber via electrospinning followed by crosslinking reaction.

To graft enzymes on nanofiber surface, the polymer used should possess reactive groups for chemical bonding [5, 6]. In some studies, polymer blends containing at least one reactive polymer were used [7, 8]. The immobilized enzymes normally showed a slightly reduced activity in aqueous environment compared with the un-immobilized native counterpart, but the activity in non-aqueous solution was much higher. For example, α- chymotrypsin was used as a model enzyme to bond chemically on the surface of electrospun PS nanofibers. The enzyme was measured to cover over 27.4% monolayer of the nanofiber surface, and the apparent hydrolytic activity of the enzyme-loaded was 65% of the native enzyme, while the activity in non-aqueous solution was over 3 orders of magnitude higher than that of its native enzyme under the same condition. In another study using PAN nanofibers to immobilize lipase, the tensile strength of the nanofiber mat was improved after lipase immobilization, and the immobilized lipase retained >90% of its initial reactivity after being stored in buffer at 30 °C for 20 days, whereas the free lipase lost 80% of its initial reactivity. Also the immobilized lipase still retained 70% of its specific activity after 10 repeated reaction cycles [9]. In addition, the immobilized enzyme also showed improved pH and thermal stabilities [10]. Ethylenediamine was used to modify PAN nanofiber mat to introduce active and hydrophilic groups, followed by a chitosan coating for improvement of biocompatibility [9].

Enzymes were incorporated into nanofibers via electrospinning, and subsequent crosslinking the enzymes incorporated effectively prevented their leaching. In the presence of PEO or PVA, casein and lipase were electrospun into ultra-thin fibers. After crosslinking with 4,4'-methylenebis(phenyl diisocyanate) (MDI), the fibers became insoluble, and the lipase en- capsulated exhibited 6 times higher hydrolysis activity towards olive oil than that of the films cast from the same solution [11]. The cross-linked enzymes in nanofibers showed very high activity and stability. For example, the immobilized α- chymotrypsin in a shaken buffer solution maintained the same activity for more than two weeks [12].

In addition to chemical bonding, the enzymes were also applied onto nanofibers simply via physical adsorption [13]. Polyacrylonitriles-2-methacryloyloxyethyl phosphoryl choline (PANCMPC) nanofiber was reported to have high biocompatibility with enzymes because of the formation of phospholipid microenvironment on the nanofiber surface. Lipase on the nanofibers showed a high immobilization rate, strong specific activity and good activity retention.

4.1.5 PHOTOCATALYSIS

The increasing industrial needs and growing urbanization have led to water scarcity issues around the globe and the wastewater produced has to be treated for re-utilization of clean water in daily activities. Among the various techniques, the heterogeneous photocatalysis system is an effective method for treating wastewater and photodegrading organic pollutants. The semiconductor metal oxides have been used as photocatalysts where upon irradiation of sunlight, create electron-hole pairs, which in turn produce radicals in different pathways as shown in Figure 4.2 [14].

The photocatalytic mechanism is as follows: upon irradiation semiconductor metal oxides eject an electron from the valance band to the conduction band, thereby leaving behind a hole in the valence band. The generated electrons and holes produce superoxide radicals to degrade the pollutants by reacting with chemisorbed oxygen on the catalyst surface and oxygen in the aqueous solution [15, 16].

Semiconducting metal-oxides such as TiO_2, ZnO, Fe_2O_3, WO_3, Bi_2WO_6, CuO and many more have been widely used as oxidative photocatalysts for effective removal of industrial pollutants and wastewater treatment [17, 18].

Shengyuan et al. [19] synthesized novel rice grain-shaped TiO_2 mesostructures by electrospinning and observed the phase change with increasing temperature from 500 °C to 1000 °C. A comparison study was performed on the photocatalytic activity of the TiO_2 rice grain and P-25. They observed

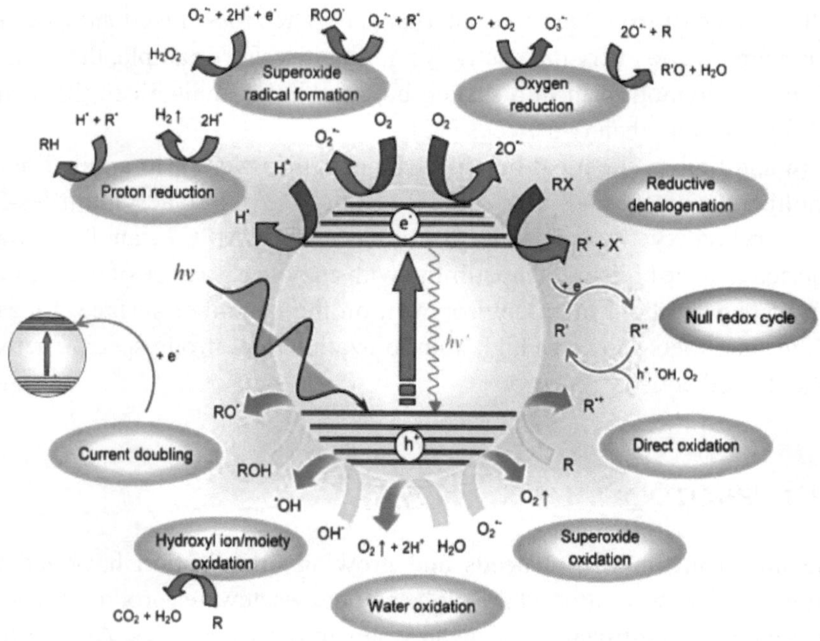

FIGURE 4.2 Possible photocatalytic mechanism of generating free radicals in the conduction band (CB) and valence band (VB) of semiconducting metal oxides.

an enhanced photocatalytic activity on alizarin red dye in rice grain-shaped TiO_2 which was due to its single crystalline nature and larger surface area than P-25. Meng et al. [20] synthesized anatase TiO_2 nanofibers by using a simple electrospinning technique and were able to grow high dense rutile TiO_2 nanorods along the fibers using hydrothermal treatment. The nanofibril-like morphology of TiO_2 nanorods/nanofibers with rutile and anatase phase was able to degrade the rhodamine-6G effectively under UV radiation. Zhang et al. [21] adopted the core/shell technique for synthesizing hollow mesoporous TiO_2 nanofibers with a larger surface area of around 118 m^2/g.

Similarly, Singh et al. [22] was able to show the synthesis of highly mesoporous ZnO nanofibers from electrospinning with high crystallinity and large surface area. The fibers had better interaction with polycyclic aromatic hydrocarbons such as naphthalene and anthracene due to the higher surface to volume ratio of the nanofibers, and thereby had higher rate constants for the UV light photodegradation of the aromatic compounds.

Ganesh et al. [23] demonstrated the superhydrophilic coating of TiO_2 films on glass substrates using electrospinning with the film acting as a self-cleaning coating for photodegradation of alizarin red dye. The self-cleaning property of the TiO_2 can be easily adopted on solid surfaces like stainless steel and also fabrics. Bedford et al. [24] synthesized photocatalytic self-cleaning textile fibers using coaxial electrospinning where cellulose acetate was used as core fiber and TiO_2 nanofiber as shell. They studied the self-cleaning photocatalytic performance of TiO_2 based textile fibers by photodegradation of dyes like keyacid blue and sulforhodamine at moderate exposure of light and were able to degrade it significantly. Neubert et al. fabricated the fibrous conductive catalytic filter membrane consisting of non-conductive polyethylene oxide blended with (±)-camphor-10-sulfonic acid doped conductive polyaniline. They were able to incorporate the TiO_2 nanoparticles on the membrane by electrospraying and achieved the photodegradtion of 2-chloroethyl phenyl sulfide pollutant upon UV exposure.

However, the usage of metal oxides such as TiO_2, ZnO, SnO_2 and Fe_2O_3 for practical applications is still limited because of the fast electron-hole recombination and broad bandgaps, which respond only to UV light. Therefore, it is essential to improve the visible light absorption and other shortcomings to ameliorate the photocatalytic property by either combining with graphene, other semiconducting metal oxides or metal nanoparticles with matching bandgap [25–27].

4.2 MEDICINAL APPLICATIONS FOR ELECTROSPUN NANOFIBERS

4.2.1 NANOTECHNOLOGY AND MEDICINAL APPLICATIONS IN GENERAL

It is obvious that the combination of nanoscience/nanotechnology with medicine makes a lot of sense for many reasons. One major reason is that the nanoscale is a characteristic biological scale, a scale related directly to Life. DNA strains, globular proteins such as ferricins, viruses are all on this scale. For instance, the tobacco mosaic virus is actually a nanotube. The dimensions of bacteria and of cells tend to be already in the micrometer range, but important subunits of these objects such as the membranes of cells have dimensions in the nm scale.

So there are certainly good reasons for addressing various types of medicinal problems on the basis of nanoscience, of nanostructures. Scaffolds used for engineering tissues such as bone or muscle tissues may be composed of nanofibers mimicking the extracellular matrix (ECM), nanoscalar carriers for drugs to be carried to particular locations within the body to be released locoregionally rather than systemic are examples. Wound healing exploiting fibrillar membranes with a high porosity and pores with diameters in the nanometer scale, thus allowing transport of fluids, gases from and to the wound yet protecting it from bacterial infections, are further examples for the combination of nanoscience and medicine with the focus here on nanofibers.

A highly interesting example along this line certainly concerns inhalation therapy. The concept is to load specific drugs onto nanorods with a given length and diameter rather than onto spherical objects such as aerosols as already done today. The reasoning is that such spherical particles tend to become easily exhaled so that frequently only a minor part of them can become active in the lung. Furthermore, the access to the lung becomes limited with increasing volume of these particles.

However, rather long fibers are known to be able to penetrate deeply into the lung. The reason is that the aerodynamic radius controls this process with the aerodynamic radius of rods being controlled mainly by the diameter and only weakly by the length, as detailed later in more detail. So, inhalation therapy based on nanorods as accessible via nanofibers offers great benefits.

In the following, different areas where nanofibers and nanorods for that matter can contribute to problems encountered in medicine will be discussed.

4.2.2 INTRODUCTION

Regenerative medicine combines the principles of human biology, materials science, and engineering to restore, maintain or improve a damaged tissue function. Regenerative medicine is divided into cell therapy or "cell transplantation" and "tissue engineering."

The National Institute of Biomedical Imaging and Bioengineering (NIBIB) defines tissue engineering as "a rapidly growing area that seeks to create, repair and/or replace tissues and organs by using combinations of cells, biomaterials, and/or biologically active molecules."

Tissue engineering has emerged through a combination of many developments in biology, material science, engineering, manufacturing and medicine. Tissue engineering involves the design and fabrication of three-dimensional substitutes to mimic and restore the structural and functional properties of the original tissue. The term 'tissue engineering' is loosely defined and can be used to describe not only the formation of functional tissue by the use of cells cultured on a scaffold or delivered to a wound site, but also the induction of tissue regeneration by genes and proteins delivered in vivo.

Cell transplantation is performed when only cell replacement is required. However, in tissue engineering, the generated tissue should have similar properties to the native tissue in terms of biochemical activity, mechanical integrity and function. This necessitates providing a similar biological environment as that in the body for the cells to generate the desired tissue. Figure 4.3 summarizes the important steps in tissue engineering. First, cells are harvested from the patient and are expanded in cell culture medium. After sufficient expansion, the cells are seeded into a porous scaffold along with signaling molecules and growth factors, which can promote cell growth and proliferation. The cell-seeded scaffold will be then placed into a bioreactor before being implanted into the patient's body. As it is evident in Figure 4.3, the three major elements of tissue engineering include [28]:

(a) cells;
(b) scaffolds; and
(c) bioreactors.

- **Cells**

Cells are the building block of all tissues. Therefore, choosing the right cell source with no contamination that is compatible with the recipient's immune system is the critical step in tissue engineering. Stem cells are employed as the main source of cells for tissue engineering and are taken from autologous,

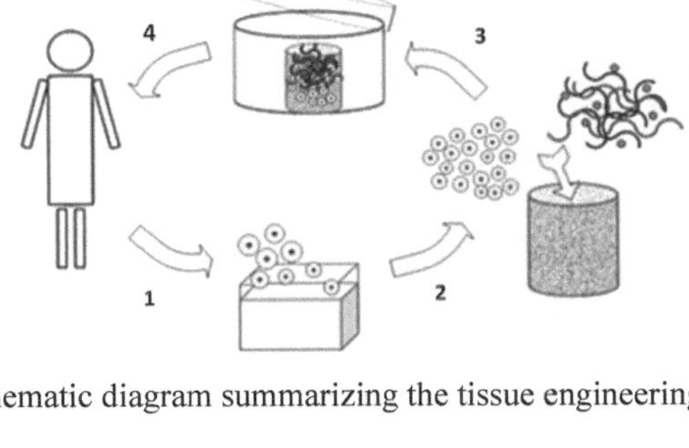

FIGURE 4.3 Schematic diagram summarizing the tissue engineering process.

allogeneic or xenogeneic sources for different applications. Stem cells are divided into the following three groups: embryonic stem cells (ESCs), induced pluripotent stem cells, and adult stem cells. ESCs are isolated from the inner cell mass of pre-implantation embryos. These cells are considered pluripotent since they can differentiate into almost any of the specialized cell types. Induced pluripotent stem cells are the adult cells that have been transformed into pluripotent stem cells through programming. Among adult stem cells, Mesenchymal stem cell is widely used as a multipotent source. Mesenchymal stem cell is derived from bone marrow stroma and can differentiate into a variety of cell types *in vitro*. Other sources of adult stem cells include the amniotic fluid and placental derived stem cells [29, 30].

- **Scaffolds**

Scaffolds are temporary porous structures used to support cells by filling up the space otherwise occupied by the natural ECM and by providing a framework to organize the dissociated cells. A biocompatible and biodegradable material is chosen for tissue engineering scaffolds, which have sufficient porosity and pore-interconnectivity to promote cell migration and proliferation, and allow for nutrient and waste exchange. The rate of degradation should be tuned with the rate of cell growth and expansion, so that as the host cells expand and produce their own ECM, the temporary material degenerates with a similar rate. Moreover, the by-products should be confirmed to be non-toxic. Mechanical properties (Figure 4.4) should also match that of the native ECM [31, 32].

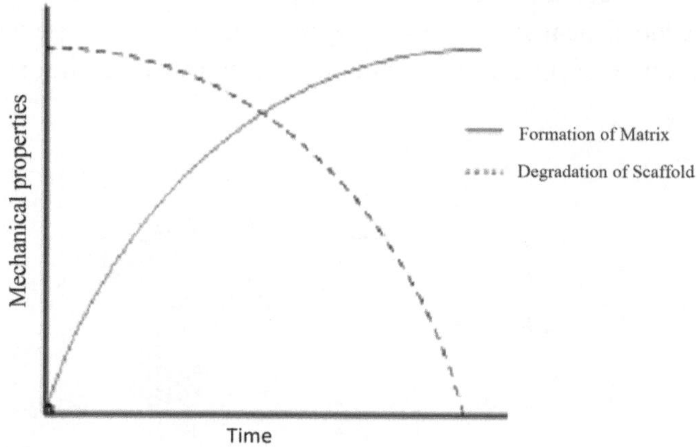

FIGURE 4.4 Mechanical properties of an ideal scaffold.

- **Bioreactors**

In-vitro tissue engineering requires bioreactors in order to provide sufficient nutrients and oxygen to the cells while removing the toxic materials left by the proliferating cells. Moreover, the essential cell-specific mechanical stimuli are also provided by bioreactors. Each type of cell (cartilage, bone, myocardium, endothelial, etc.) has different requirements in terms of pH, oxygen tension, mechanical stimulation, temperature, etc. As a result, it is necessary to use cell specific bioreactors for generation of different tissues [33].

For the treatment of tissues or organs in malfunction in a human body, one of the challenges to the field of tissue engineering/biomaterials is the design of ideal scaffolds/synthetic matrices that can mimic the structure and biological functions of the natural ECM. Human cells can attach and organize well around fibers with diameters smaller than those of the cells. In this regard, nanoscale fibrous scaffolds can provide an optimal template for cells to seed, migrate, and grow. A successful regeneration of biological tissues and organs calls for the development of fibrous structures with fiber architectures beneficial for cell deposition and cell proliferation [34].

In general, tissue engineering involves the fabrication of three dimensional scaffolds that can support cell in-growth and proliferation. In this context, the generation of scaffolds with tailored, biomimetic geometries (across multiple scales) has become an increasingly active area of research [35].

4.2.3 TISSUE ENGINEERING: BACKGROUND INFORMATION

There is without any doubt a constantly increasing need for tissues and organs to replace those that have been damaged by sickness or accidents. It is also without any doubt that this need cannot be covered by allogeneic transplants.

The shortage of donor organs, immunological problems, and possibly contamination of the donor tissue limit the use of organ transplants. This is a good reason for having a closer look at the emerging science of tissue engineering.

Tissue engineering involves the cultivation of various types of tissues to replace such damaged tissues or organs. Cartilage, bone, skin tissue, muscle, blood vessels, lymphatic vessels, lung tissue, and heart tissue are among the target tissues. In vitro, in vivo as well as combination approaches are known. In vitro approaches taken in tissue engineering rely on the seeding of specific cells on highly porous membranes as scaffold. Both homologous and autologous cells have been used for this purpose. Autologous cells, that is, cells that are harvested from an individual for the purpose of being used on that same individual, have the benefit of avoiding an immunologic response in tissue engineering. The bodies of the patients will not reject the engineered tissue because it is their own tissue and the patients will not have to take immunosuppressive drugs.

The concept is that such scaffolds will mimick to a certain extent the extracellular matrix surrounding cells in living tissue. The extracellular matrix is known to have a broad range of tasks to accomplish. It embeds the cells of which the particular tissue is composed, it offers points of contacts to them, provides for the required mechanical properties of the tissue. So, the expectation is that cells seeded on adequate porous scaffolds experience an enhanced proliferation and growth, covering finally the whole scaffold. A further task is to define the three-dimensional shape of the tissue to be engineered. Ideally, such a scaffold may then be reimplanted into the living body provided that an appropriate selection of the nature of the seeded cells was done.

So, as an example, to replace muscles, muscle cells might be chosen for the seeded cells. Yet, frequently rather than choosing specific cell lineages stem, cells such as, for instance, mesenchymal stem cells, to be discussed below in more detail, are seeded for various reasons. The proliferation of such cells is, in this case, just one step, the next involving the differentiation of the cells along specific target cell lines depending of the target tissue. To induce such differentiation various types of biological and chemical signals have been developed. To enhance proliferation it will in general be necessary to include functional compounds such as growth factors, etc., into the scaffold membranes, as discussed below.

Another approach used currently less often (that is, an in vivo approach) in tissue engineering consists in implanting the original scaffold directly into the body to act as nucleation sites for self-healing via seeding of appropriate cells.

Because the tissue engineering technology is based essentially on the seeding of cells into three-dimensional matrices, the material properties of the matrix as well as its architecture will fundamentally influence the biological functionality of the engineered tissue. One has to keep in mind that the carrier matrix has to fulfill a diverse range of requirements with respect to biocompatibility, biodegradability, morphology, sterilizability, porosity, ability to incorporate and release drugs, and mechanical suitability. Also, the scaffold architecture, porosity and relevant pore sizes are very important. In general, a high surface area and an open and interconnected 3D pore system are required for scaffolds.

These factors affect cell binding, orientation, mobility, etc. The pores of scaffolds are, furthermore, very important for cell growth as nutrients diffuse through them. The minimum pore size required is decided by the diameter of cells and therefore varies from one cell type to another. Inappropriate pore size can lead to either no infiltration at all or nonadherence of the cells. Scaffolds with nanoscalar architectures have bigger surface areas, providing benefits for absorbing proteins and presenting more binding sites to cell membrane receptors.

Biological matrices are usually not available in sufficient amounts, and they can be afflicted with biological-infection problems. It is for this reason that during the last decades man-made scaffolds composed of a sizable number of different materials of synthetic or natural, that is biological, origins have been used to construct scaffolds characterized by various types of architectures.

These include powders, foams, gels, porous ceramics and many more. However, powders, foams, and membranes are often not open-pored enough to allow cell growth in the depth of the scaffold; consequently, the formation of a three-dimensional tissue structure is frequently restricted. Even loose gel structures (e.g., of polypeptides) may fail. Furthermore, smooth walls and interfaces, which occur naturally in many membranes and foams, are frequently unfavorable for the adsorption of many cell types.

From a biological viewpoint, almost all of the human tissues and organs are deposited in nanofibrous forms or structures. Examples include: bone, dentin, collagen, cartilage, and skin. All of them are characterized by well-organized hierarchical fibrous structures realigning in nanometer scale. Nanofibers are defined as the fibers whose diameter ranges in the nanometer range. These have a special property of high surface area and increased porosity which makes it favorable for cell interaction and hence it makes its a potential platform for tissue engineering. The high surface area to volume ratio of the nanofibers combined with their microporous structure favors cell adhesion, proliferation, migration, and differentiation, all of which are highly desired properties for tissue engineering applications. There are mainly three techniques involved in synthesizing nanofibers, namely: phase separation, self-assembly, and electrospinning [36].

(a) Phase Separation

In this technique water-polymer emulsion is formed which is thermodynamically unstable. At low gelation temperature, nanoscale fibers network is formed, whereas high gelation temperature leads to the formation of platelet-like structure. Uniform nanofiber can be produced as the cooling rate is increased. Polymer concentration has a significant effect on the nanofiber properties, as polymer concentration is increased porosity of fiber decreased and mechanical properties of fiber are increased. The final product obtained is mainly porous in nature but due to controlling the key parameters we can obtain a fibrous structure. The key parameters involved are as follows [36]:

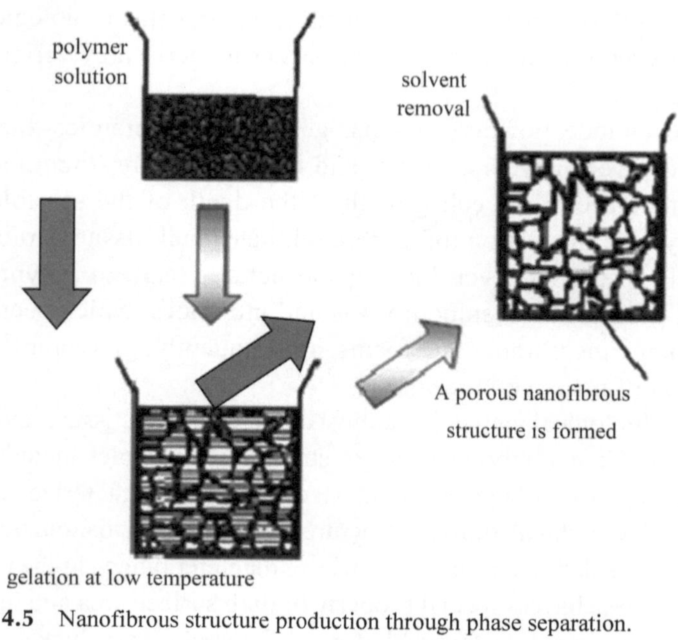

FIGURE 4.5 Nanofibrous structure production through phase separation.

a) type of polymers and their viscosity;
b) type of solvent and its volatility;
c) quenching temperature; and
d) gelling type.

(b) Self Assembly

It is a powerful approach for fabricating supra molecular architectures. Self-assembly of peptides and proteins is a promising route to the fabrication of a variety of molecular materials including nanoscale fibers and fiber network scaffolds. The main mechanism for a generic self-assembly is the intermolecular forces that bring the smaller unit together [36, 37].

(c) Electrospinning

It is a term used to describe a class of fibers forming processes for which electrostatic forces are employed to control the production of the fiber. Electrospinning readily leads to the formation of continuous fibers ranging from 0.01 to 10 μm. Electrospinning is a fiber forming processes by which electrostatic forces are employed to control the production of fibers. It is closely related to the more established technology of electrospraying, where the droplets are formed. "Spinning" in this context is a textile term that derives from the early use of spinning wheels to form yarns from natural fiber. In both electrospinning and electrospraying, the role of the electrostatic forces is to supplement or replace the conventional mechanical forces (e.g., hydrostatic, pneumatic) used to form the jet and to reduce the size of the fibers or

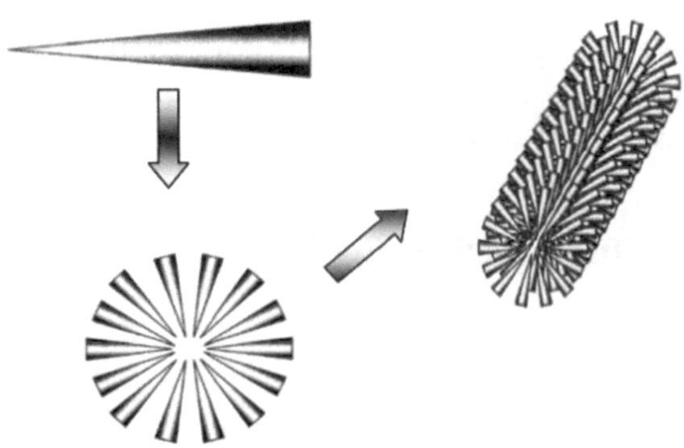

FIGURE 4.6 Schematic presentation of self-assembled nanofiber production.

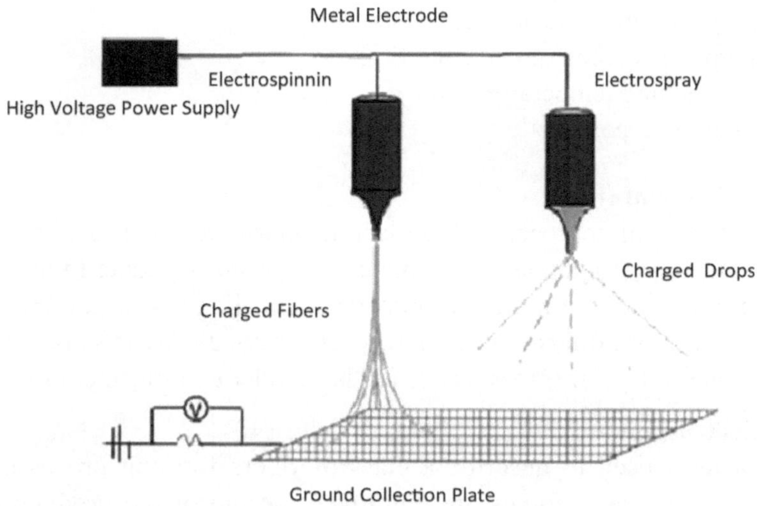

FIGURE 4.7 Electrospinning and electrospraying principles.

droplets, hence the term "electrohydrodynamic jetting." Polymer nanofibers fabricated via electrospinning have been proposed for a number of soft tissue prostheses applications such as blood vessel, vascular, breast, etc. In addition, electrospun biocompatible polymer nanofibers can also be deposited as a thin porous film onto a hard tissue prosthetic device designed to be implanted into the human body. This method will be discussed in detail, later.

4.2.4 NANOFIBERS IN TISSUE ENGINEERING APPLICATIONS

A rapidly growing field of application of polymer nanofibers is their use in tissue engineering. The main areas of intensive research are nerve, blood vessel, skeletal muscle, cartilage, bone and skin tissue engineering.

4.2.4.1 Nerve Tissue Engineering

Application of electrospun polymeric nanofibers for nerve tissue regeneration is a very significant issue. The most important observation is the elongation and neurite growth of cells parallel to fiber direction, and the effect of fiber diameter is not so significant. They have found that aligned nanofibrous scaffolds are good scaffolds for neural tissue engineering [38].

4.2.4.2 Skin Tissue Engineering

Nanofibers exhibit higher cell attachment and spreading, especially when coated with collagen than microfibers. The results, which researchers were obtained, prove the potential of electrospun nanofibers in wound healing and regeneration of skin and oral mucosa [39].

4.2.4.3 Blood Vessel Tissue Engineering

A number of studies have shown that the biodegradable polymers mimic the natural ECM and show a defined structure replicating the in vivo-like vascular structures and can be ideal tools for blood vessel tissue engineering [40].

4.2.4.4 Skeletal Muscle Tissue Engineering

Skeletal muscle is responsible for maintenance of structural contours of the body and control of movements. Extreme temperature, sharp traumas or exposure to myotoxic agents are among the reasons of skeletal muscle injury. Tissue engineering is an attractive approach to overcome the problems related to autologous transfer of muscle tissue. It could also be a solution to donor shortage and reduction in surgery time. The studies demonstrate the absence of toxic residuals and satisfactory mechanical properties of the scaffold [41].

4.2.4.5 Cartilage Tissue Engineering

There are three forms of cartilage in the body that vary with respect to structure, chemical composition, mechanical property and phenotypic characteristics of the cells. These are hyaline cartilage, fibrocartilage and elastic cartilage. Cells capable of undergoing chondrogenic differentiation upon treatment with appropriate factors and a 3-D scaffold that provides a suitable environment for chondrogenic cell growth are the two main requirements for successful cartilage tissue engineering. In addition, there are some other conditions to fulfill. First, the matrix should support cartilage-specific matrix production; second, it should allow sufficient cell migration to achieve a good bonding to the adjacent host tissue and finally, the matrix should provide enough mechanical support in order to allow early mobilization of the treated joint [42].

4.2.4.6 Bone Tissue Engineering

Bone engineering has been studied for a long time to repair fractures and in the last decades, used in preparation of dental and orthopedic devices and bone substitutes. Bone tissue engineering is a more novel technique, which deals with bone restoration or augmentation. The matrix of bone is populated by osteogenic cells, derived of mesenchymal or stromal stem cells that differentiate into active osteoblasts. Several studies have demonstrated that it is possible to culture osteogenic cells on 3-D scaffolds and achieve the formation of bone. They designed a novel 3-D carrier composed of micro and nanofibers and have observed that cells used these nanofibers as bridges to connect to each other and to the microfibers. Furthermore, a higher ability for enhancement of cell attachment and a higher activity was observed in the nano/microfiber combined scaffolds compared to the microfibrous carrier. The fibrous scaffolds improved bone formation [43, 44].

4.3 TISSUE ENGINEERING: CELLS AND SCAFFOLDS

At this point in history tissue engineering is largely an Edisonian exercise in which the scaffold provides mechanical support while host-appropriate cells populate the structure and the deposit ECM components specific to the organ targeted for replacement. The current goal is a 'neotissue' that the body can "work with" and eventually adapt to carry out the full range of expected biological activities. The primary constituent of the various ECM's involved is typically collagen; the ratios of collagen type and hierarchical organization define the mechanical properties and organization of the evolving neotissue. In addition, the ECM provides cells with a broad range of chemical signals that regulate cell function [45–47]. Cells have been mainly cultured at the surface of the electrospun materials instead of in the bulk material. 2D monolayer culture models are easy and convenient to set up with good viability of cells in culture. Although, cells on electrospun surfaces have shown 3D matrix adhesion, considerations must be made at a 3D level to truly assess the potential of electrospun biomaterials for tissue engineering by providing cells with the 3D environment found in natural tissues [48, 49].

In recent years, there have been a large number of patients who suffered from the bone defects caused by tumor, trauma or other bone diseases. Generally, autogenetic and allogenetic bones are used as substitutes in treatment of bone defects. However, secondary surgery for procuring autogenetic bone from patient would bring donor site morbidity and allogenetic bone would cause infections or immune response [50, 51]. Thus, it is necessary to find new approach for the bone regeneration. As a promising approach, the tissue engineering develops the viable substitutes capable of repairing or regenerating the functions of the damaged tissue. For the bone tissue engineering, it requires a scaffold system to temporarily support the cells and direct their growth into the corresponding tissue in vivo [52, 53].

There are three basic tissue-engineering strategies that are used to restore, maintain, or improve tissue function, and they can be summarized as cell transplantation, scaffold-guided regeneration, and cell loaded scaffold implantation [54, 55] (Figure 4.8).

1. Cell transplantation involves the removal of healthy cells from a biopsy or donor tissue and then injecting the healthy cells directly into the diseased or damaged tissue. However, this technique does not guarantee tissue formation and generally has less than 10% efficiency.
2. Scaffold-guided regeneration involves the use of a biodegradable scaffold implanted directly into the damaged area to promote tissue growth.
3. Cell-loaded scaffold implantation involves the isolation of cells from a patient and a biodegradable scaffold that is seeded with cells and then implanted into the defect location.

Prior to implantation, the cells can be subjected to an in vitro environment that mimics the in vivo environment in which the cell/polymer constructs can develop into functional tissue. This in vitro environment is generally the result of a bioreactor, which provides growth factors and other nutrients

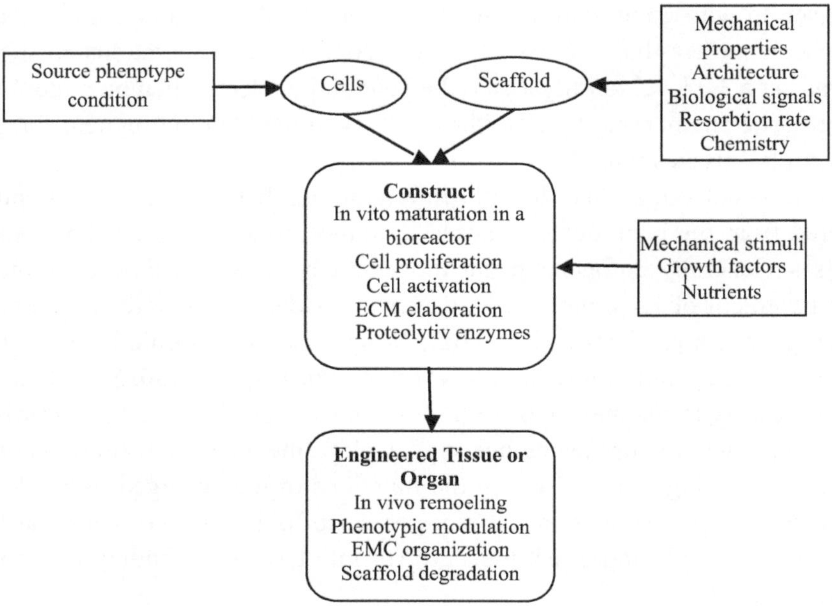

FIGURE 4.8 Tissue engineering paradigm.

while also providing mechanical stimuli to facilitate tissue growth. The first phase is the in vitro formation of a tissue construct, by placing the chosen cells and scaffold in a metabolically and mechanically supportive environment with growth media [35, 56]. The key processes occurring during the in vitro and in vivo phases of tissue formation and maturation are:

(i) cell proliferation, sorting and differentiation; (ii) extracellular matrix production and organization; (iii) degradation of the scaffold; and (iv) remodeling and potentially growth of the tissue [57].

It is generally accepted that electrospinning has the potential to fabricate scaffolds as it results in a material with sufficient strength, nanostructure, biocompatibility, and economic attractiveness. Structures composed of the thin fibers generated by the electrospinning fall into this category as demonstrated by the widespread use of the process. Electrospinning produces non-woven meshes containing fibers ranging in diameter from tens of microns to tens of nanometers [58, 59].

A scaffold design based on nanofibers can successfully mimic the structure and components of ECM component in the body and therefore properties of other native tissues. Specifically, the ECM consists of a cross-linked network of collagen and elastin fibrils (mechanical framework), interspersed with glycosaminoglycans (biochemical interactions). In spite of its remarkable diversity due to the presence of various biomacromolecules and their organization, a key feature of native ECM is the nanoscale dimension of its internal components [60].

4.4 EXTRACELLULAR MATRIX

4.4.1 COLLAGEN

Collagen types II, VI, IX, X, and XI are found in articular cartilage, although type II accounts for 90–95% of the collagen in the matrix. Type II collagen has a high amount of bound carbohydrate groups, allowing more interaction with water than some other types. Types IX and XI, along with type II, form fibrils that interweave to form a mesh. This organization provides tensile strength as well as physically entrapping other macromolecules. Although the exact function of types IX and XI are unknown, type IX has been observed to bind superficially to the fibers and extending into the inter-fiber space to interact with other type IX molecules, possibly acting to stabilize the mesh structure. Type X is found only near areas of the matrix that are calcified [61, 62].

4.4.2 PROTEOGLYCANS

Proteoglycans are composed of about 95% polysac-charide and about 5% protein. The protein core is associated with one or more varieties of glycosaminoglycan (GAG) chains. GAG chains are unbranched polysaccharides made from disaccharides of an amino sugar and another sugar. At least one component of the disaccharide has a negatively charged sulfate or carboxylate group, so the GAGs tend to repel each other and other anions while attracting cations and facilitating interaction with water. Hyaluronic acid, chondroitin sulfate, keratan sulfate, dermatan sulfate and heparan sulfate are some of the GAGs generally found in articular cartilage [63].

There are both large aggregating monomers and smaller proteoglycans present in articular cartilage. The aggregating proteoglycans, or aggregans, are composed of monomers with keratan sulfate and chondroitin sulfate GAGs attached to the protein core. In most aggregan molecules, link proteins connect many (up to 300) of these monomers to a hyaluronic acid chain. Aggregans fill most of the interfibrillar space of the ECM and are thought to be responsible for much of the resilience and stress distribution in articular cartilage through their ability to attract water. There are no chemical bonds between the proteoglycans and collagen fibers; aggregation prevents diffusion of the proteoglycans out of the matrix during joint loading [64, 65].

The smaller proteoglycans include decorin, biglycan and fibromodulin. They have shorter protein cores and fewer GAG chains than their larger counterparts. Unlike aggregans, these molecules do not affect physical properties of the tissue, but are thought to play a role in cell function and organization of the collagen matrix [64].

4.4.3 NONCOLLAGENOUS PROTEINS

In contrast to proteoglycans, glycoproteins have only a small amount of oligosaccharide associated with the protein core. These polypeptides help to stabilize the ECM matrix and aid in chondrocyte-matrix interactions. Both anchorin CII and cartilage oligomeric protein anchor chondrocytes to the surrounding matrix. Other noncollagenous proteins commonly found in most tissues, such as fibronectin and tenascin, are also observed in articular cartilage and are believed to perform similar functions as the glycoproteins [66].

4.4.4 TISSUE FLUID

Tissue fluid is an essential part of hyaline cartilage, comprising up to 80% of the wet weight of the tissue. In addition to water, the fluid contains gasses, metabolites and a large amount of cations to balance the negatively charged GAG's in the ECM. It is the exchange of this fluid with the synovial fluid that provides nutrients and oxygen to the avascular cartilage. In addition, the entrapment of this fluid though interaction with ECM components provides the tissue with its ability to resist compression and return to normal shape after deformation [28].

4.5 SCAFFOLDS

It is evident that scaffolds mimicking the architecture of the extracellular should offer great advantages for tissue engineering. The extracellular matrix surrounds the cells in tissues and mechanically supports them, as discussed above. This matrix has a structure consisting of a three-dimensional fiber network, which is formed hierarchically by nanoscale multifilaments. An ideal scaffold should replicate the structure and function of the natural extracellular matrix as closely as possible, until the seeded cells have formed a new matrix.

The use of synthetic or natural nanofibers to build porous scaffolds, therefore, seems to be especially promising and electrospinning seems to be the fabrication method of choice for various reasons. Electrospinning first of all allows construction of nanofibers from a broad range of materials of synthetic and natural origin. The range of accessible diameters of the nanofibers is extremely broad covering the range from a few nm up to several micrometers. Secondly, the nonwovens composed of nanofibers and produced by electrospinning have a total porosity of up to 90%, which is highly favorable in view of the requirements defined above. By controlling the diameter of the nanofibers one is able to control directly the average pore sizes within the nonwovens.

When constructing the scaffold by electrospinning the material choice for nanofibers is important. One often chooses degradable polymers designed to degrade slowly in the body, disappearing as the cells begin to regenerate. The degradation rate must therefore match the regeneration rate of tissue in this case.

Biocompatible and biodegradable natural and synthetic polymers such as polyglycolides, polylactides, polycaprolactone (PCL), various copolymers, segmented polyurethanes, polyphosphazenes, collagens, gelatin, chitosans, silks, and alginates are used as the carrier materials. Mixtures of gelatin and chitosans or synthetic polymers like PCL and PEO (polyethyleneoxide) are also employed, as are PCL modified by grafting, and copolymers coated or grafted with gelatin. The material choice for the applications depends upon the type of scaffold required, nature of the tissues to be regenerated and their regeneration time. The correct material helps in fulfilling the requirement of specific mechanical properties and degradation times for the particular application.

The highly porous nature of the scaffolds is apparent in all cases and the fiber diameter and thus the pore dimensions can be controlled over a sizable range in all cases, as discussed above. Functional compounds such as growth factors can be introduced in large quantities of up to 50% and more into the fibers if required, by adding these compounds to the spinning solution.

A variety of cells (e.g., mesenchymal stem cells, endothelial cells, neural stem cells, keratinocytes, muscle cells, fibroblasts, and osteoblasts) have been seeded onto carrier matrices for the generation of target tissues (such as skin tissue, bone, cartilage, arteries, and nerve tissue). The diameters of the fibers used generally conform to the structural properties of the extracellular matrix and are of the order of 100 nm. However, in some cases, fibers with diameters of less than 100 nm or of the order of 1 nm were used. The observation is that the number of cells located on the scaffold increase in time due to proliferation processes, if the nature of the scaffold is chosen appropriately.

A frequent requirement for the growth of cells in tissue engineering is that the cells are not oriented randomly within the scaffold but are oriented planar or even uniaxial. Tissue engineering of bones or muscles is an example.

In several studies, the proliferation behavior of cells in such fiber structures was compared with that on films cast from the same polymer material. The results showed that the fiber architecture generally affects cell growth positively. For endothelial cells, however, it was reported that a smoother surface can be beneficial for cell adhesion and proliferation. Another conclusion made was that the biocompatibility of a material improves with decreasing fiber diameter. Porosity also seems to have a favorable influence on cell growth. For instance, it was observed that mesenchymal stem cells form branches to the pores on porous nanofibers.

Another important requirement is that the scaffolds are porous enough to allow cells to grow in their depths, while being provided with the necessary nutrients and growth factors. The degree of porosity and the average pore dimensions are significant factors for cell proliferation and the formation of three-dimensional tissues. Depending on the cell type, the optimal pore diameters are 20–100 nm; pore diameters larger than 100 nm are in general not required for optimal cell growth. It was also found that cells can easily migrate to a depth of about 100 nm, but encounter problems at greater depths.

One solution to this problem involves a layer-by-layer tissue-generation procedure. In this approach, cells are uniformly assembled into multilayered three-dimensional (3D) structure with the assistance of electrospun nanofibers. This approach offers lot of flexibility in terms of varying cell seeding density and cell type for each cell layer, the composition for each nanofiber layer, precise control of fiber layer thickness, fiber diameter, and fiber orientation. A further answer to this problem consists in introducing the cells directly during the preparation step of the scaffold via electrospinning. The concept is to combine the spinning process of the fibers with an electrospraying process of the cells.

Combination of electrospinning of fibers and electrospraying of cells Scaffolds based on nanofiber nonwovens offer a lot of further advantages. One important prerequisite for a scaffold is a sufficient mechanical compatibility. Cartilage, for example, is characterized by a Young's modulus of about 130 MPa, a maximum deformation stress of about 20 MPa, and a maximum deformation of 20–120%; the corresponding values for skin tissue are 15–150 MPa, 5–30 MPa, and 35–115%. These ranges of values can be achieved with electrospun nanofibers. For instance, for scaffolds composed of electrospun collagen fibers with diameters of about 100 nm, a Young's modulus of 170 MPa and maximum deformation stress of 3.3 MPa were found. However, the maximum elongation is usually less than 10%. Another important finding is that the fibers can impart mechanical stress to the collective of growing cells. It was reported that the production of extracellular tissue is greater if oriented rather than unoriented matrix fibers are employed. This production can be significantly increased by the application of a periodical mechanical deformation (typically 6%).

Mimicking functional gradients (one of the important characteristic features of living tissue) that is gradients in composition, microstructure and porosity in scaffolds is also possible in a simple way by electrospinning. For example layer-by-layer electrospinning with composition gradients via controlled changes in the composition of the electrospinning solutions provide functional gradient scaffolds. The incorporation of bioactive agents in electrospun fibers will lead to advanced biofunctional tissue-engineering scaffolds. The biofunctionalization can alter the efficiency of these fibers for regenerating biological functional tissues. The bioactive agents can be easily incorporated onto fibers just by mixing them in electrospinning solution or by covalent attachment.

Scaffolds fabricated from electrospun nanofibers have definitely several advantages. However, considerable room for optimization remains with respect to architecture, surface properties, porosity, mechanical and biomechanical properties and functional gradient, and also with respect to the seeding of cells in the three-dimensional space and the supply of nutrients to the cells. It is often observed that the cells preferentially grow on the surfaces or that they initially adhere to the carrier fibers, but then detach after differentiation. Toxicity of the organic solvents used for electrospinning is another issue for in vivo applications. The solution is to use either water-soluble polymers for electrospinning with subsequent cross-linking after scaffold formation or to make use of water-based polymeric dispersions. One final remark: early investigations on the co-growth of different types of cells on scaffolds are very promising. For instance, the co-growth of fibroblasts, keratinocytes, and endothelial cells was reported; the astonishing result is that co-growth enhances cell growth. So, a lot can be expected from scaffolds for tissue engineering based on electrospun nanofibers.

Ideal scaffolds probably approximate the structural morphology of the natural collagen found in the target organ. The ideal scaffold must satisfy a number of often conflicting demands: (i) appropriate levels and sizes of porosity allowing for cell migration; (ii) sufficient surface area and a variety of surface chemistries that encourage cell adhesion, growth, migration, and differentiation; and (iii) a degradation rate that closely matches the regeneration rate of the desired natural tissue [35]. While a broad range of tissue engineering matrices has been fabricated, a few types of synthetic scaffolding show special promise. Synthetic or natural materials can be used that eliminate concerns regarding unfavorable immune responses or disease transmission [58, 59].

Synthetic tissues help stimulate living tissues to repair themselves in various parts of the human body, such as cartilage, blood vessels, bones and so forth, due to diseases or wear and tear. Victims whose skin are burned or scalded by fire or boiling water may also find an answer in synthetic tissues. The newest generation of synthetic implant materials, also called biomaterials, may even treat diseases such as Parkinson's, arthritis and osteoporosis. The uses of synthetic tissues are numerous. And there are several methods available to create them. One method is to make use of scaffold fabrication technology. Under this technology, synthetic tissues are cultured and placed on the scaffold that is shaped accordingly to, say a tendon or ligament, and then grafted onto the damaged part of the body. Once the new tissues grow over the damaged part of the organ or achieved sufficient structural integrity, the scaffolds would eventually degrade until only the tissues remain. It is also possible to use biocompatible materials that do not degrade, in which case, the scaffolds remain harmlessly in the body.

Synthetic polymers typically allow a greater ability to tailor mechanical properties and degradation rate. Clearly, the electrospinning process can eventually be developed to achieve successful utilization in vivo on a routine basis. Electrospinning offers the ability to fine-tune mechanical properties during the fabrication process, while also controlling the necessary biocompatibility and structure of the tissue engineered grafts [67]. The ability of the electrospinning technique to combine the advantages of synthetic and natural materials makes it particularly attractive, where a high mechanical durability, in terms of high burst strength and compliance (strain per unit load), is required. Advances in processing techniques, morphological characteristics and interesting, biologically relevant modifications are underway [68, 69].

However, there is one big drawback. This is the unstable dynamical behavior of the liquid jet, which is formed during the electrospinning process. This instability inhibits the fibers to be aligned in a regular way, which is crucial to satisfy the scaffolds specifications. Many researches have been done to study this jet behavior. When the instability of the jet can be controlled, electrospinning cannot only be adapted to produce high quality scaffolds for tissue engineering, but also for many other applications.

4.6 WOUND HEALING

An interesting application of electrospun nanofibers is the treatment of large wounds such as burns and abrasions [70–72]. It is found that these types of wounds heal particularly rapidly and without complications if they are covered by a thin web of nanofibers, in particular, of biodegradable polymers. Such nanowebs have suitable pore size to assure the exchange of liquids and gases with the environment, but have dimensions that prevent bacteria from entering. Mats of electro-spun nanofibers generally show very good adhesion to moist wounds. Furthermore, the large specific surface area of up to 100 m^2g^{-1} is very favorable for the adsorption of liquids and the local release of drugs on the skin, making these materials suitable for application in hemostatic wound closure. Further, multifunctional bioactive nanofibrous wound healing dressings can be made available easily simply by blending with bioactive therapeutic agents (like antiseptics, antifungal, vasodilators, growth factors, etc.) or by coaxial electrospinning. Compared to conventional wound treatment, the advantage is also that scarring is prevented by the use of nanofibers.

The nanofibrillar structure of the nanoweb promotes skin growth, and if a suitable drug is integrated into the fibers, it can be released into the healing wound in a homogeneous and controlled manner. The charging of biodegradable nanofibers with antibiotics was realized with the drugs cefazolin and mefoxin. Generally, different drugs with antiseptic and antibiotic effects, as well as growth and clotting factors, are available for wound healing. Polyurethane (PU) is widely used as the nanoweb material because of its excellent barrier properties and oxygen permeability. Electrospun mats of PU nanofibers as wound dressings were successfully tested on pigs. Histological investigations showed that the rate of epithelialization during the healing of wounds treated with nanofiber mats is higher than that of the control group. Another promising and, in contrast to PU, biodegradable material is collagen. The wound healing properties of mats of electrospun fibers of type I collagen can be investigated on wounds in mice. It was found that especially in the early stages of the healing process better healing of the wounds was achieved with the nanofiber mats compared to conventional wound care. Blends of collagen or silk and PEO were also electrospun into fibers and used in wound dressings.

Numerous other biodegradable polymers that can be electrospun can be applied in wound healing, for example, PLA and block-copolymer derivatives, PCL, chitin, and chitosan. Using tetracycline hydrochloride as a model drug, it was shown that the release kinetics can be adjusted by varying the polymer used for the fabrication of the nanofibers. Poly[ethylene-co-(vinyl acetate)] (PEVA), PLA, and a 50,50 mixture of the two polymers were investigated. With PEVA, faster drug release was observed than with PLA or the blend. With PLA, burst release occurred, and the release properties of the blend are intermediate between those of the pure polymers. The morphology of the fibers and their interaction with the drug are critical factors. The concentration of the drug in the fibers also affects the release kinetics. The higher the concentration, the more pronounced the burst, evidently because of an enrichment of the drug on the surface.

Handheld electrospinning devices have been developed for the direct application of nanofibers onto wounds. In such a device, a high voltage is generated with the voltage supplied by standard batteries. The device has a modular construction, so that different polymer carriers and drugs can be applied, depending on the type of wound, by exchanging containers within the spinning device.

4.7 TRANSPORT AND RELEASE OF DRUGS

Nanostructured systems for the release of drugs (or functional compounds in general) are of great interest for a broad range of applications in medicine, including among others tumor therapy, inhalation and pain therapy [72–74].

Nanoparticles (composed of lipids or biodegradable polymers, for example) have been extensively investigated with respect to the transport and release of drugs. Such nanostructured carriers must fulfill diverse functions. For example, they should protect the drugs from decomposition in the bloodstream, and they should allow the controlled release of the drug over a chosen time period at a release rate that is as constant as possible. They should also be able to permeate certain membranes (e.g., the blood/brain barrier), and they should ensure that the drug is only released in the targeted tissue. It may also be necessary for the drug release to be triggered by a stimulus (either external or internal) and to continue the release only as long as necessary for the treatment. A variety of methods have been used for the fabrication of such nanoparticles, including spraying and sonification, as well as self-organization and phase-separation processes. Such nanoparticles are primarily used for systemic treatment. Experiments are currently being carried on the targeting and enrichment of particular tissues (vector targeting) by giving the nanoparticles specific surface structures (e.g., sugar molecules on the surface).

A very promising approach is based on the use of anisometric nanostructures that is, of nanorods, nanotubes, and nanofibers for the transport and release of drugs. In the focus of such an approach will, in general, be a locoregional therapy rather than a systemic therapy. In a locoregional therapy the drug carriers are localized at the site where the drug is supposed to be applied. Such anisometric carriers can be fabricated by electrospinning with simultaneous incorporation of the drugs via the spinning solution. Another approach envisions the preparation of core-shell objects via coaxial electrospinning where the drug is incorporated in the core region of the fibers with the shell being composed of a polymer.

Nanofibers with incorporated super paramagnetic Fe_3O nanoparticles serve as an example the carrier should be possible with the application of an external magnetic field.

An interesting property of super paramagnetic systems is that they can be heated by periodically modulated magnetic fields. This feature allows drug release to be induced by an external stimulus.

A broad set of in vitro experiments on the release kinetics of functional molecules has been performed among others by fluorescence microscopy. The experiments often have demonstrated that the release occurs as a burst, that is, in a process that is definitely nonlinear with respect to time. It was, however, found that the release kinetics, including the linearity of the release over time and the release time period, can be influenced by the use of core-shell fibers, in which the core immobilizes the drugs and the shell controls their diffusion out of the fibers.

In addition to low molecular weight drugs, macromolecules such as proteins, enzymes, growth factors and DNA are also of interest for incorporation in transport and release systems. Several experimental studies on this topic have been carried out. The incorporation of plasmidic DNA into PLA-b-PEG-b-PLA block copolymers and its subsequent release was investigated, and it was shown that the released DNA was still fully functional. Bovine serum albumin (BSA) and lysozyme were also electrospun into polymer nanofibers, and their activities after release were analyzed, again yielding positive results. In the case of BSA, is was shown that the use of core-shell fibers fabricated by the chemical vapor deposition (CVD) of poly (p-xylylene) PPX onto electrospun nanofibers affords almost linear release over time. Further investigations deal with the incorporation and release of growth factors for applications in tissue engineering. In the following, some specific applications of nanofibers in drug release are described in more detail.

4.8 APPLICATION IN TUMOR THERAPY

Nanofibers composed of biodegradable polymers were investigated with respect to their use in local chemotherapy via surgical implantation. A selection of approaches will be discussed in the following. The water-insoluble antitumor drug paclitaxel (as well as the antituberculosis drug rifampin) was electrospun into PLA nanofibers. In some cases, a cationic, anionic, or neutral surfactant was added, which influenced the degree of charging of the nanofibers. Analysis of the release kinetics in the presence of proteinase K revealed that the drug release is nearly ideally linear over time. The release is clearly a consequence of the degradation of the polymer by proteinase. Analogous release kinetics were found when the degree of charging was increased to 50%. Similar investigations were also carried out with the hydrophilic drug doxorubicin.

To obtain nanofibers with linear release kinetics for water-soluble drugs like doxorubicin, water–oil emulsions were electrospun, in which the drug was contained in the aqueous phase and a PLA-co-PGA copolymer (PGA: polyglycolic acid) in chloroform was contained in the oil phase. These electrospun fibers showed bimodal release behavior consisting of burst kinetics for drug release through diffusion from the fibers, followed by linear kinetics for drug release through enzymatic degradation of the fibers by the proteinase K. In many cases, this type of bimodal behavior may be desired. Furthermore, it was shown that the antitumor drug retained its activity after electrospinning and subsequent release. The drug taxol was also studied with respect to its release from nanofibers. These few examples show that nanofibers may in fact be used as drug carrier and release agents in tumor therapy.

4.9 INHALATION THERAPY

Finally, a unique application for anisometric drug carriers, inhalation therapy, will be discussed. The general goal is to administer various types of drugs via the lung. One key argument is that the surface of the lung is, in fact, very large, of the order of 150 m^2, so that this kind of administration should be very effective.

Indications for such treatments are tumors, metastases, pulmonary hypertension, and asthma. But these systems are also under consideration for the administration of insulin and other drugs through the lung.

Further advantages of anisometric over spherical particles as drug carriers for inhalation therapy are that a significantly larger percentage of anisometric particles remain in the lung after exhalation and that the placement of the drug carriers in the lung can be controlled very sensitively via the aerodynamic radius. To produce rod-shaped carriers with a given aerodynamic diameter, nanofibers were electrospun from appropriate carrier polymers that were subsequently cut to a given length either by mechanical means or by laser cutting.

Further progress in inhalation therapy will mainly depend first of all on finding biocompatible polymer systems that do not irritate the lung tissue and on the development dispensers able to dispense such rod-like particles.

4.10 CONCLUDING REMARK

Enzyme-loaded nanoparticles as well as some other nanodispersed biocatalysts are remarkable from several perspectives. It appeared that the use of nanofibers, typically, electrospun nanofibers, provide a large surface area for the attachment or entrapment of enzymes and enzyme reaction. In the case of porous nanofibers, they can reduce diffusion path of substrates from reaction medium to enzyme active sites, due to the reduced thickness.

Tissue engineering has provided a new medical treatment as an alternative to traditional transplantation methods. It is a promising area to repair or replace task of damaged tissues or organs. Natural polymers offer the advantage of being very analogous, often identical to macromolecular substances existing in the human body. Thus, the biological environment is prepared to distinguish and interact with natural polymers. Some of the natural polymers applied as scaffolds in the nervous system are collagen, gelatin hyaluronic acid, chitosan and elastin. Synthetic polymers can be tailored to develop a wide range of mechanical features and degradation rates. They can also be processed to minimize immune response.

Polymeric materials can lead to great development in tissue engineering of damaged nervous system, but there are still many questions to be answered before their application, such as type and characteristic of polymer and the complementary methods which are appropriate for specific neurological dysfunctions and further investigation is needed to promote them as ideal scaffolds for nervous tissue engineering.

KEYWORDS

- **catalyst**
- **enzyme**
- **nanofibers**
- **tissue engineering**

REFERENCES

1. Fang, J., X. Wang, and T. Lin, *Functional applications of electrospun nanofibers.* Nanofibers-production, properties and functional applications, 2011, 287–326.
2. Su, L., et al., *Free-standing palladium/polyamide 6 nanofibers for electrooxidation of alcohols in alkaline medium.* The Journal of Physical Chemistry, 2009, 113(36), 16174–16180.
3. Kim, H. J., et al., *Pt and PtRh nanowire electrocatalysts for cyclohexane-fueled polymer electrolyte membrane fuel cell.* Electrochemistry Communications, 2009, 11(2), 446–449.
4. Stasiak, M., et al., *Design of polymer nanofiber systems for the immobilization of homogeneous catalysts–Preparation and leaching studies.* Polymer, 2007, 48(18), 5208–5218.
5. Wang, Y. and Y. L. Hsieh, *Enzyme immobilization to ultra-fine cellulose fibers via amphiphilic polyethylene glycol spacers.* Journal of Polymer Science Part A: Polymer Chemistry, 2004, 42(17), 4289–4299.
6. Stoilova, O., et al., *Functionalized electrospun mats from styrene–maleic anhydride copolymers for immobilization of acetylcholinesterase.* European Polymer Journal, 2010, 46(10), 1966–1974.

7. Jia, H., et al., *Enzyme-carrying polymeric nanofibers prepared via electrospinning for use as unique biocatalysts.* Biotechnology progress, 2002, 18(5), 1027–1032.
8. Kim, B. C., et al., *Preparation of biocatalytic nanofibers with high activity and stability via enzyme aggregate coating on polymer nanofibers.* Nanotechnology, 2005, 16(7), p. S382.
9. Li, S.-F., J.-P. Chen, and W.-T. Wu, *Electrospun polyacrylonitrile nanofibrous membranes for lipase immobilization.* Journal of Molecular Catalysis B: Enzymatic, 2007, 47(3), 117–124.
10. Huang, X.-J., D. Ge, and Z.-K. Xu, *Preparation and characterization of stable chitosan nanofibrous membrane for lipase immobilization.* European Polymer Journal, 2007, 43(9), 3710–3718.
11. Xie, J. and Y.-L. Hsieh, *Ultra-high surface fibrous membranes from electrospinning of natural proteins: casein and lipase enzyme.* Journal of Materials Science, 2003, 38(10), 2125–2133.
12. Herricks, T. E., et al., *Direct fabrication of enzyme-carrying polymer nanofibers by electrospinning.* Journal of Materials Chemistry, 2005, 15(31), 3241–3245.
13. Wang, G., et al., *Fabrication and characterization of polycrystalline WO3 nanofibers and their application for ammonia sensing.* The Journal of Physical Chemistry B, 2006, 110(47), 23777–23782.
14. Teoh, W. Y., J. A. Scott, and R. Amal, *Progress in heterogeneous photocatalysis: from classical radical chemistry to engineering nanomaterials and solar reactors.* The Journal of Physical Chemistry Letters, 2012, 3(5), 629–639.
15. Bhatkhande, D. S., V. G. Pangarkar, and A. A. Beenackers, *Photocatalytic degradation for environmental applications–a review.* Journal of Chemical Technology and Biotechnology, 2002, 77(1), 102–116.
16. Han, F., et al., *Tailored titanium dioxide photocatalysts for the degradation of organic dyes in wastewater treatment: a review.* Applied Catalysis A: General, 2009, 359(1), 25–40.
17. Alves, A., et al., *Photocatalytic activity of titania fibers obtained by electrospinning.* Materials Research Bulletin, 2009, 44(2), 312–317.
18. Szilágyi, I. M., et al., *Photocatalytic Properties of WO_3/TiO_2 Core/Shell Nanofibers prepared by Electrospinning and Atomic Layer Deposition.* Chemical Vapor Deposition, 2013, 19(4–6), 149–155.
19. Shengyuan, Y., et al., *Rice grain-shaped TiO_2 mesostructures—synthesis, characterization and applications in dye-sensitized solar cells and photocatalysis.* Journal of Materials Chemistry, 2011, 21(18), 6541–6548.
20. Meng, X., et al., *Growth of hierarchical TiO2 nanostructures on anatase nanofibers and their application in photocatalytic activity.* Cryst. Eng. Comm., 2011, 13(8), 3021–3029.
21. Zhang, X., et al., *Novel hollow mesoporous 1D TiO_2 nanofibers as photovoltaic and photocatalytic materials.* Nanoscale, 2012, 4(5), 1707–1716.
22. Singh, P., K. Mondal, and A. Sharma, *Reusable electrospun mesoporous ZnO nanofiber mats for photocatalytic degradation of polycyclic aromatic hydrocarbon dyes in wastewater.* Journal of Colloid and Interface Science, 2013, 394, 208–215.
23. Ganesh, V. A., et al., *Photocatalytic superhydrophilic TiO2 coating on glass by electrospinning.* RSC Advances, 2012, 2(5), 2067–2072.

24. Bedford, N. and A. Steckl, *Photocatalytic self cleaning textile fibers by coaxial electrospinning.* ACS Applied Materials and Interfaces, 2010, 2(8), 2448–2455.
25. Bao, N., et al., *Adsorption of dyes on hierarchical mesoporous TiO2 fibers and its enhanced photocatalytic properties.* The Journal of Physical Chemistry C, 2011, 115(13), 5708–5719.
26. Pant, B., et al., *Carbon nanofibers decorated with binary semiconductor (TiO_2/ZnO) nanocomposites for the effective removal of organic pollutants and the enhancement of antibacterial activities.* Ceramics International, 2013, 39(6), 7029–7035.
27. Su, Q., et al., *Effect of the morphology of V_2O_5/TiO_2 nanohetero structures on the visible light photocatalytic activity.* Journal of Physics and Chemistry of Solids, 2013, 74(10), 1475–1481.
28. Temenoff, J. S. and A. G. Mikos, *Review: Tissue Engineering for Regeneration of Articular Cartilage.* Biomaterials, 2000, 21(5), 431–440.
29. Bianco, P. and P. G. Robey, *Stem Cells in Tissue Engineering.* Nature, 2001, 414(6859), 118–121.
30. Griffith, L. G. and G. Naughton, *Tissue Engineering–current Challenges and Expanding Opportunities.* Science, 2002, 295(5557), 1009–1014.
31. Raghunath, J., et al., *Biomaterials and Scaffold Design: Key to Tissue-engineering Cartilage.* Biotechnology and Applied Biochemistry, 2007, 46(2), 73–84.
32. Stocum, D. L., *Stem Cells in CNS and Cardiac Regeneration*, in *Regenerative Medicine I.* 2005, Springer. p. 135–159.
33. Ratcliffe, A. and L. E. Niklason, *Bioreactors and Bioprocessing for Tissue Engineering.* Annals of the New York Academy of Sciences, 2002, 961(1), 210–215.
34. Hutmacher, D. W., *Scaffold Design and Fabrication Technologies for Engineering Tissues—State of the Art and Future Perspectives.* Journal of Biomaterials Science, Polymer Edition, 2001, 12(1), 107–124.
35. Lannutti, J., et al., *Electrospinning for Tissue Engineering Scaffolds.* Materials Science and Engineering: C, 2007, 27(3), 504–509.
36. Smith, L. A. and P. X. Ma, *Nano-Fibrous Scaffolds for Tissue Engineering.* Colloids and Surfaces B: Biointerfaces, 2004, 39(3), 125–131.
37. Ma, Z., et al., *Potential of Nanofiber Matrix as Tissue-Engineering Scaffolds.* Tissue Engineering, 2005, 11(1–2), 101–109.
38. Yang, F., et al., *Electrospinning of Nano/Micro Scale Poly (L-lactic acid) Aligned Fibers and their Potential in Neural Tissue Engineering.* Biomaterials, 2005, 26(15), 2603–2610.
39. Noh, H. K., et al., *Electrospinning of Chitin Nanofibers: Degradation Behavior and Cellular Response to Normal Human Keratinocytes and Fibroblasts.* Biomaterials, 2006, 27(21), 3934–3944.
40. Kwon, K., S. Kidoaki, and T. Matsuda, *Electrospun Nano-to Microfiber Fabrics Made of Biodegradable Copolyesters: Structural Characteristics, Mechanical Properties and Cell Adhesion Potential.* Biomaterials, 2005, 26(18), 3929–3939.
41. Riboldi, S. A., et al., *Electrospun Degradable Polyesterurethane Membranes: Potential Scaffolds for Skeletal Muscle Tissue Engineering.* Biomaterials, 2005, 26(22), 4606–4615.
42. Hwang, N. S., S. Varghese, and J. Elisseeff, *Cartilage Tissue Engineering*, in *Stem Cell Assays*. 2007, Springer. 351–373.

43. Tuzlakoglu, K., et al., *Nano-and Micro-fiber Combined Scaffolds: A New Architecture for Bone Tissue Engineering.* Journal of Materials Science: Materials in Medicine, 2005, 16(12), 1099–1104.
44. Santos, M. I., et al., *Endothelial Cell Colonization and Angiogenic Potential of Combined Nano-and Micro-fibrous Scaffolds for Bone Tissue engineering.* Biomaterials, 2008, 29(32), 4306–4313.
45. Holzwarth, J. M. and P. X. Ma, *Biomimetic Nanofibrous Scaffolds for Bone Tissue Engineering.* Biomaterials, 2011, 32(36), 9622–9629.
46. Sill, T. J. and H. A. Recum, *Electrospinning: Applications in Drug Delivery and Tissue Engineering.* Biomaterials, 2008, 29(13), 1989–2006.
47. Gautam, S., A. K. Dinda, and N. C. Mishra, *Fabrication and Characterization of PCL/gelatin Composite Nanofibrous Scaffold for Tissue Engineering Applications by Electrospinning Method.* Materials Science and Engineering: C, 2013, 33, 1228–1235.
48. Mouthuy, P. A. and H. Ye, *Biomaterials: Electrospinning* in *Comprehensive Biotechnology*. 2011, Elsevier: Oxford. p. 23–36.
49. Shen, Q., et al., *Progress on Materials and Scaffold Fabrications Applied to Esophageal Tissue Engineering.* Materials Science and Engineering: C, 2013, 33, 1860–1866.
50. Jang, J. H., O. Castano, and H. W. Kim, *Electrospun Materials as Potential Platforms for Bone Tissue Engineering.* Advanced Drug Delivery Reviews, 2009, 61(12), 1065–1083.
51. Schneider, O. D., et al., *In vivo and* in vitro *Evaluation of Flexible, Cotton wool-like Nanocomposites as Bone Substitute Material for Complex Defects.* Acta Biomaterialia, 2009, 5(5), 1775–1784.
52. Meng, Z. X., et al., *Electrospinning of PLGA/gelatin Randomly oriented and Aligned Nanofibers as Potential Scaffold in Tissue Engineering.* Materials Science and Engineering: C, 2010, 30(8), 1204–1210.
53. Chahal, S., F. S. J. Hussain, and M. B. M. Yusoff, *Characterization of Modified Cellulose (MC)/Poly (Vinyl Alcohol) Electrospun Nanofibers for Bone Tissue Engineering.* Procedia Engineering, 2013, 53, 683–688.
54. Kramschuster, A. and L. S. Turng, *Fabrication of Tissue Engineering Scaffolds.* Handbook of Biopolymers and Biodegradable Plastics: Properties, Processing and Applications. Vol. 17. 2013, Elsevier.
55. Shoichet, M. S., *Polymer Scaffolds for Biomaterials Applications.* Macromolecules, 2009, 43(2), 581–591.
56. Orlando, G., et al., *Regenerative Medicine as Applied to General Surgery.* Annals of Surgery, 2012, 255(5), 867–880.
57. Lakshmipathy, U. and C. Verfaillie, *Stem Cell Plasticity.* Blood Reviews, 2005, 19(1), 29–38.
58. Yoshimoto, H., et al., *A Biodegradable Nanofiber Scaffold by Electrospinning and its Potential for Bone Tissue Engineering.* Biomaterials, 2003, 24(12), 2077–2082.
59. Barnes, C. P., et al., *Nanofiber Technology: Designing the next Generation of Tissue Engineering Scaffolds.* Advanced Drug Delivery Reviews, 2007, 59(14), 1413–1433.
60. Janković, B., et al., *The Design Trend in Tissue-engineering Scaffolds Based on Nanomechanical Properties of Individual Electrospun Nanofibers.* International Journal of Pharmaceutics, 2013, 455(1), 338–347.

61. Hay, E. D., *Cell Biology of Extracellular Matrix*. 1991, Springer. 468.
62. Khadka, D. B. and D. T. Haynie, *Protein-and Peptide-based Electrospun Nanofibers in Medical Biomaterials*. Nanomedicine: Nanotechnology, Biology and Medicine, 2012, 8(8), 1242–1262.
63. Kjellen, L. and U. Lindahl, *Proteoglycans: Structures and Interactions*. Annual Review of Biochemistry, 1991, 60(1), 443–475.
64. Rosso, F., et al., *From Cell–ECM Interactions to Tissue Engineering*. Journal of Cellular Physiology, 2004, 199(2), 174–180.
65. Schaefer, L. and R. M. Schaefer, *Proteoglycans: From Structural Compounds to Signaling Molecules*. Cell and Tissue Research, 2010, 339(1), 237–246.
66. Roughley, P. J., *Articular Cartilage and Changes in Arthritis: Noncollagenous Proteins and Proteoglycans in the Extracellular Matrix of Cartilage*. Arthritis Res, 2001, 3(6), 342–347.
67. Liang, D., B. S. Hsiao, and B. Chu, *Functional Electrospun Nanofibrous Scaffolds for Biomedical Applications*. Advanced Drug Delivery Reviews, 2007, 59(14), 1392–1412.
68. Hasan, M. D., et al., *Electrospun Scaffolds for Tissue Engineering of Vascular Grafts*. Acta Biomaterialia, 2013.
69. Metter, R. B., et al., *Biodegradable Fibrous Scaffolds with Diverse Properties by Electrospinning Candidates from a Combinatorial Macromere Library*. Acta Biomaterialia, 2010, 6(4), 1219–1226.
70. Greiner, A. and J. H. Wendorff, *Electrospinning: a fascinating method for the preparation of ultrathin fibers*. Angewandte Chemie International Edition, 2007, 46(30), 5670–5703.
71. Reneker, D., et al., *Electrospinning of nanofibers from polymer solutions and melts*. Advances in applied mechanics, 2007, 41, 43–346.
72. Agarwal, S., J. H. Wendorff, and A. Greiner, *Use of electrospinning technique for biomedical applications*. Polymer, 2008, 49(26), 5603–5621.
73. Zeng, J., et al., *Poly (vinyl alcohol) nanofibers by electrospinning as a protein delivery system and the retardation of enzyme release by additional polymer coatings*. Biomacromolecules, 2005, 6(3), 1484–1488.
74. Xie, J. and C.-H. Wang, *Electrospun micro-and nanofibers for sustained delivery of paclitaxel to treat C6 glioma* in vitro. Pharmaceutical Research, 2006, 23(8), 1817–1826.

CHAPTER 5

HEAT AND MOISTURE TRANSFER IN CLOTHING SYSTEM

A. AFZALI and SH. MAGHSOODLOU

University of Guilan, Rasht, Iran

CONTENTS

ABSTRACT

The apparel products with high thermal quality have become increasingly attractive to clothing industry as it can provide people comfort wearing feeling thus improve the quality of life. A multi-structural computational scheme for textile thermal engineering design plays an important role in translating complex mathematical models into computational algorithms and to implement computational simulation of clothing wearing system consisting of human body, clothing and external environment.

5.1 THERMAL BEHAVIORS IN THE CLOTHING WEARING SYSTEM

Clothing is one of the most intimate objects in people's daily life since it covers most of the human body most of the time. People may keep having subjective psychological feelings of the clothing and consciously judging the warm/cold/comfort sensation during the wearing time. On the basis of wearing experience, people can make a rough evaluation of the thermal function of clothing and choose suitable clothing for their daily activities. However, as projected in the thermoscopic world, the wearing situation of people can be regard as a complex and interactive multi-component system. Figure 5.1 shows the components of the clothing wearing system. The thermal behaviors involved in the wearing situation may be categorized as [1]:

(i) Heat and moisture transfer in the textile materials. This is the physical behavior mainly deciding the thermal performance of clothing. It can be regarded as the following physical process:
- The heat transfer process in the textile material in terms of conduction, convection and radiation;
- The vapor moisture transfer process in the textile material in terms of diffusion and convection;
- The liquid water transfer process in the textile material;
- Phase change process in the textile material. It is an approach allowing the heat and moisture transfer in a coupled way, including moisture condensation/evaporation in the fabric air void volume, moisture absorption/desorption of fibers, and micro-encapsulated phase change materials;

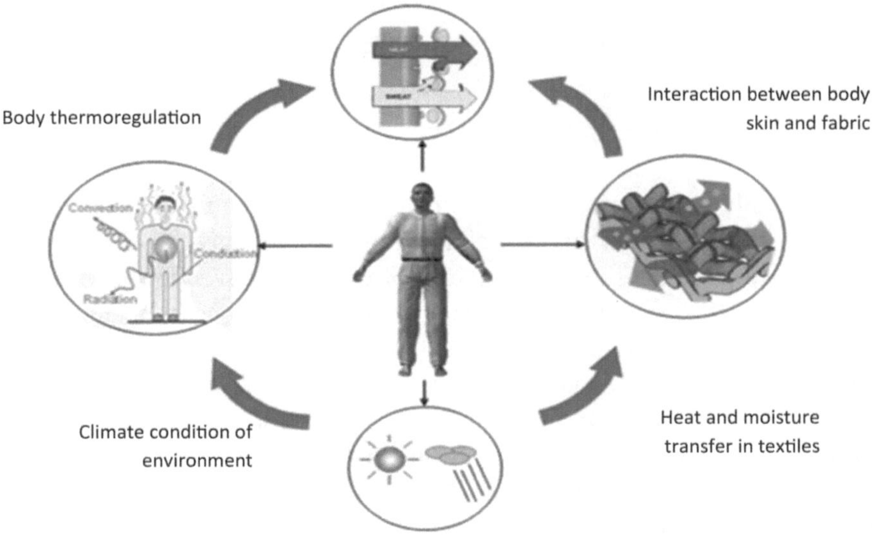

FIGURE 5.1 Components of the clothing wearing system.

- Influence of functional treatments of textiles on the heat and moisture transfer process, such as waterproof fabric, moisture management treatment, PCM coating and heating fabrics.

(ii) Thermoregulatory behaviors of the human body, such as sweating, shivering and biological metabolism, and heat and moisture exchange of body skin and environment.

(iii) Interactions between the inner clothing and body skin.

(iv) The climatic conditions of wearing situation in terms of temperature, relative humidity and wind velocity, which influence the heat and moisture behaviors of textile materials and the human body.

5.2 THEORETICAL MODELING OF THE CLOTHING WEARING SYSTEM

5.2.1 HEAT AND MOISTURE TRANSFER IN THE TEXTILE MATERIALS

Normally, the heat and moisture transfer processes in the textile material occur when there are gradients of temperature and water vapor pressure across textile structures, and these two processes are often coupled accompanying with the appearance of phase change process.

5.2.2 HEAT TRANSFER PROCESS

The overall heat transfer in textile materials is the sum of contributions through the fiber and interstitial medium, which may involve multiple transfer mechanisms in terms of conduction, convection and radiation. Theoretically, conduction heat transfer always occurs in the solid fiber material and the medium trapped in the spaces between the fibers as long as a temperature gradient is presented. Convection heat transfer will be obviously observed if the medium is gaseous and if the space is large enough, that is to say, the more porous the textile material is, the more effectively convection takes place. Radiation heat transfer can be ignored when the temperature gradient is small. Consequently, the heat transfer via conduction is the most dominant transfer mechanism.

In the engineering applications, thermal conductivity is usually adopted to express the thermal properties of materials because thermal conduction is better documented and mathematically analyzed [2]. Unlike other porous materials, in textiles the air filled in the space between fibers has substantially bigger proportion than that of the fibers and the thermal conductivity of fiber is much smaller than air. The heat flow by thermal conduction at any position (x) inside the textile structure can be expressed by the Fourier's law:

$$Q_k(6.x)_c = -k\frac{dT(6.x)}{dx} \tag{5.1}$$

$$k = (1-\epsilon)k_f + k_a \tag{5.2}$$

where, Q_k is conductive heat flux, T is the temperature and K is the effective thermal conductivity of the textile structure, which is the combination of the conductivities of the air (K_a) and the solid fiber (K_f) is the porosity of textile material.

The radioactive heat transfers in the ways of emitting or absorbing electromagnetic waves when the standard temperature of the textile material is above zero. The intensity of radioactive is depended on the ratio between the radiation penetration depth and the thickness of textile material, besides the temperature difference between the two surfaces of textile material. Farnworth reported that when the radiation depth is similar to the thickness of fabric, the amount of radiation should be taken into account as it is comparable to the amount of the conduction heat flow, and the radiation heat flux in fabric can be expressed by [3]]:

$$\frac{\partial F_R}{\partial x} = -\beta F_R + \beta \sigma T^4 \frac{\partial F_L}{\partial x} = \beta F_L - \beta \sigma T^4 \tag{5.3}$$

$$\beta = \frac{(6.1-\varepsilon)}{r}\varepsilon_r \tag{5.4}$$

where, β is the radiation absorption constant for textile materials, σ is the Boltzmann constant, F_R and F_L, respectively is the total thermal radiation incident traveling to the right and left way.

5.2.3 MOISTURE TRANSFER PROCESS

Thanks to the porosity of the fabric, the interstices between fibers provide the space for moisture to flow away. There are four ways of moisture (in the phase of water vapor or liquid) transfer occurring in textile materials as summarized by Mecheels [4]:

(i) diffusion through the space between the fibers,
(ii) absorption/desorption by the fiber materials,
(iii) transfer of liquid water through capillary interstices in yarns/fibers, and
(iv) migration of liquid water on the fiber surface.

There are many similarities between heat conduction and moisture diffusion in the textile material. When the system scale, material properties and initial and boundary conditions are similar, the governing equations, analysis methods and results would be analogous for these two processes [5]. When there is a difference between the water vapor concentration on the fibers' surface and that of the air in the fiber interstices, there will be a net exchange of moisture. The water vapor diffusion through the textile material can be described by the First Fick's law [5]:

$$Q_w = D_a \frac{\Delta C}{L} \tag{5.5}$$

where Q_w is the moisture transfer rate, D_a is the diffusion coefficient of water vapor through the textiles, L is the thickness of the fabric sample, and ΔC is the vapor concentration gradient of two fabric sides.

Since the fiber has a small proportion of volume in the fabric, the main contribution of moisture flux is from the diffusions process through the air in the fiber interstices. However, it was identified by Wenhner et al. [6] that absorption of moisture by the fiber also importantly affected the response of fabric to the moisture gradient. The water vapor concentration on the fiber surface, theoretically, depends on the amount of absorbed moisture onto the surface and the local temperature of the fiber. The fiber will keep absorbing as much moisture as it can until it reaches a saturated status with respect to the absorption rate. And when the fiber becomes saturated, additional vapor moisture may condense into liquid phase onto the fiber surface. With regard to the physic nature of fibrous materials, condensate water may be held on the surface of the fiber and be relative immobile, or may be transferred across the textile structure by capillary actions.

The moisture absorption capacity of fibers is described by the property of hygroscopicity (also called moisture regain), which means the amount of moisture that the fiber contains when placed in an environment at certain temperature and relative humidity. In 1967, Nordonb land David proposed an exponential relationship to describe the change rate of water content of the fibers, and developed a numerical solution with computer technology at that time. Li and Holcombe, in 1992, devised a new absorption rate equation by analyzing the two-stage sorption kinetics of wool fibers and incorporating it with more realistic boundary condition [7]. They assumed the water vapor uptake rate of fiber is composed by two components associated with the two stages of sorption:

$$\frac{\partial C_f}{\partial t} = (1-\alpha)R_1 + \alpha R_2 \quad 0 \le \alpha \le 1 \tag{5.6}$$

$$R_1 = \frac{\partial C_f}{\partial t} = \frac{1}{r}\frac{\partial}{\partial r}\frac{(6.rD_f \partial C_f)}{\partial r} \tag{5.7}$$

where, R_1 and R_2 are the moisture sorption rates in the first and second stages respectively, α is the proportion of uptake occurring during the second stage. R_1 can be obtained by regarding the sorption/desorption process as Fickian diffusion. R_2 is identified by experimental data and related to the local temperature, humidity and sorption history of the fibers.

5.2.4 LIQUID TRANSFER PROCESS

When liquid water transfers across the textile material, it will experience wetting and wicking stages, in which wetting of textile is prerequisite for the wicking process [8]. Both wetting and wicking are determined by surface tensions between the solid-vapor-liquid interfaces. In view of the macroscopic world, these tensions are the energy that must be supplied to increase the surface/interface area by one unit. The liquid water put in touch with fibrous material comes to an equilibrium state with regard to minimization of interfacial free energy on the surface. The force involved in the equilibrium can be expressed with the well-known Young's equation:

$$\gamma_{LV}.\cos\theta = \gamma_{SV} - \gamma_{SL} \tag{5.8}$$

where, γ_{LV}, γ_{SV}, γ_{SL}, respectively, represents the interfacial tensions that exists between the solid-vapor, solid-liquid and liquid-vapor interfaces. And the term γ_{LV} is also usually regarded as the surface tension of the liquid. θ is equilibrium contact angle, which is the consequence of wetting instead of the cause of it. The term $\gamma_{LV} \cdot \cos\theta$ is defined as adhesion tension or specific wettability of textile material. This equation shows that wettability increases with the decreased equilibrium contact angle θ. The equilibrium contact angle is an intrinsic value described by the Young equation for an ideal system. However, precise measurement of surface tension is not commonly possible. The experimentally measured contact angle between the fiber and the liquid can be observed on a macroscopic scale and is an apparent physical property.

When textiles surface is fully wet by the liquid, the wicking process will happen spontaneously, where the liquid water transports into the capillary space formed between fibers and yarns by capillary force. The fibrous textile assembly is a complex nonhomogeneous capillary system due to irregular capillary spaces. These spaces have various dimensions and discontinuous radius distributions. The practical engineering field, an indirectly determined effective capillary radius, is adopted to represent the no-uniform capillary spaces in yarns and fabrics. If the penetration of liquid is limited to the capillary space and the fiber does not absorb the liquid, the wicking process is called capillary penetration, and the penetration is originally driven by the wettability of the fiber, which is decided by the chemical nature and geometry structure of its surface [8]. Ito and Muraoka pointed out that the wicking process will be suppressed with the decreased number of fibers in the textiles [9]. When the number of fibers becomes greater, water moves along the void spaces even between the untwisted fibers, which indicates that sufficient number of continuity of pores is very important to the wicking process.

5.3 COUPLED HEAT AND MOISTURE TRANSFER

The research of the heat transfer and moisture diffusion in textile materials was initially regarded as independent under the assumption that the temperature and moisture concentration of the clothing is steady over a period of time. In this steady-state condition, there is no need to address the interactions between heat and moisture transfer process [10].

However, under some transient situations where some phase change processes happen, such as moisture sorption/desorption and evaporation/condensation, these two processes are coupled and interact significantly.

The moisture absorption/desorption capability of the solid fiber depends on the relative humidity of the enclosed air in the microclimate around the fiber and the type of fiber material. When fibers absorb moisture, heat is generated and released. Consequently, the temperature of fiber will rise and thus results in an increase of dry heat flow and a decrease in latent heat flow across the fabric [11]. The absorbed/desorbed moisture of fibers and the water vapor in the enclosed microclimate in textiles compose the water content of the textile material, which can originate from the wicking process or result from condensation in case of the fully saturated water vapor in fibrous materials [12].

Similar to the phase change process of moisture sorption/desorption, liquid condensation/evaporation pose an impact on the flow of heat and moisture across the textiles by acting as a heat source or be merged into the heat transfer process. Condensation is a physical phenomenon, which commonly takes place when the fibrous material is exposed to a large temperature gradient and high humid source, both of which cause the local relative humidity to attain 100% or full saturation. And provided that there is still extra moisture diffusing into the fibrous material, condensation continues. The condition of condensation is different from the transient process of moisture sorption/desorption. When the relative humidity of surrounding microclimate is less than 100%, evaporation occurs.

The first model describing the transient heat and moisture transfer process in porous textile material was developed by Henry (1939) in terms of two differential equations respectively for mass and heat governing formulation [13], as listed in the following. A simple linearity assumption was made in this model to describe the moisture absorption/desorption of fibers to obtain analytical solution. This modeling work established a basic framework of modeling the complicated coupled process of heat and moisture transfer through the textiles material.

$$\epsilon.\frac{\partial C_a}{\partial t}=\frac{D_a.\epsilon}{\tau_a}.\frac{\partial^2 C_a}{\partial x^2}-(1-\epsilon).\Gamma_f \tag{5.9}$$

$$C_v.\frac{\partial T}{\partial t}=K_{fab}.\frac{\partial^2 T}{\partial x^2}+(1-\epsilon).\lambda_v.\Gamma_f \tag{5.10}$$

$$\Gamma_f=\frac{\partial C_f}{\partial t}=const+a_1C_a+a_2T \tag{5.11}$$

where, a_1 and a_2 are coefficients. Henry assumed the moisture sorption rate Γ_f to be a linear relationship between temperature and moisture concentration, allowing him to solve the equations analytically.

Ogniewicz and Tien [13] proposed a model considering the heat transport that happened by the ways of conduction, convection and condensation in a pendular state. That model ignored moisture sorption and was lack of a clear definition of the volumetric relationship between the gas phases and the liquid phase. Motakef [14] extended this analysis to describe mobile condenses, in which the moisture condensation was taken into account with simultaneous mass and heat transfer process.

$$\frac{\partial C_a}{\partial t} = D_a \frac{\partial C_a}{\partial x} + \Gamma_{1g} \tag{5.12}$$

$$C_v.\frac{\partial T}{\partial t} = K\frac{\partial^2 T}{\partial x^2} - \lambda_{1g}.\Gamma_{1g} \tag{5.13}$$

$$\rho_c\epsilon \frac{\partial \theta}{\partial t} = \rho_c\epsilon \frac{\partial}{\partial t}(6.D_l(\theta)\frac{\partial \theta}{\partial x} - \Gamma_{1g} \tag{5.14}$$

In Motakef's model, a concept of critical liquid content (CLC) was introduced to address the liquid diffusivity. When the liquid content θ is below the CLC, the liquid is in the pendular state and has no tendency to diffuse. When the liquid content θ is beyond the CLC, a liquid diffusivity D_l *(θ)* was introduced to describe the liquid transfer by surface tension force from regions of higher liquid content to the drier regions. D_l *(θ)* is a complicated function of the internal geometry and structure of the medium.

Fan and Luo [12] incorporated the new two-stage moisture sorption/desorption model of fibers into the dynamic heat and moisture transfer model for porous clothing assemblies. They considered the radiation heat transfer and the effect of water content of fibers on the thermal conductivity of fiber material. Further, Fan and his co-workers improved the model by introducing moisture bulk flow, which was caused by the vapor-pressure gradients and supersaturation state [12]. This improvement made up for the ignorance of liquid water diffusion in the porous textile material in previous models. The equations of the model are listed as follows:

$$\epsilon \frac{\partial C_a}{\partial t} = -\epsilon \frac{\partial C_a}{\partial x} + \frac{D_a\epsilon}{\tau}\frac{\partial^2 C_a}{\partial x^2} - \Gamma(6.x,t) \tag{5.15}$$

$$\rho(6.1-\varepsilon).\frac{\partial(6.W-W_f)}{\partial t}=\rho(6.1-\varepsilon).D_l\frac{\partial^2(6.W-W_f)}{\partial x}+\Gamma(6.x,t) \quad (5.16)$$

$$C_v(x,t)\frac{\partial T}{\partial t}=-\varepsilon\mu C_{va}(x,t)\frac{\partial T}{\partial x}+\frac{\partial}{\partial x}\left((x,t)\frac{\partial T}{\partial x}\right)-\frac{\partial F_R}{\partial x}+\frac{\partial F_L}{\partial x}+\lambda(6.x,t)\Gamma(6.x,t) \quad (5.17)$$

where, $\Gamma(x, t)$ accounts for moisture change due to absorption/desorption of fibers and the water condensation/evaporation; $W - W_f$ is the free water content in the fibrous material; D_l is the diffusion coefficient of free water in the fibrous batting, which is assumed with a constant value with reference to some previous work. C_v is the effective volumetric heat capacity of fibrous material.

In 2002 year, Li and Zhu reported a new model for simulation of coupled heat and moisture transfer processes, considering the capillary liquid diffusion process in textile [15], which developed the liquid diffusion coefficient as a function of fiber surface energy, contact angle, and fabric pore size distribution. Based on this new model, Wang et al. [16] considered more the radiative heat transfer and moisture sorption and condensation in the porous textile, achieving more accurate simulation for the realistic situation. The governing equations of the model are shown as follows:

$$\epsilon\,\frac{\partial(6.C_a\varepsilon_a)}{\partial t}=-\frac{1}{\tau_a}\frac{\partial}{\partial x}\left(D_a\frac{\partial(6.C_a\varepsilon_a)}{\partial t}\right)-\varepsilon_f\xi_1\Gamma_f+\Gamma_{1g} \quad (5.18)$$

$$\frac{\partial(6.\rho_l\varepsilon_l)}{\partial t}=-\frac{1}{\tau_l}\frac{\partial}{\partial x}\left(D_l\frac{\partial(6.\rho_l\varepsilon_l)}{\partial t}\right)-\varepsilon_f\xi_2\Gamma_f+\Gamma_{1g} \quad (5.19)$$

$$C_v\frac{\partial T}{\partial t}=\frac{\partial}{\partial x}\left(K_{min}(x)\frac{\partial T}{\partial x}\right)+\frac{\partial F_R}{\partial x}-\frac{\partial F_L}{\partial x}+\varepsilon_f\Gamma_f(\xi_1\lambda_v+\xi_2\lambda_l)-\lambda_{1g}\Gamma_{1g} \quad (5.20)$$

The previous research about the heat and moisture transfer in textile material limited their focus on a single layer of porous textiles. Li and Wang extended the model for coupled heat and moisture transfer to multilayer fabric assemblies [16]. They described the geometrical features, layer relationships and blend fibers of the multilayer fabric assemblies by the following definitions:

$$l_{(6.i-1)} = l_{i1}(6.2 < i < n) \tag{5.21}$$

$$\text{Contact}_{\text{in}} = \begin{matrix} 1 & l_{ij} = 0 & (6.1 \le i \le n, j = 0,1) \\ 0 & l_{ij} \ne 0 & \end{matrix} \tag{5.22}$$

$$\bar{P} = \sum_{ti=1}^{tn} f_{ti} P_{ti} \qquad (6.\text{tn} \ge n) \tag{5.23}$$

where, l_{i0} and l_{il} are defined as the thickness of the left and right gap between neighbored layers. Contact is defined to express the contact situation at boundaries between layers, which determines the heat and moisture transfer behavior at the layer boundary. The symbol of is the weighted mean property of all blend fibers in the fabric based on their fractions f_{ti}. In addition, they individually developed boundary conditions for different fabric layers to achieve the numerical solutions.

5.4 WATERPROOF FABRIC

Waterproof fabric, which is laminated or coated with micro-porous or hydrophilic films, is frequently used in the design of functional clothing for the weather of low temperature, wind, rain, and even more extreme situations. With waterproof fabric, the clothing can effectively protect the body from the wind and water; as well as reduce the heat loss from the body to the environment. These functions of waterproof fabric, scientifically, are achieved by significantly affecting the processes of heat and moisture transfer through the textile products.

The term 'water vapor permeability' (WVP, $g.m^{-2}.day^{-1}$) is commonly used to measure the breathability of the fabric, which indicates the moisture transfer resistance in the heat and mass transfer processes. This property can be obtained by the experimental measurement with the Evaporation and Desiccant methods. The calculation formulation is expressed according to the first Fick's law of diffusion [7]:

$$\mathrm{WVP} = \frac{Q_W}{tA} = \frac{\Delta C}{W_n} \tag{5.24}$$

where, Q_w is the weight loss/gain in grams over a period time t through an area A, W_n is the water vapor resistance and ΔC is the difference of water vapor concentration on the two surfaces of the fabric sample.

Meanwhile, R_n is employed to express the thermal resistance of the waterproof fabric. Thus, the simulation of the thermal effect of the waterproof fabric can be realized by specifying the heat and moisture transfer coefficients at the boundaries, as shown by the following formula:

$$H_{mn} = \frac{1}{W_n + \dfrac{1}{h_{mn}}} \tag{5.25}$$

$$H_{cn} = \frac{1}{R_n + \dfrac{1}{h_{nc}}} \tag{5.26}$$

where, h_{mn} and h_{cn} are respectively the mass and heat transfer coefficients of the inner and outer fabric surface. When the fabric is laminated or coated with microporous or hydrophilic films as being waterproof, the combined mass and heat transfer coefficients H_{mn} and H_{cn} (n = 0.1) can be obtained by integrating the water vapor resistance W_n and the thermal resistance R_n of the waterproof fabric.

5.5 PHASE CHANGE MATERIAL FABRIC

The PCM which has the ability to change its phase state within a certain temperature range, such as from solid to liquid or from liquid to solid, is microencapsulated inside the textile fabrics in the functional clothing design in recent years to improve the thermal performance of clothing when subjected to heating or cooling by absorbing or releasing heat during a phase change at their melting and crystallization points.

With PCM technology, the temperature of clothing is able to gain a change delay due to the energy released/absorbed from the PCM when exposed to a very hot/cold environment.

In the textile application, the PCM is enclosed in small plastic spheres with diameters of only a few micrometers. These microscopic spheres containing PCM are called PCM microcapsules, and are either embedded in the fibers or coated on the surface of the fabric.

Research on qualifying the effect of the PCM fabric in clothing on the heat flow from the body was experimentally conducted by Shim. He measured the effect of PCM clothing on heat loss and gain from the manikin, which moved from a warm environment to a cold environment and back again. Ghali et al. [17] analyzed the sensitivity of the amount of PCM inside the textile material on the fabric thermal performance. The percentage of PCM in the fabric was found to influence the length of time period during which the phase change occurs.

Also, they drew a conclusion that under steady-state environmental conditions, PCM has no effect on the thermal performance; while when there is a sudden change in the ambient temperature, PCM can delay the transient response and decrease the heat loss from the human body.

In order to investigate the mechanisms of thermal regulation of the PCM on the heat and moisture transfer in textiles, Li and Zhu developed a mathematical model to describe the energy loss rate from the microspheres which is considered to be a sphere consisting of solid and liquid phases [18], as shown in the following equations:

$$\frac{\partial T_{ms}(6.x,r,t)}{\partial t} = a_{ms}\frac{1}{r^2}\frac{\partial}{\partial r}\left(6.r^2\frac{\partial T_{ms}(6.x,r,t)}{\partial r}\right) \quad \text{Spherical core (5.27)}$$

$$\frac{\partial T_{ms}(6.x,r,t)}{\partial t} = a_{ms}\frac{1}{r^2}\frac{\partial}{\partial r}\left(6.r^2\frac{\partial T_{ms}(6.x,r,t)}{\partial r}\right) \quad \text{Spherical shell (5.28)}$$

where, T_{ms} and T_{ml} are respectively the temperature distributions in a sphere containing solid and liquid PCM. These energy-governing equations are developed on the radial coordinate. r denotes the radius of the latest phase interface in micro-PCM. The smaller is the radius of the micro-spheres, the more significant is the effect of the PCM.

5.6 HUMAN BODY THERMOREGULATORY SYSTEM

To sustain life in various environments, the human body must have the ability to keep the temperature of core and skin at a reasonable range under a variety of external conditions, that is, the core body temperature should be maintained at 37.0 ± 0.5 °C, and the skin temperature should be managed at approximately 33 °C. In the human body, the regulation of the body temperature is implemented by the thermoregulation system, which responds to produce/dissipate heat when the body core temperature drops/rises. The basic mechanism of the body thermoregulation system involves two processes:

(i) when the body feels warm, the blood vessels react with vasodilatation and the glands begin to perform the sweating process;
(ii) when the body feels cold, the blood vessels reduce the blood flow to the skin and increase the heating production by muscle shivering. The schematics of thermoregulatory system of the human body.

The thermal mathematical models for human body can be classified for a single part and/or for the entire body. The models for a single part of the body, which was usually developed by physiologists, most of them are too complicated to simulate the physiological and anatomical details of the specific parts. It seems that these partial models theoretically can be added together to form a complete representation of the thermal exchange of the whole body. However, Fu claimed that such methods were not practical due to the fact that the connection between these models is very difficult, and adding the models together would need a supercomputer even if the connection between models were feasible. The thermal models for the entire body, reviewed by Fu, can be characterized by the following classifications:

(i) one-node models,
(ii) two-node thermal models,
(iii) multi-node models, and
(iv) multi-element models.

Though most of these models are likely to produce acceptable simulation results under the condition that the temperature is relatively uniform throughout the body, the multi-node and multi-element models perform better when large temperature gradients exist within the body because of the greater amount of details provided about the body temperature fields.

5.7 ONE-NODE THERMAL MODELS

One-node thermal models, in which the human body is represented by one node, are also called empirical models. They usually depend on experiments to determine the thermal response of the human body, and therefore, are not mathematical models in a phenomenological sense. A well-known empirical prediction model for the entire human body was reported by Givoni and Goldman. It was derived by fitting curves to the experimental data obtained from the subjects exposed to various environments.

5.8 TWO-NODE THERMAL MODELS

Two-node thermal models tend to divide the entire human body into two concentric shells of an outer skin layer and a central core representing internal organs, bone, muscle and tissue. The temperature of each node is assumed to be uniform. The energy balance equations are usually developed for each node and solved to produce the skin and core temperature and other thermal responses.

The early two-node models were not widely used by people due to the lack of sufficient consideration of the complicated physiological phenomena of the human body. Gagge et al. [19] introduced a more complete two-node model for the entire body, which includes the unsteady-state energy balance equation for the entire human body and two energy balance equations for the skin node and core node, as listed in the following:

$$S = M - E_{res} - C_{res} - E_{sk} - R - C - W \tag{5.29}$$

$$S_{cr} = M - E_{res} - C_{res} - W - \left(K_{min} + c_{pbl}V_{bl}\right)(6.T_{cr} - T_{sk}) \tag{5.30}$$

$$S_{sk} = \left(K_{min} + c_{pbl}V_{bl}\right)\left(T_{cr} - T_{sk}\right) - E_{sk} - R - C \tag{5.31}$$

where, S_{cr} and S_{sk} are respectively the heat storage of core and skin shell, M is the metabolic heat, E_{res} and C_{res} are respectively latent and dry respiration heat loss, W is moisture transfer resistance, R is radioactive heat loss, C is convective heat loss, V_{bl} is skin blood flow rate, and c_{pbl} is specific heat at constant pressure of blood.

Gagge et al. [19] later improved their two-node model by the development of the thermal control functions for the blood flow rate (V_{bl}), the sweat rate (RSW), and the shivering metabolic rate (M).

$$V_{bl} = [6.3 + 200(6.WARM_{cr})]/[1 + 0.1(6.\mathrm{COL}\,D_{cr})] \tag{5.32}$$

$$RSW = 4.72\times10^{-5}.WARM_{bm}.e^{\left(6.\frac{WARM_{cr}}{10.7}\right)} \tag{5.33}$$

$$M = 58.2 + 19.4.COLD_{cr}.COLD_{sk} \tag{5.34}$$

Due to the two-node nature of this model, it is able to be applied easily and simply with the straightforward numerical solution. Smith pointed out that Gagge's model was applicable for situations with moderate levels of activity and uniform environment conditions. However, due to the limitation imposed by only two nodes, Gagge's model can only be applied under uniform environmental circumstances.

5.9 THE MULTI-NODE MODELS

Multi-node models divided the entire human body into more than two nodes and developed energy balance equation for each node as well the control functions for blood flow rate, shivering metabolic rate, and so on. Stolwijk et al. [20] presented a more complex multi-node mathematical thermal model of the entire human body, in which many efforts are made to the statement of the thermal controller. This model firstly divided the body into six cylindrical parts of head, trunk, arm, hands, legs and feet and a spherical body part comprising the head. Each part is further divided into four concentric shells of core, muscle, fat and skin tissue layers. Specifically, in this model all the blood circulation in the human body is regarded as a node called the central blood pool, which is the only communication connecting each body part. Therefore, Stolwijk's model is also called a 25-node model, and energy balance equations are developed for each node with the assumption of uniform temperature in each layer, which includes heat accumulation, blood convection, tissue conduction, metabolic generation, respiration and heat transfer to the environment. The descriptions for these equations are shown as follow:

$$\text{Core layer}: C(6.i,1)\frac{dT(6.i,1)}{dt} = \mathrm{Q(6.i,1) - B(6.i,1) - D(6.i,1) - RES(6.i,1)} \tag{5.35}$$

$$\text{Muscle layer}: C(6.i,2)\frac{dT(6.i,2)}{dt} = \mathrm{Q(6.i,2) - B(6.i,2) + D(6.i,1) - D(6.i,2)} \tag{5.36}$$

$$\text{Fat layer}: C(6.i,3)\frac{dT(6.i,3)}{dt} = \mathrm{Q(6.i,3) - B(6.i,3) + D(6.i,2) - D(6.i,3)} \tag{5.37}$$

$$\text{Skin layer}: C(6.i,4)\frac{dT(6.i,4)}{dt} = \mathrm{Q(6.i,4) - B(6.i,4) + D(6.i,3) - E(6.i,4) - Q(6.I,4)} \tag{5.38}$$

where, $C(i, j)$ is the heat capacity of each node, $T(i, j)$ is the node temperature, $Q(i, j)$ is the sum of the basic metabolic rate, $B(i, j)$ is the heat exchanged between each node and central blood compartment, $D(i, j)$ is the heat transmitted by conduction to the neighboring layer with the same segment. $E(i, j)$ is the evaporative heat loss at skin surface.

Similar to Gagge's model [19], Stolwijk et al. [19] also developed the thermal control functions in terms of tissue temperature signals, in which the warm signal (WARMS) and cold signal (COLDS) are corresponding to warm and clod receptors of the skin and are calculated by the error signal (ERROR). These controller equations produce the signals to drive the regulator, including total efferent sweat command

(SWEAT), total efferent shivering command (CHILL), total efferent skin vasodilation command (DILAT), and total efferent skin vasoconstriction command (STRICT).

Stolwijk's model [19] has made much advancement compared to previous multi-node models as it is not only capable of calculating the spatial temperature distribution for each node, but also has improved the representation of the human's circulatory system since the blood circulation is the most important function of human body. This model has been validated with the good agreements between the experimental and predicted results of most cases. The limitation of this model is that it cannot be used for the highly non-uniform environmental situations caused from the negligence of spatial tissue temperature gradients.

5.10 MULTI-ELEMENT MODELS

The greater difference between the multi-element thermal models and the two-node or multi-node models is that it divides the human body into several parts or elements without further division, and the temperature of each part or elements is no longer assumed as uniform. With the lifting of node uniform assumption, the mathematical descriptions of thermal functions, circulation, respiration etc. have also become more detailed to correspond with the detailed temperature filed.

Wissler [10] developed a multi-element model for the entire body by dividing the human body into six elements: head, torso, two arms and two legs, which were connected by the heart and lung where venous streams were mixed. Later on, Wissler improved his model and divided the human body into 15 elements to represent the head, thorax, abdomen, and the proximal, medical, and distal segments of the arms and legs, which were connected by the vascular system composed of arteries, veins and capillaries. Energy balance equations for each element and the arterial and venous pools were developed with the assumption that the blood temperature of arteries and veins in each element were uniform, and the thermal control equations for blood flow rates, the shivering metabolic rate and the sweat rate were built up.

The limitations of this model are that it is not applicable to the situations where a large internal temperature gradient or highly nonuniform environmental conditions exist. Additionally, the effect of the vasodilation and vasoconstriction was not included in the model. Finally, the parameters and constants used in the control equation are not easy to determine.

Smith developed a 3-D, transient, multi-element thermal model for the entire human body with detailed control functions for the thermoregulation system. Compared to the previous models, the improvements were that he:

(i) developed a 3-D temperature description of the human body;
(ii) provided a detailed description of the circulatory system, the respiratory system and the control system;
(iii) employed the finite-element method to get the numerical solutions of the model, which made a 3-D transient model for the entire human body possible.

The model divided the human body in 15 cylindrical body parts: head, neck, torso, upper arms, thighs, forearms, calves, hand and feet. Each body part is connected only by the blood flow and without tissue connection. The simulation results showed this model works well for situations of human thermal response during sedentary conditions in both uniform and non-uniform environments for either hot or cold stress conditions. However, the behaviors of the model during cold or hot exercising conditions were less satisfactory.

Fu summarized the limitations of the previous models and developed a 3-D transient, mathematical thermal model for the clothed human to simulate the clothed human thermal response under different situations. The main improvement of this model is the addition of the subcutaneous fat layer, the accumulation of moisture on the skin, and the blood perfusion and blood pressure to Smith's model. The development of the human model includes the thermal governing equations of the passive and control systems. Fu's 3-D transient model has a good ability of simulating the human body thermoregulatory system in situations where there exists high temperature vibration and even in the extremely atrocious weather conditions.

Though the multi-elements models can predict the thermal status of the human body in more detail, however, it should be noticed that there are many difficulties in applying the multi-element models into clothing engineering design due to the following considerations:

(i) the multi-element model requires the clothing to be 3-D meshed and modeled, which may cause great complexity in the integration of the models of clothing and human body, and the computation load is very intensive;
(ii) the many parameters involved in the multi-element models have high demanding on the data availability in the engineering application.

5.11 INTERFACES BETWEEN THE BODY SKIN AND CLOTHING

In the daily life, the clothing acts as an important barrier for heat and vapor transfer between the skin and the environment, protecting against extreme heat and cold, but meanwhile hampering the loss of superfluous heat during physical effort. This barrier is formed by the clothing materials themselves and by the air they enclose as well as the still air bound to the outer surfaces of the clothing.

Some researchers experimentally observed the phenomena of heat and moisture that exchange actively between the clothing and skin, and found the exchanged amount is considerable compared to the total increase/ decrease volume. The maximum heat flow from the skin to clothing depends on the heat conductivity of the inner layer of clothing and covering area of skin. Also, the heat exchange between the human body and clothing is dependent on the external parameters, such as air temperature, air humidity, and wind speed.

The clothing with the least porous, greatest thickness and lowest permeability will provide the greatest protection to heat and perspiration from the skin to environment provided with least porous and thickest thickness.

Li and Holcombe [21] reported a new model by interfacing the model for a naked body with a heat and moisture transfer model of a fabric. They developed the boundary condition between the body and clothing by quantifying the heat and mass flow. At the fabric-skin interface,

Heat:

$$M_t = h_{ti}(T_{sk} - T_{fi}) \tag{5.39}$$

Mass:

$$M_d = h_{ci}(C_{sk} - C_{fi}) + L_{sk}\frac{\partial C_{sk}}{\partial t} \tag{5.40}$$

At the interface between fabric and ambient air
Heat:

$$K\frac{\partial T}{\partial x} = -h_t\left(T - T_{ab}\right) \tag{5.41}$$

Heat:

$$D_a\frac{\partial C_a}{\partial x} = -h_c(6.C_a - C_{ab}) \tag{5.42}$$

where, M_d is the moisture flow from the skin and M_t is the heat flow from the skin. Since the air spacing between the skin and the fabric continuously varies in time depending on the level of activity and the location, the thermal transfer processes are influenced by the ventilating motion of air through the fabric initialed from the relative motion between the body and surrounding environment, such as in the walking situations.

Ghali et al. [22] developed a 1D model of the human body. The oscillating trapped air layer gap width and the periodically ventilated fabric predicts the effect of walking on exchanges of heat and mass. Murakami et al. [23] presented a numerical simulation of the combined radiation and moisture transport for heat release from a naked body in a house where continuous slight air flow exists by considering the thermal interaction between the human body and the environment and the intrinsic complex air situation in the real world.

Though much attention has been paid to the simulation of the thermal behaviors in the integrated system of human body, clothing and environment, and some numerical algorithms were reported for the simulation, these research studies put their focuses only on the scientific exploration and investigation. Few are developed systemically with a user-oriented purpose and used for clothing functional design.

5.12 CLOTHING THERMAL ENGINEERING DESIGN

5.12.1 CLOTHING THERMAL FUNCTIONAL DESIGN

The clothing thermal functional design, if following the traditional way of the clothing design and production, begins from the conception design and the prototypes making. As a result, a series of testing configured with experimental protocols will be performed by employing wearing subjects or by thermal manikins, and related thermal data will be measured using various equipment's during the experiments. Based on the analysis of experimental data, designers attempt to find the difference of measured thermal functions of clothing and their design concepts to obtain feedback to improve their design. After the iterative trial and error process, the final products can be put on the market. This traditional design process is very expensive, time-consuming and tedious due to the real prototypes making, experimental testing and burdensome data analysis.

These shortcomings of the traditional design method make it difficult to satisfy the requirements of designers and manufacturers. People come to resort to the powerful capacity of the computer in the design process of thermal functional clothing. Antunano et al. [24] employed a computer model and heat-humidity index to evaluate the heat stress in protective clothing. Schewenzferier et al. [25] optimized thermal protective clothing by using a knowledge bank concept and a learning expert system. Computer has acted a role in the clothing thermal functional design. However, it still cannot directly help the designer to preview the thermal performance of clothing, which is a crucial function for clothing thermal functional design.

James et al. [26] applied the commercial software of computational fluid dynamics (CFD) in their strategy to simulate the heat and moisture diffusive and convective transport as well as effect of sweating to predict the performance of chemical and steam/fire protective clothing. Kothari et al. [27] simulated the convective heat transfer through textiles with the help of CFD to observe the effects of convection on the total heat transfer of the fabric. The software tools like CFD provide a possible pathway for the user to simulate the heat and fluid distribution in the clothing. However, these tools do not take into account the structural features of the textile materials and the special features of the heat and moisture transfer process in textile materials that are related to the physical properties and chemical compositions of the textile materials. They cannot reflect the practical wearing situation and preview the true complex thermal behaviors in clothing.

In order to obtain scientific simulation of clothing thermal performance, some researchers have made efforts to apply the theoretical models describing the complex heat and moisture behaviors in clothing wearing system to the clothing thermal functional design. Parsonin [28] adopted thermal models for the clothed body including human thermoregulation and clothing to work as tools for evaluating clothing risks and controls. Prasad et al. in 2002 [29] constructed a detail mathematical model to study transient heat and moisture transfer through wet thermal liners and evaluated the thermal performance of fire fighter protective clothing.

Their research has made good exploration in clothing thermal functional design with the computer tools. Design and evaluation models, experts systems, applications of CFD software, and theoretical simulation models have been used to help to carry out clothing thermal functional designs. They either simply focus on evaluating certain thermal properties of clothing or need specialized knowledge to understand. They are very difficult to be applied as engineering tools for the general designers.

5.12.2 CAD SYSTEMS FOR CLOTHING THERMAL ENGINEERING DESIGN

The application of CAD technologies for clothing design is a significant sign of revolutionary advancement in the development of computerization and automation in the clothing industry. Clothing designers/engineers are offered a number of flexibilities in their design with CAD systems, such as the usage of textile material, exploration of functional design and products display.

Currently many CAD packages are available, targeting at pattern design, garment construction, fashion design and physical fitting simulation, which have made many achievements in catering for different requirements of the clothing industry [30, 31]. They are helpful to shorten the design cycle and save time and money on the prototypes preparation as well as improve productivity considerably. Recently, 3D clothing design and visualizations have been developed to simulate and visualize the physical performance of the clothing wear on the body in 3D virtual ways [30, 32], which enable the designs to be more realistic and make detailed analysis and evaluation of the clothing mechanical performance. However, these pioneer achievements are mainly focused on the mechanical behaviors of clothing.

The newest interest in the CAD system for clothing functional design places the focus on the thermal behaviors of clothing. Clothing thermal engineering design with CAD systems and tools is an effective and economical solution of designing clothing with superior thermal performance for wearing in various environments with a feeling of comfort. The CAD system for clothing thermal engineering design aims to create a virtual platform, which offers designers the ability to conceive their products using the engineering method. Thermal engineering design of clothing is the application of a systematic and quantitative way of designing and engineering of clothing with the inter-disciplinary combination of physical, physiological, mathematical, computational and software science, and engineering principles to meet with the thermal biological needs of protection, survival and comfort of human body.

The research in these fields discussed in the above sections lays substantial foundation to achieve this strategy. With this engineering design system, designers can simulate the thermal behaviors of the clothing, human body and environment system in specified scenarios, preview and analyze the thermal performance of the clothing, and iteratively improve their designs for desirable thermal functions of clothing.

In order to help the designers and manufacturers to quickly carry out clothing thermal functional design, Mao et al. [33] have developed two software systems respectively for multi-layer and multi-style thermal functional design of clothing. With these tools, the designers and manufacturers may have a quick preview of the different thermal performance when using different textile materials to have a comparison and make decision; or they can consider the different style (long, media or short) of the clothing and its thermal performance on the different body parts. The thermal performance of the sportswear can be quickly simulated and validated with the CAD system before the physical pattern making to reduce the design cycle and lower the design cost.

This new application of CAD technologies demonstrates great potentials in the clothing thermal engineering design, because the capacity of simulating and predicting the thermal performance of clothing is indispensable for designing clothing for thermal protection and comfort. While physical fit and a good-looking fashion style are crucial aims of clothing design, the thermal performance of clothing is another critical aspect that relates to the survival, health and comfort of human beings living in various environmental conditions. As more and more consumers want to wear clothing with higher functional and comfort performance, there is an urgent need for a CAD system to design clothing and analyze its thermal performance effectively and efficiently.

desirable functions, and then produce real products. The theoretical research of physics, chemistry related to textile and clothing, physiological thermoregulations of the human body, and the dynamic interactions between the clothing and body lead to thermal engineering design of clothing to achieve desirable thermal functions. The computational simulation can be enabled by related mathematical models, which is the substantial foundation for the engineering design, and the CAD systems provide a friendly tool for users through a series of functionalities to quickly carry out engineering design of clothing for desirable thermal functions. This strategy of thermal engineering design of clothing can hopefully speed up the design cycle and reduce the design cost.

5.13 CONCLUDING REMARKS

Traditionally, the comfort performance associated with the heat and moisture transfer behaviors of clothing are evaluated by subjective wearer trials. However, the subjective wearer trials may fail to simulate the practical environmental conditions accurately, which might lead to the inconsistent results. On the other hand, the costs of some subjective wearer trials are high not only in terms of the required environmental simulation, and clothing but also because the subjects have to be paid to participate. Therefore, the objective simulation tests are developed as a more acceptable option. The mathematical modeling and numerical simulation of the heat and moisture transfer in clothing materials and the human body have been extensively reported. More recently, a mathematical model to describe the complicated and coupled physical mechanisms concerning the heat and mass transfer in vivo worn facemasks during breathing cycles has been developed. With this model, the theoretical predictions are compared with experimental data of in-vivo protective performance of facemasks, and good agreement is observed between the two, indicating that the model is satisfactory. However, these simulations are mainly expressed using mathematical models, relevant computational algorithms and numerical solution methods, which limits the use of the model to a few professionals in the field.

KEYWORDS

- **thermal behaviors**
- **clothing wearing system**
- **thermal model**

REFERENCES

1. Pan, N. and P. Gibson, *Thermal and moisture transport in fibrous materials. 2006,* Cambridge: Woodhead Pub.
2. Martin, J. R. and G. E. Lamb, *Measurement of thermal conductivity of nonwovens using a dynamic method.* Textile Research Journal, 1987, 57(12), 721–727.
3. Farnworth, B., *Mechanisms of heat flow through clothing insulation.* Textile Research Journal, 1983, 53(12), 717–725.
4. Mecheels, J. *Concomitant heat and moisture transmission properties of clothing.* in *3rd Shirley Institute Seminar, Textiles in Comfort.* 1971.
5. Crank, J., *The mathematics of diffusion.* 1979, Oxford University Press.
6. Wehner, J. A., B. Miller, and L. Rebenfeld, *Dynamics of water vapor transmission through fabric barriers.* Textile Research Journal, 1988, 58(10), 581–592.
7. Li, Y. and B. Holcombe, *A two-stage sorption model of the coupled diffusion of moisture and heat in wool fabrics.* Textile Research Journal, 1992, 62(4), 211–217.
8. Kissa, E., *Wetting and wicking.* Textile Research Journal, 1996, 66(10), 660–668.
9. Ito, H. and Y. Muraoka, *Water transport along textile fibers as measured by an electrical capacitance technique.* Textile Research Journal, 1993, 63(7), 414–420.

10. Wissler, E. H., *Steady-state temperature distribution in man.* Journal of Applied Physiology, 1961, 16(4), 734–740.
11. Downes, J. and B. Mackay, *Sorption kinetics of water vapor in wool fibers.* Journal of Polymer Science, 1958, 28(116), 45–67.
12. Li, Y. and Z. Luo, *An improved mathematical simulation of the coupled diffusion of moisture and heat in wool fabric.* Textile Research Journal, 1999, 69(10), 760–768.
13. Ogniewicz, Y. and C. Tien, *Analysis of condensation in porous insulation.* International Journal of Heat and Mass Transfer, 1981, 24(3), 421–429.
14. Motakef, S. and M. A. El-Masri, *Simultaneous heat and mass transfer with phase change in a porous slab.* International Journal of Heat and Mass Transfer, 1986, 29(10), 1503–1512.
15. De Vries, D., *Simultaneous transfer of heat and moisture in porous media.* Transactions, American Geophysical Union, 1958, 39, 909–916.
16. Wang, Z., et al., *Radiation and conduction heat transfer coupled with liquid water transfer, moisture sorption, and condensation in porous polymer materials.* Journal of Applied Polymer Science, 2003, 89(10), 2780–2790.
17. Schlangen, H., *Experimental and Numerical Analysis of Fracture Processes in Concrete.* 1993.
18. Sarhadov, I. and M. Pavluš, *Models of Heat and Moisture Transfer in Porous Materials.*
19. Gagge, A., *An effective temperature scale based on a simple model of human physiological regulatory response.* Ashrae Trans., 1971, 77, 247–262.
20. Stolwijk, J. A., *A mathematical model of physiological temperature regulation in man.* Vol. 1855, 1971, National Aeronautics and Space Administration.
21. Li, Y. and B. Holcombe, *Mathematical simulation of heat and moisture transfer in a human-clothing-environment system.* Textile Research Journal, 1998, 68(6), 389–397.
22. Ghaddar, N., K. Ghali, and B. Jones, *Integrated human-clothing system model for estimating the effect of walking on clothing insulation.* International Journal of Thermal Sciences, 2003, 42(6), 605–619.
23. Murakami, S., S. Kato, and J. Zeng, *Combined simulation of airflow, radiation and moisture transport for heat release from a human body.* Building and Environment, 2000, 35(6), 489–500.
24. Antunano, M. and S. Nunneley, *Heat stress in protective clothing: validation of a computer model and the heat-humidity index (HHI).* Aviation, Space, and Environmental Medicine, 1992, 63(12), 1087–1092.
25. Schwenzfeier, L., et al. *Optimization of the thermal protective clothing using a knowledge bank concept and a learning expert system.* in *The sixth biennial conference of the European Society for Engineering and Medicine.* 2001.
26. Barry, J. J. and R. W. Hill, *Computational modeling of protective clothing.* Int Nonwovens J, 2003, 12, 25–34.
27. Bhattacharjee, D. and V. Kothari, *Prediction Of Thermal Resistance of Woven Fabrics. Part II: Heat transfer in natural and forced convective environments.* Journal of the Textile Institute, 2008, 99(5), 433–449.
28. Parsons, K., *Computer models as tools for evaluating clothing risks and controls.* Annals of Occupational Hygiene, 1995, 39(6), 827–839.

29. Prasad, K., W. H. Twilley, and J. R. Lawson, *Thermal Performance of Fire Fighters' Protective Clothing: Numerical Study of Transient Heat and Water Vapor Transfer*. 2002, US Department of Commerce, Technology Administration, National Institute of Standards and Technology.
30. Choi, K.-J. and H.-S. Ko, *Research problems in clothing simulation.* Computer-Aided Design, 2005, 37(6), 585–592.
31. Breen, D. E., D. H. House, and M. J. Wozny. *Predicting the drape of woven cloth using interacting particles.* in *Proceedings of the 21st annual conference on Computer graphics and interactive techniques*. 1994, ACM.
32. Wang, C. C. and M. M. Yuen, *CAD methods in garment design.* Computer-Aided Design, 2005, 37(6), 583–584.
33. Yi, L., et al., *P-smart—a virtual system for clothing thermal functional design.* Computer-Aided Design, 2006, 38(7), 726–739.

CHAPTER 6

ELECTROSPINNING OF NANOFIBERS AND POROSITY

A. AFZALI and SH. MAGHSOODLOU

University of Guilan, Rasht, Iran

CONTENTS

ABSTRACT

The nanofiber mat has extremely high specific surface area, adequate porosity and small pores due to their small diameters. The fiber size and morphology depends on various parameters involved in the preparation method. Electrospun fibrous membranes are highly porous structures that can be produced from a number of polymer/solvent combinations. Pore sizes ranging from 0.1 to 6 nm in diameter can be produced from solvent electrospinning. In this study different ways of producing porous structures are investigated. Then porosity measurement relationships are reviewed.

6.1 INTRODUCTION

This study exploits fundamental physics, chemistry and engineering principles to understand how liquids wet, permeate/flow, and reside in nanometer size porous fibrous structures and to develop user friendly computer model packages for industry applications.

Ultrafine fibers, called "nanofibers" are a unique nanomaterial because of the nanoscaled dimensions in the cross-sectional direction and the macroscopic length of the fiber axis (*see* Figure 6.1). Therefore, nanofibers have both the advantages of functionality due to their nanoscaled structure and the ease of manipulation due to their macroscopic length. In addition, three-dimensional nanofiber network assemblies (nanofibrous membranes or fabrics) provide good mechanical properties and good handling characteristics [1].

Many of the potential uses for nonwoven fabrics comprised of nanofibers are expected to take advantage of the large specific surface area, high porosity and small pore size of these fabrics. Both theoretical models and experimental studies have indicated that fiber diameter strongly influences the pore diameter in the fabric, with smaller fiber diameters resulting in smaller pores. For some applications, however, it would be interesting to

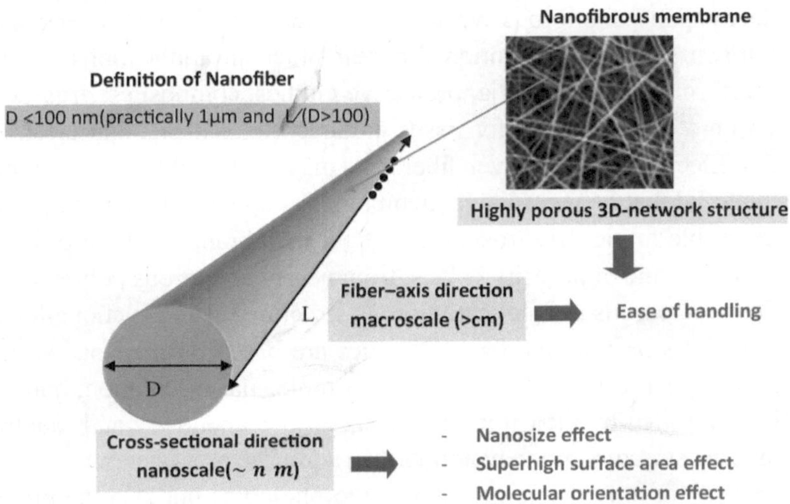

FIGURE 6.1 Characteristics of nanofiber and nanofibrous web.

combine the increased surface area associated with fine fibers with larger pores for fluid or cell transport. To accomplish this, the pore size must be de-linked from the fiber diameter [2].

Electrospinning is a simple technique that has garnered much attention recently because of its capability and feasibility in the generation of large quantities of nanofibers. The standard setup for electrospinning consists of a spinneret with a metallic needle, a syringe pump, a high-voltage power supply, and a grounded collector. A polymer, sol-gel, or composite solution (as well as melt) is loaded into the syringe and this viscous liquid is driven to the needle tip by a syringe pump, forming a droplet at the tip. When a voltage is applied to the metallic needle, the droplet is first stretched into a structure called a Taylor cone and finally into an electrified jet. The jet is then elongated and whipped continuously by electrostatic repulsion until it is deposited on the grounded collector [3].

Electrospinning results in submicrometer size fibers lay down in a layer that has high porosity but very small pore size. For fibers spun from polymer solutions, the presence of residual solvent in the electrospun fibers facilitates bonding of intersecting fibers, creating a strong cohesive porous structure [4]. Porous areas have the potential to stretch during fiber formation to form longitudinal striations along the axis of the fiber and then develop axial cracks that cause catastrophic failure.

Electrospun nanofibrous webs or membranes with high surface areas have drawn significant attention for their practical applications, such as high-performance filter media, protective clothes, composites, drug delivery systems, and scaffolds for tissue engineering, sensors, and electronic devices. Electrospun nonwoven fiber mats may be thought of as a microporous material that behaves like a membrane, as opposed to a more porous, air-permeable fabric. Electrospun nanofiber membranes may be produced over a wide range of porosity values, from nearly nonporous polymer coatings, to very porous and delicate fibrous structures. The functionalities of the nanofibers or nanofibrous membranes are based on their nanoscaled-size, high specific surface area, and high molecular orientation, and they can be controlled by their fiber diameter, surface chemistry and topology, and internal structure of the nanofibers. In addition, processing innovations to improve not only the controlling of morphologies but also the production capacity of electrospun nanofibers and nanofibrous membranes are in progress. In particular, the high-throughput electrospinning systems are an ongoing development (multi-needle and needleless processes) [1, 4].

6.2 PRODUCING POROUS NANOFIBERS

Porous fibers of variety polymers can be prepared through different methods, with different pore structure. Here, some of the more useful methods are reviewed.

6.2.1 MULTIPLE POLYMERS ELECTROSPINNING

Fibers can be selectively dissolved to increase the void volume and associated pore size. Electrospinning multiple polymers from multiple spinnerets into a single nonwoven fabrics has been approached in several ways. Side-by-side arrays of spinnerets have been used to increase the rate of electrospun nonwoven production. Fibers produced from adjacent spinnerets are overlap somewhat, but are not intimately mixed in the resulting fabrics. The tendency of fibers formed in side-by-side charged jets to resist mingling is easily explained by electrostatic repulsion of similarly charged materials. Other strategies for mingling materials in an electrospun

nonwoven fabric has included side-by-side or core/sheath arrangements of two materials from a single spinneret. Many interesting structures have been produced by these methods, but spinning conditions for the two materials are not completely independent. Frequently, both polymers are soluble in the same solvent. The applied voltage is necessarily the same for both materials [2].

To spin two dissimilar polymers from different solvents at different applied voltages, a simple electrospinning set up was arranged as shown in Figure 6.2. Each material is spun from a separate spinneret with separate high voltage sources. Since the charge drops with the distance squared to the grounded collector, no electrical field interference is expected at the rotating collector and none was observed. The collector rotates rapidly enough to create an intimate mixture of the two different fibers, but not rapidly enough to impart anisotropy or alignment in the resulting fabric. Porosity of the fabrics is measured before and after one of the fibers is removed by selective dissolution. Using this method, an intimately mixed fabric of two dissimilar polymers is created and porosity and pore size distribution is de-linked from fiber diameter [2].

6.3 THERMALLY INDUCED PHASE SEPARATION (TIPS)

A novel method producing porous nanofibers is developed by Xia et al. In his work, TIPS happen between the solvent-rich and solvent-poor regions in the fiber during electrospinning by immersing the collector in a bath of liquid nitrogen. Finally with removing the solvent in vacuum, porous poly(styrene) nanofibers are produced. In this method, phase separation into solvent-rich

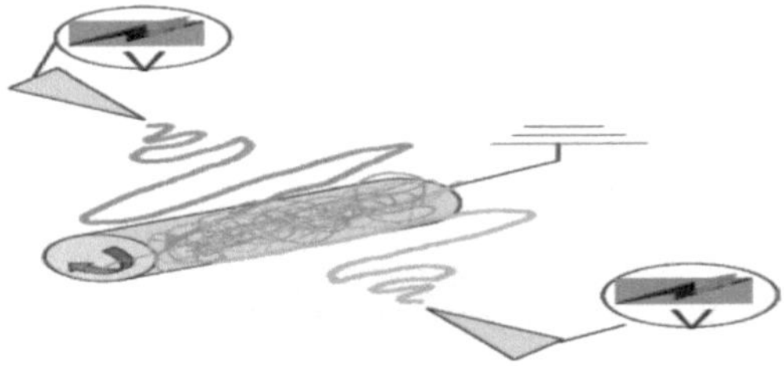

FIGURE 6.2 Two electrospinning units are positioned on opposite sides of a spinning collector.

and solvent-poor regions is induced when the remaining solvent is frozen in nitrogen bath before reaching the collector. These porous fibers are promising for use in the encapsulation of substance, as supports of catalysts, lightweight reinforcement, and hydrophobic coating. The electrospinning setup with the modified collector used in this method is shown in Figure 6.3. Nanofibers produced by this method have larger diameters than those collected without the use of a liquid nitrogen bath, since the fibers are drawn in shorter distance. The SEM image shows that the fibers are not interconnected, and that the morphology of individual fibers is not degraded by this method. This method can be readily used with nonvolatile solvents and does not require selective dissolution of phase-separated polymers [5, 6].

6.4 SELECTIVE DISSOLUTION

A facile method for the preparation of porous ultrafine nanofibers is selective dissolution technique that was used by number of groups. Scientists [5–7] produced highly porous PAN nanofibers, via this technique, PAN/ $NaHCO_3$ composite nanofibers were electrospun, and $NaHCO_3$ was removed by a selective dissolution and reaction with the solution of hydrochloric acid, Different content of $NaHCO_3$ had significant effect on the morphology of the final porous fibers. In the other work, You et al. [7] used this technique to produce porous ultrafine poly (glycolic acid)(PGA) fibers, thus ultrafine (PGA)/poly(L-lactic acid)(PLA) blend fibers were electrospun and then the PLA was removed by dissolving with chloroform.

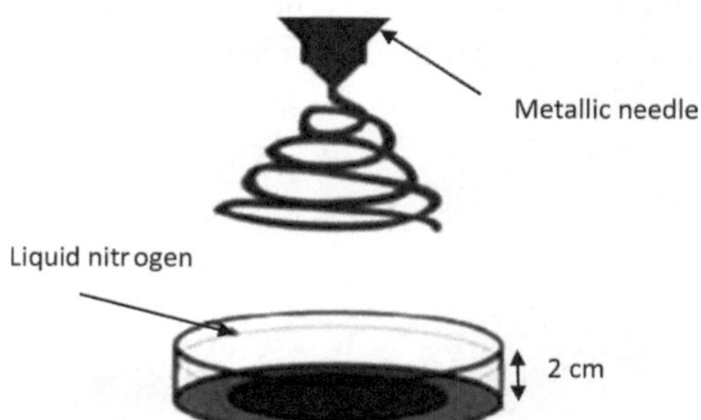

FIGURE 6.3 Electrospinning set up for thermally induced phase separation method.

The resulting PGA fibers had three dimensionally interconnected pores with a circular shape and the very narrow pore size distribution [7].

6.5 SELECTIVE PYROLYZATE COMPOSITE FORMATION

Researchers [8–9] used a facile method to obtain carbon nanofibers with a nanoporous structure using selective pyrolyzate composite formation. A blend of polyacrylonitrile and a copolymer of acrylonitrile and methyl methacrylate in dimethylformamide was electrospun into nano fibers with a microphase-separated structure. With pyrolyzing the copolymer domains in oxidation process, nanoporous structure was obtained, and preserved after carbonization.

Rapid solvent evaporation and solidification of the electrospun fibers during electrospinning, decreases the domain coarsening, so there are fine and stretched phase domains in the nanofibers, and carbon nanofibers with a large amount of nanopores throughout the surface and the interior of the fibers were obtained. The nanopores in the fibers at about several tens of nanometers in widths are continuous. Great potential of the nanoporous carbon fiber obtained in this study make many interesting applications for them in different fields [8].

6.6 ELECTROSPINNING A TERNARY SYSTEM OF NONSOLVENT/SOLVENT/POLYMER

In the work, presented in Refs. [8–10], facile method for fabricating fibers with micro-and nano-porous structure by electrospinning was introduced. Poly L-lactic acid was dissolved in a mixture of dichloromethane (solvent) and butanol (nonsolvent) with a certain ratio. Phase separation happens during the electrospinning, due to different evaporation ratio of solvent and nonsolvent. Thus the jet yielded to different phase separated structures, and further evaporation of the residual nonsolvent would lead to porous fibers. Porous biodegradable polymer fibers with large surface area and rough surface structure are so ideal candidates for drug delivery and tissue engineering [8].

6.7 POROUS TEMPLATE WITH HIERARCHICAL STRUCTURE

Recently, a novel method (wetting porous templates) has been combined with electrospinning technique to construct novel hierarchical fiber by Chen and groups.

Figure 6.4 shows the whole and simple process. The mechanism lies in the thermal annealing above the glass transition temperature of polymer nanofibers. At this time, the wetting of polymer chains within the polymer nanofiber will have enough mobility to move into the template. Finally, controllable hierarchical structures can be constructed on the outer surface of electrospun polymer fibers [9].

6.8 INTERIOR STRUCTURE WITHIN THE FIBER INTERIOR PORES THROUGH THE POLYMER BLENDS

In 2001, Wendorff and groups firstly build interior porous structures based on electrospun fibers through polymer blends with polylactide (PLA) and polyvinylpyrrolidone (PVP) as model. In this system, different polymeric phase separations, solubilities, and the decomposition temperature are the key for porous structures within the fibers. (i) Different polymeric phase separations form the PVP-rich and PLA-rich regions within the fibers. (ii) The lower decomposition temperature of PLA makes the removal of PLA easy. (iii) The good water solubility of PVP makes the removal of PVP easy. Finally, porous structures can be easily obtained in polymer fibers (Figure 6.5) [10].

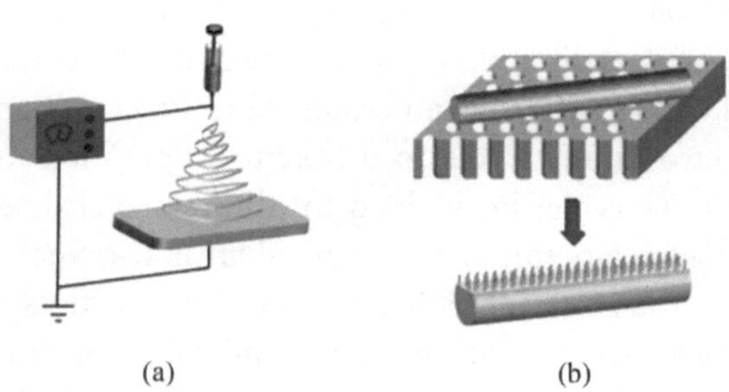

FIGURE 6.4 Schematic illustration of the whole process to make hierarchical polymer structures based on electrospinning and wetting of porous templates.

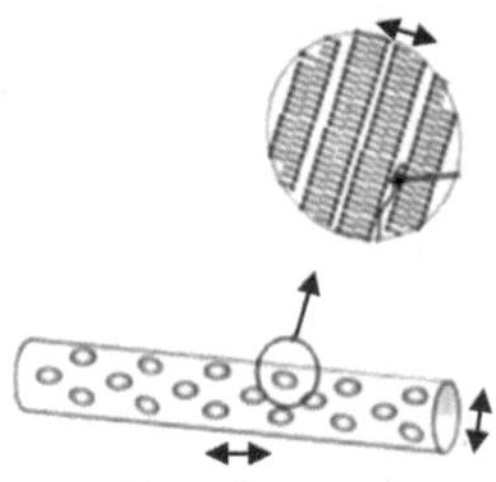

FIGURE 6.5 Schematic illustration of the electrospun fibers containing the internal periodic structures (self-assembled region).

It is important to emphasize that this chapter is the pioneer work to fabricate interior porous structured polymeric fibers through polymer blends. Inspired by this work, a number of interior porous structured polymeric fibers have been fabricated.

In 2005, Ruotsalainen et al. [11] designed a very interesting route for internal periodic structures within the electrospun fibers by hierarchical self-assembly of polymeric comb-shaped supramolecules.

6.9 CONTROL OF STRUCTURE OF NANOFIBERS AND NANOFIBROUS WEBS

6.9.1 CONTROL OF SIZE, INTERNAL STRUCTURE, AND SURFACE OF NANOFIBER

Many material parameters and process parameters have important effects on both the electrospinning process and the resulting fiber morphology. The fiber diameter depends on the solution properties (e.g., viscosity, conductivity, surface tension, permittivity, and boiling point) and/or operating conditions (e.g., applied voltage, spinneret-to-collector distance, and flow rate) summarized in Table 6.1.

Practically, the viscosity and electric conductivity of the spinning solutions are crucial factors for controlling the fiber diameter. The solution viscosity is changed by polymer concentration and the more viscous solution tends to form a thicker fiber. The increase in the entanglement of polymer chains caused by increasing the polymer concentration and/or molecular

TABLE 6.1 Control Parameters for Morphology and Diameter of Electrospun Fibers

Solution properties
Viscosity (molecular weight of polymer and concentration)
Electric conductivity
Solvent properties (surface tension, boiling point, polarity, and permittivity)
Operating conditions
Applied voltage (typically from several kV to several 10 kV) distance between spinneret and collector (typically from several cm to 50 cm)
Feeding rate of polymer solution
Spinneret (inner diameter, shape, and material)
Surrounding conditions
Temperature
Humidity

weight of the polymer contributes to the formation of the fibrous structure. In addition, higher-molecular-weight polymers improve the electrospinnability from the lower viscosity solution and are effectively fabricated into thinner and homogenous fibers by electrospinning. The fiber diameter also decreases with an increase in the solution conductivity due to the addition of a small number of electrolytes. The adequate conductivity solution enhances electrostatic repulsion force on the surface of the jet during electrospinning, and consequently, the fiber diameter decreases. However, the high-conductivity solution prevents electric-field induced charging of the solution, and consequently show a low electrospinnability [1].

Porosity is one of the important parameter in filter design and filter performance. Past studies have shown that the thickness and porosity of the nanofiber mats can be controlled by changing the deposition rate of nanofibers. Adequate porosity and surface area of the nanofiber mat has turned nanofiber coatings as an important candidate for high performance filters.

6.10 POROSITY MEASUREMENT

Pore sizes can be measured by saturating the porous material with a wetting liquid of known surface tension. Gas pressure on one side of the sample is increased until liquid from the largest pores was expelled. As the pressure increased, smaller pores opened up and the flow rate of gas through the sample increased until all the accessible pores are emptied. A plot of the pressure versus flow rate through the wetted sample, when compared with the equivalent pressure/flow rate curve for a dry sample, gave an estimate of pore size distribution in the material. In the standard mode, the pores measured with this method only include those pores that provide a continuous path from one side of the material to the other. Dead-end pores are not measured with this method. If the mean pore size measurements are plotted with the corresponding air flow resistance measurements, a general correlation between these two properties. This is in general agreement with models explaining fluid flow phenomena based on nominal length scales related to pore or fiber diameters in fibrous media. In fact, if a sample's porosity can be determined from sample thickness, sample weight, and nominal density of the solid polymer, this correlation can also serve as a basis for measuring the mean fiber diameter of the porous structure, in addition to the mean pore size.

There are few methods for measuring the porosity such as conventional methods using apparent density and bulk density, image analysis and mercury porometer. However, till date, an accurate estimation of porosity in these grades of materials nanofiber mat (NFM) is a difficult task [12].

Generally in conventional measurement porosity is estimated with the following relation.

$$NFM\ Apparent\ Density = \frac{NFM\ Mass}{NFM\ Area \times NFM\ Thickness} \tag{6.1}$$

TABLE 6.2 Common Method Used for Measuring the Porosity

Method	Information yielded	Pore width range(nm)
Mercury porosimetry	Porosity, pore diameter and pore size distribution	10 to 10,000
Image analysis by FESEM, TEM	Porosity, pore diameter and pore size distribution	0.1 to 1000
Capillary flow porometry	Median pore diameter, pore size distribution and permeability	10 to 1000
Adsorption and condensation/Brunauer, Emmett and Teller (BET)	Pore size and pore size distribution of the fiber but not mat	0.1 to 10

$$Porosity = 1 - \frac{NFM\ Apparent\ Density}{Bulk\ Density} \tag{6.2}$$

In conventional method the mat thickness is measured with micrometer as the fiber mats are relatively thick for biomaterial applications and total volume (V) is calculated as product of area and thickness [12].

6.11 SPECIFIC SURFACE AREA AND POROUS STRUCTURE

It is well known that the difference in fiber size leads to change in the material structures such as specific surface area and pore size distribution.

Each value of BET surface area is much higher than the traditional fibrous materials, though these values are slightly lower than that of untreated nanofiber membrane.

BET most common method used to describe specific surface area:

$$\frac{1}{W((P_0/P)-1} = \frac{1}{W_m C} + \frac{C-1}{W_m C}\left(\frac{P}{P_0}\right) \tag{6.3}$$

$$S = \frac{C-1}{W_m C} \tag{6.4}$$

$$i = \frac{1}{W_m C} \tag{6.5}$$

$$W_M = \frac{1}{S+i} \tag{6.6}$$

Total surface area (S_t) can then be derived:

$$S_t = \frac{W_m N A_{cs}}{M} \tag{6.7}$$

Specific surface area (S) is then determined by total surface area by sample weight

$$S = S_t / w \tag{6.8}$$

IUPAC classification on pores is:

Macroporous >50 nm
Mesoporus 2–50 nm
Microporus <2 nm

Total pore volume is derived from the amount of vapor adsorbed at a relative temperature close to unity (assuming pores are filled with liquid adsorbate).

$$V_{liq} = \frac{P_a V_{ads} V_m}{RT} \tag{6.9}$$

Average pore size is estimated from the pore volume. Assuming cylindrical pore geometry (type A hysteresis) average pore radius (r_p) can be expressed as:

$$r_p = \frac{2V_{liq}}{S} \tag{6.10}$$

Other pore geometry models may require further information on the isotherm hysteresis before applying appropriate model [13, 14].

6.12 CONTACT ANGLES

The surface contact angles were firstly measured on a Drop Shape Analysis System (DSA100) (KRUSS, Germany). Deionized water was dropped onto the sample from a needle on a micro syringe during the test. A picture of the drop was captured after the drop set onto the sample. The contact angles could be calculated by the software through analyzing the shape of the drop. The contact angle θ was an average of 20 measurements [15].

The method for measuring the contact angle relies on the test fluids penetrating the porous sample, which can be expressed by the theory of Washburn [15]:

$$\frac{l^2}{l} = \frac{\sigma r \cos\theta}{2\eta} \tag{6.11}$$

where l, σ, r, θ and η represent rising height, surface tension of liquid, radius of capillary, advancing contact angle and viscosity of liquid, respectively. For the nanofiber membrane, this equation can be modified such that they are seen as a bundle of capillaries with a mean radius of capillary, r. A modified Washburn equation can be used [15]

$$\frac{l^2}{l} = \frac{(cr)\sigma\cos\theta}{2\eta} \tag{6.12}$$

In this equation, c is a constant to estimate the tortuous path of the capillaries. The relationship becomes non-linear at higher σ values, that is, the rise height decreases. Once (cr) is determined for a given sample, the advancing angle of liquids with cos θ < 1. Equation (6.11) can be calculated by replacing the rise height of the liquid with the weight gain of the sample. Hence the Eq. (6.12) becomes [15]

$$\frac{W^2}{t} = \frac{\left[(cr)w^2(\pi R^2)\right]\rho^2\sigma\cos\theta}{2\eta} = \frac{K\rho^2\sigma\cos\theta}{2\eta} \tag{6.13}$$

where *W*, ρ, ω, *R*, represent weight of penetrating liquid, density of measuring liquid, relative porosity and inner radius of the measuring tube. For a given sample, nanofiber membrane in this study, *K*, the geometric factor was found by conducting a preliminary test on the nanofiber membrane using n-hexane as a totally wetting liquid (cos θ = 1), which can be characterized by the

$$\mathrm{K} = \frac{2\mathrm{W}^2\eta}{t\rho^2\sigma} \tag{6.14}$$

Thus, the Washburn contact angle of nanofiber membrane can be calculated by Eqs. (6.13) and (6.14).

The membranes obtained with different electrospinning voltages and grafted under the same conditions present varied contact angles Roughness, however, is so complicated that it is difficult to develop a general method for the roughness measurement.

In this study, specific surface area based on BET method is used for roughness characterization [see Eqs. (6.15) and (6.16)] [15]:

$$B = \frac{S_r}{m} \tag{6.15}$$

$$R = \frac{S_r}{S} = \frac{mB}{S} \tag{6.16}$$

It can be seen from the formula that the contact angle, θ' of the membrane depends on the ideal contact angle, θ and surface area, and the specific surface area plays an important role in the contact angle of a material.

Porous contact angle is an important fact influencing the wetting behavior of fibrous membrane, which can be calculated by a modified Washburn equation.

6.13 NANOWEB POROSITY

The porosity ϕ of the nanoweb cab be estimated by mass of liquid saturation with a gas such as hexadecane [16]:

$$m_W - m_d = r_1 \phi V_T \tag{6.17}$$

where m_w and m_d are the mass of the wet and dry strips, respectively; r_1 is the density of gas and V_T is the apparent volume of the sample. The area is derived from a digital image of the sample analyzed with the Image program. The thickness d is measured with a Lasico ocular grid mounted on a Zeiss microscope by the gap between two slides sandwiching the samples. The measurements agreed very well with independent data obtained from the cross-sectional view, which can be obtained from SEM images.

Assuming that the Lucas-Washburn law works for the nanowebs, the velocity of front propagation (dL/dt) is expressed as:

$$\frac{dL}{dt} = \frac{k}{\eta L}(P_C - \rho g L) \tag{6.18}$$

where L is the front position, k is the permeability, P_c is the capillary pressure, g is the gravity constant. The weight of liquid column is important only if it is comparable with the capillary pressure P_c. Otherwise, we can ignore this effect and consider a simple theory, expressing the absorbing power of the nanoweb as $L = \alpha t^{1/2}$. The characteristic coefficient $\chi = \sqrt{\frac{2P_c k}{\eta}}$ shows the effect of capillary pressure, nanoweb permeability and fluid viscosity on the rate of liquid absorption. The capillary pressure is estimated as $P_c \approx 2\sigma / r$.

6.14 DETERMINATION OF WEIGHT REDUCTION AND MESH DENSITY FOR SCAFFOLDS

Many medical scaffolds need to allow cell infiltration and tissue ingrowths. Scaffold porosity is a crucial parameter. Lack of pore interconnectivity of electrospun fiber scaffolds is a major drawback of the technique, since the scaffolds are very dense allowing only poor cell infiltration and tissue ingrowths. Recently, several solutions have been proposed; among those, cryoelectrospinning-using ice-crystals as templates for enlarged pore formation leading to better cell infiltration in vitro and in vivo [17].

The polymer to compose the scaffold is spun simultaneously with a water-soluble template polymer that can be removed by extensive rinsing after production of the scaffold. The removed co-spun polymer leaves behind voids throughout the scaffold thus providing improved porosity and interfiber spacing. Therefore the porosity of the final scaffold can be tuned by carefully selecting the water-soluble polymer (type and molecular weight), which deposits fibers with variable fiber diameters leading to increased pore sizes after its removal [18].

In addition the ratio between water-soluble and water-nonsoluble polymer can be adjusted leading to variations in porosity and pore interconnectivity.

Fiber fleeces were weighed as-spun and after overnight rinse in water. The scaffolds were dried in a vacuum oven and the mass loss was determined. The bulk densities ρ of the electrospun polymer meshes were determined gravimetrically using the weights of precisely cut mesh samples of defined area and thickness. The scaffold dimensions were measured using SEM micrographs of the scaffold. The overall mesh porosity P was calculated according to the following equation [17]:

$$P = (1 - \frac{\rho_0}{\rho}) \times 100[\%] \quad (6.19)$$

6.15 CONCLUSION

Electrospinning is a simple and convenient method to produce nanofibers. With modifying nanofibers surface morphology, by introducing pores, their surface are to volume ratio increases greatly, so nanofiber potential for application in chemical filtration, fuel cell membrane, tissue engineering, and catalyst sensor would be improved. There are several methods for producing porous nanofibers, of a variety of polymers, with different pore structures. Thermally induced phase separation can be readily used with nonvolatile solvents and does not require selective dissolution of phase-separated polymers, however Some approaches considered electrospinning of mixtures composed of two immiscible polymers and a common solvent and selective dissolution could be employed to produce highly porous nanofibers, in the other method when a highly volatile solvent is applied in the electrospinning process, porous fibers or nanoscaled structures can be obtained. Nanofibers are produced in different methods have different pore structures. So porosity measurement is a critical step. For this purpose besides considering the structure surface area and contact angle should be investigated.

KEYWORDS

- **electrospinning**
- **nanofibers**
- **porosity**

REFERENCES

1. Matsumoto, H. and A. Tanioka, *Functionality in Electrospun Nanofibrous Membranes Based on Fiber's Size, Surface Area, and Molecular Orientation.* Membranes, 2011, 1(3), 249–264.

2. Frey, M. W. and L. Li, *Electrospinning and Porosity Measurements of Nylon-6/poly (Ethylene Oxide) Blended Non-Wovens.* Journal of Engineered Fibers and Fabrics, 2007, 2, 31–37.
3. McCann, J. T., D. Li, and Y. Xia, *Electrospinning of Nanofibers with Core-sheath, Hollow, or Porous Structures.* Journal of Materials Chemistry, 2001, 15(7), 735–738.
4. Gibson, P., H. S. Gibson, and D. Rivin, *Transport Properties of Porous Membranes Based on Electrospun Nanofibers.* Colloids and Surfaces A: Physicochemical and Engineering Aspects, 2001, 187, 469–481.
5. Kim, H. D., et al., *Effect of PEG–PLLA Diblock Copolymer on Macroporous PLLA Scaffolds by Thermally Induced Phase Separation.* Biomaterials, 2004, 25(12), 2319–2329.
6. Rowlands, A. S., et al., *Polyurethane/Poly (lactic-co-glycolic) Acid Composite Scaffolds Fabricated by Thermally Induced Phase Separation.* Biomaterials, 2007, 28(12), 2109–2121.
7. You, Y., et al., *Preparation of Porous Ultrafine PGA Fibers via Selective Dissolution of Electrospun PGA/PLA Blend Fibers.* Materials Letters, 200 60(6), 757–760.
8. Esfandarani, M. S. and M. S. Johari, *Producing Porous Nanofibers.* Textile Engineering Department, 2010, 12, 1-
9. Chen, J. T., W. L. Chen, and P. W. Fan, *Hierarchical Structures by Wetting Porous Templates with Electrospun Polymer Fibers.* ACS Macro Letters, 2011, 1(1), 41–4.
10. Bognitzki, M., et al., *Preparation of Fibers with Nanoscaled Morphologies: Electrospinning of Polymer Blends.* Polymer Engineering and Science, 2001, 41(6), 982–989.
11. Ruotsalainen, T., et al., *Towards Internal Structuring of Electrospun Fibers by Hierarchical Self-Assembly of Polymeric Comb-Shaped Supramolecules.* Advanced Materials, 200 17(8), 1048–1052.
12. Sreedhara, S. S. and N. R. Tata, *A Novel Method for Measurement of Porosity in Nanofiber Mat using Pycnometer in Filtration.* Journal of Engineered Fibers and Fabrics, 2013, 8(4), 132–137.
13. Allen, T., *Particle Size Measurement: Volume 2, Surface Area and Pore Size Determination.* 1997, Springer, 252.
14. Dullien, F. A. L. and H. Brenner, *Porous Media: Fluid Transport and Pore Structure.* 1991, Elsevier Science, 574.
15. Huang, F., et al., *Dynamic Wettability and Contact Angles of Poly (Vinylidene Fluoride) Nanofiber Membranes Grafted with Acrylic Acid.* Express Polymer Letters, 2010, 4(9), 551–558.
16. Hsieh, Y. L., N. Pan, and A. V. Neimark, *Liquid Wetting and Flow in Nano-Fibrous Systems: Multi-scale and Heterogeneous.* National Textile Center Annual Report, 2007.
17. Milleret, V., et al., *Tuning Electrospinning Parameters for Production of 3D-fiber-fleeces with Increased Porosity for Soft Tissue Engineering Applications.* Eur Cell Mater, 2011, 21, 286–303.
18. Baker, B. M., et al., *The Potential to Improve Cell Infiltration in Composite Fiber-Aligned Electrospun Scaffolds by the Selective Removal of Sacrificial Fibers.* National Center for Biotechnology Information, 2008, 29(15), 2348–2358.

CHAPTER 7

SYNTHESIZE OF NANOCOMPOSITES: NEW ACHIEVEMENTS

D. S. DAVTYAN, A. O. TONOYAN, A. Z. VARDERESYAN, and S. P. DAVTYAN

State Engineering University of Armenia, 105 Teryana Str., Yerevan, 375009, Armenia, E-mail: atonoyan@mail.ru

CONTENTS

ABSTRACT

Characteristics of the frontal copolymerization of acrylamide with methyl methacrylate in the presence of single-wall carbon nanotubes in different amounts are studied. It is shown that adding of bentonite, which represents the natural lamellar nanomaterial with nanodimensional layers, results in the formation of polyacrylamide-bentonite hydrogels. It is shown that the filling by nanotubes by more than 20% (of the initial weight of comonomers) causes the loss of stability of copolymerization thermal waves with occurrence of periodical, spin and chaotic modes. The mechanism of periodical modes formation is offered. Physical and mechanical, dynamic and mechanical and thermochemical properties of obtained polymer nanocomposites are studied. On the basis of analysis of the data on the influence of amounts of single-wall nanotubes on the properties of copolymer nanocomposites the conclusion is drawn relative to the intercalation of copolymer macromolecules into the inner surface of nanotubes.

7.1 INTRODUCTION

Poor compatibility of carbon nanotubes with many polymer binders, organic and aqueous solutions considerably restrict their application as nanofillers. Therefore, there are many papers (e.g., see [1–3] and cited references) devoted to the research of capabilities of considerable enhancement of interaction of single-wall (SWCNT) and multi-wall (MWCNT) carbon nanotubes surface with polymer macromolecules.

High physical and mechanical performance of carbon nanotubes (tensile strength ~100 GPa, modulus of elasticity ~1000 GPa and elongation up to ~0.4%) are good preconditions for the enhancement of properties of nanocomposites – polymer/carbon nanotubes. However, as it was mentioned in the paper [4], and as analysis of other papers shows [5–11], the data of physical and mechanical properties of nanocomposites (polymer/carbon nanotubes) is inconsistent. Most probably, first of all such status is connected with the uneven distribution of nanotubes in the polymer volume. Also the methods of nanocomposites generation [4, 12], which influence on the morphology of binder macromolecules, directly on the surfaces of phases of nanotube-polymer matrix are very important factors. Reliable comprehension of the results of many papers is complicated also due to the fact that often they do not give data on the thermal and temperature conditions of nanocomposites synthesis. Therefore, development of new methods of polymer nanocomposites synthesis which will provide the even distribution of carbon nanotubes in the binder volume as well as the enhancement of reliability and reproducibility of their generation process are the topical tasks for the obtaining of nanomaterials – polymer/carbon nanotubes.

The purpose of this chapter is to synthesize nanocomposites using the method of frontal copolymerization of acrylamide (AAM) with methylmethacrylate (MMA) in the presence of SWCNT and to distribute them evenly in the polymer matrix; to investigate their physical and mechanical, dynamic and mechanical, thermochemical and electroconductive properties; to determine the boundaries of stable frontal modes depending on the nanotubes filling degree, considering the direct dependence of obtained nanocomposite properties on the capability of setting the stationarity of frontal process thermal wave propagation. It is also interesting to investigate the geometric shapes and constitution of nonlinear structures which are formed as a result of non-stationary front wave propagation.

7.2 EXPERIMENTAL PART

Sigma Aldrich AAMs and MMAs were used as co-monomers. MMAs were purified according to the methods [12]. AAMs were purified via double recrystallization from the saturated solutions of ethyl alcohol.

Initiating agent of copolymerization is dicyclohexylperoxydicarbonate (DCPC) which was used after the double recrystallization from ethyl alcohol and drying in vacuum cabinet at the room temperature until the constant weight was obtained. SWCNTs (of Sigma Aldrich brand) and aluminum nanopowder with the particles size of 40 nm (of Sigma Aldrich brand) were used as the nanofillers during the copolymerization process.

For the frontal copolymerization of AAM with MMA in the presence of SWCNT the initial mixtures were prepared as follows. In the beginning, the powdery AAM was carefully mixed with the necessary amount of nanotubes. Then, in order to ensure the stationary modes of frontal copolymerization [13, 14] the AAM mixture with nanoparticles was put into the reaction glass ampoules in individual portions and compacted. Further, MMA in quantity of 20% (of AAM weight) together with initiating agent was added to the prepared mixture. DCPC concentration in all experiments was 2% (of weight) of co-monomer amounts.

The frontal copolymerization of AAM with MMA adding nanotubes in the appropriate amounts was accomplished according to the methods which were described by us before in the Refs. [14, 15]. The reaction was carried out in the vertically established glass ampoules with diameter of 5 mm and length of 100 mm. Polymerization front was initiated from the top of reaction ampoules by application of hot (~200°C) metallic surface to the edges of reaction ampoules [16]. Temperature profiles of the frontal copolymerization were determined through the copper-constantan thermocouples performance. Thermocouple junctions were located in the middle part of ampoules. And the velocity of front propagation was determined visually according to the dependence of front coordinate on time.

Physical and mechanical (under conditions of elongation), dynamic and mechanical properties of nanocomposite samples were determined on the Perkin-Elmer Diamond DSA device.

Thermal-oxidative degradation of polymer binders was investigated via derivatographic method on MOM device with the heating-up velocity of 3.2°/min. Electroconductive properties of nanocomposite samples (cross-section 0.2 cm^2, length 1 cm) were determined via impedance measurements (frequency 1000 Hz, amplitude 5 mv) on Electrochemical workstation CHI 660D.

7.3 INFLUENCE OF AMOUNTS OF SWCNT ON THE CHARACTERISTICS OF FRONTAL COPOLYMERIZATION

Data on the influence of SWCNT amounts on the temperature profiles (Figure 7.1a) and propagation velocity (Figure 7.1b) of copolymerization front of AAM with MMA are given in Figure 7.1.

Comparison of the data in Figures 7.1a and 7.1b with analogous results obtained in the Ref. [12] displays their considerable difference. In this case, the limiting temperature (Figure 7.1a) of thermal waves and velocity of copolymerization front (Figure 7.1b) decrease practically simultaneously with adding of nanotubes. Herewith, as is seen in Figure 7.1a, the structure of temperature profiles also changes.

Observed changes of typical values of the AAM-MMA frontal copolymerization can be explained by two factors. From one hand, this is the joint effect of nanofiller quantities and intensity of thermal loss from the reaction zone to the environment on the thermal conditions of frontal copolymerization. But on the other hand, there is absence of the chemical

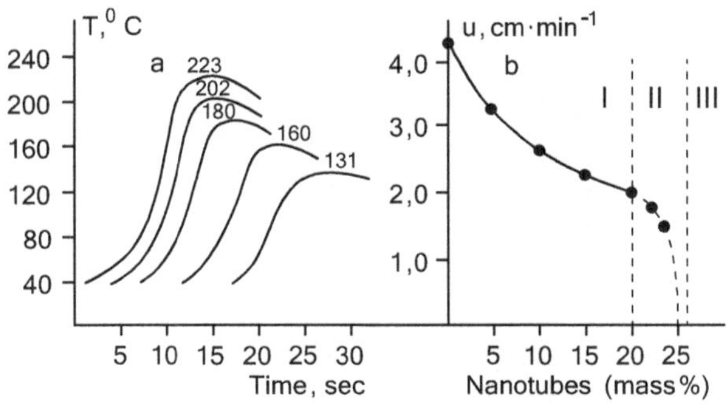

FIGURE 7.1 Influence of amounts of SWCNT on the behavior of temperature profiles (a) and front velocity (b). Ratio of AAM co-monomers: MMA= 80:20, amount of SWCNT (% of co-monomers weight): 1 – 0, 2 – 5, 3 – 10, 4 – 5 and 5 – 20.

interaction of binder macromolecules with outer and inner surfaces of SWCNT. Proof of the first factor is found in such aspects as quite significant decrease of the limiting temperatures (Figure 7.1a) and data on the influence of nanotubes amounts on the stationary status of frontal modes and their stability. And indeed, as is seen from the data in Figure 7.1b, depending on the amounts of SWCNT added, three regions of copolymerization front velocity change with different nature are observed. Region I (amount of nanoadditives up to 20%) corresponds to the stationary stable statuses of thermal copolymerization waves. In region II (amount of nanotubes is 20–25%) when amount of SWCNT grows the front velocity quite sharply reduces and the stability of frontal modes is lost. And in region III when amounts of nanotubes are higher than ~25–26% the frontal copolymerization modes do not exist. Let us consider unstable frontal modes in region II in details.

It is known [17–27] that the loss of frontal modes stability as a rule is followed by occurrence of oscillatory, periodical, single-, two-, three- and multiple-start spin modes. Herewith, on the surfaces of polymerized samples in the specified papers [17–27] the spiral hollows which are typical for unstable modes of frontal polymerization are revealed. In this case, as is seen from the data in Figure 7.1b and Figure 7.2 (samples 1–6), the stability of co-polymerization thermal waves is lost when adding the nanotubes in amount of 20% and higher. Indeed, when filling the polymerized medium with nanotubes up to 19%, the frontal

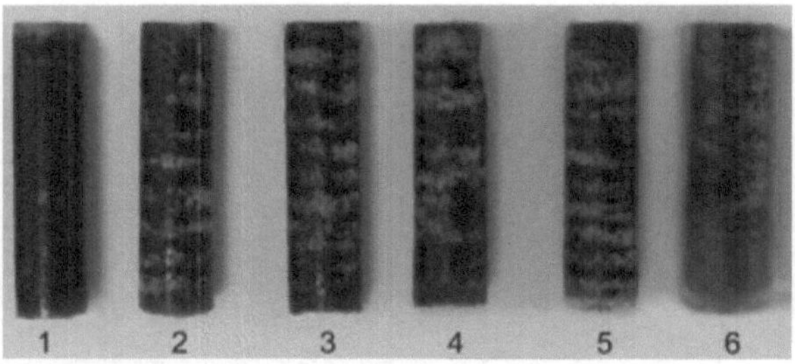

FIGURE 7.2 Samples of copolymer nanocomposites obtained under conditions of the frontal copolymerization of AAM with MMA. Filling degree (% wt. of AAM and MMA amounts): 15 – 1, 20 – 2, 22 – 3, 23 – 4, 24 – 5, 25 – 6. Stationary stable – 1 and unstable modes: periodical – 2, single-start – 3, two-start – 4, three-start – 5, spin, chaotic – 6.

modes are stable and samples have smooth surface with black color. In Figure 7.2, the photo of one nanocomposite sample (sample 1) with 15% filling is displayed.

Stability loss (Figure 7.2) is followed by the formation of periodical (sample 2), single- (sample 3), two- (sample 4), three- (sample 5) start spin and chaotic (sample 6) modes. Formation of the specified nonlinear phenomena is displayed in the form of white colorings against the background of smooth surfaces of obtained samples.

In order to reveal the sequence of unstable modes occurrence with filling up to 20% and higher the amount of SWCNT additives was increased in small portions or by 1% of total weight of co-monomers.

For the processes of frontal copolymerization of AAM with MMA in the presence of SWCNT the loss of stability of stationary thermal waves starts from the occurrence of periodical modes (Figure 7.2, sample 2). Periodical modes of the frontal copolymerization are characterized by the fluctuation [20] of front velocity about its stationary value (Figure 7.3).

Study of the mechanism of periodical modes formation and geometrical shape of front is of interest. For this purpose the polymerization was stopped by the freezing of reaction ampoules using liquid nitrogen in preassigned time intervals (points specified in Figure 7.3 by Figures 7.1–7.4) which correspond to the half of front velocity fluctuation period.

After, the reaction ampoules were independently heated up to the room temperature, without causing damage to the samples, then the glass housing was removed and photos of the obtained samples were taken (Figure 7.4).

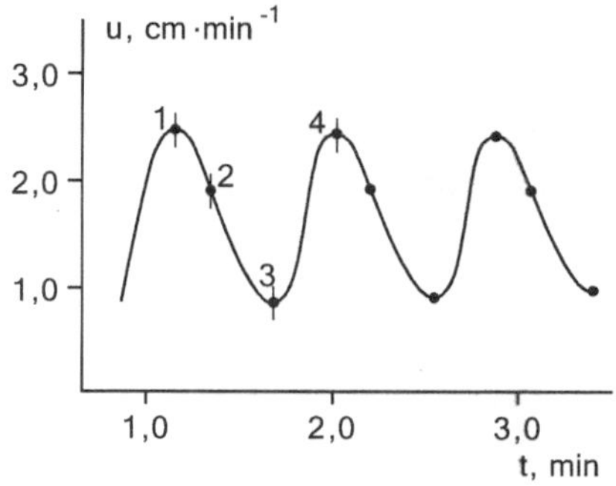

FIGURE 7.3 Oscillatory mode of the frontal copolymerization.

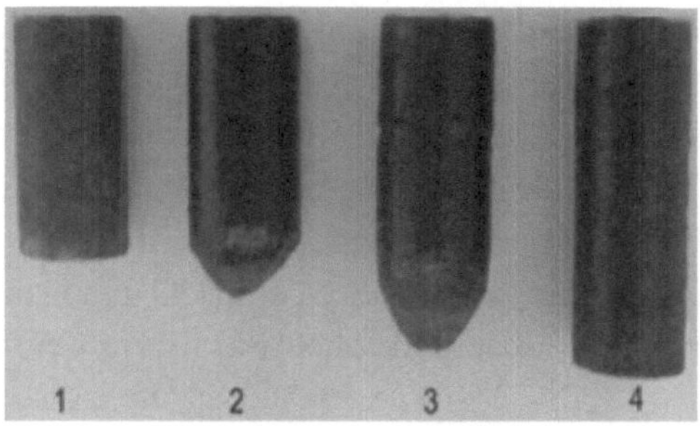

FIGURE 7.4 Change of geometrical shape of front caused by time. Time in sec. (time reference corresponds to the point 1 in Figure 7.3): 0 – 1, 20 – 2, 40 – 3 and 50 – 4.

It is seen in Figure 7.4 that in point 1 (Figure 7.3) the front shape is flat but then its incurvation occurs (Figure 7.4, 2) together with forming of "tongue" (point 2 in Figure 7.3). Then (Figure 7.4, 3), the formed "tongue" becomes longer (point 3 in Figure 7.3) and in point 4 (Figure 7.3) the front shape becomes flat again (Figure 7.4, 4). Dynamics of the front geometrical shape change shown in Figure 7.4 is connected with the heat loss from the reaction zone to environment and effect of inert SWCNT additives on the reaction mixture heating up (Figure 7.1a). Most probably, at the moment of maximum incurvation of the front geometrical shape (which corresponds to the minimum temperature in reaction zone) the nucleation site occurs which closes on itself. Or by analogy with results of the Ref. [21], thermal wave propagates not only in axial but also in radial directions of reaction ampoules. Both considered mechanisms can result in the front shape alignment and increase of the temperature in reaction zone, respectively. Therefore, velocities of thermal copolymerization waves (Figure 7.3) have maximum values for the flat (Figure 7.4, 1 and 4, 4) and vice versa, minimum values (Figure 7.3) for the most curved front shapes (Figure 7.4, 3).

Further increase of amounts of nanotubes at first results in the formation of single-start (Figure 7.2, sample 3), then two-start (Figure 7.2, sample 4), multiple-start (Figure 7.2, sample 5) spin and at the end chaotic (Figure 7.2, sample 6) modes. Stability loss of thermal waves of chemical nature with formation of spin modes for the processes of burning and SHS (self-propagating high-temperature synthesis) is considered in the papers [25–27] in details.

It should be noted that degeneration of copolymerization frontal modes of AAm with MMA is observed when adding nanotubes in amounts of 26% and higher (Figure 7.1b, region III).

Also under the conditions of frontal polymerization of MMA [12] or upon the frontal copolymerization of AAM with MMA [24, 28] (in the presence of spherical nanoparticles SiO_2 and TiO_2) the stability loss of stationary frontal modes is observed when the filling degrees are 25–30%. This phenomenon [12, 24, 28] is explained by the existence of additional heat generation source in reaction zone at the expense of exothermic interaction of binder macromolecules with nanoparticles surface.

7.4 PHYSICAL AND MECHANICAL, DYNAMIC AND MECHANICAL AND THERMOCHEMICAL PROPERTIES OF NANOCOMPOSITES

Influence of the filling degree on tensile strength (s), modulus of elasticity (E) and elongation (e) is displayed in Table 7.1. Increase of amounts of SWCNT additives in nanocomposites leads to the increase of the values s and E and decrease of deformability of samples. A 20% filling causes the growth of limiting tensile strength by ~30%, modulus of elasticity – by ~20% and the decrease of deformability by ~50%.

When the amounts of SWCNT additives increase, there is the notable growth of the tensile strength and modulus of elasticity which indicates their even distribution in the copolymer binder volume. Even distribution

TABLE 7.1 Influence of Amounts of SWCNT Additives on the Values σ, E and ε

SWCNT. % of binder weight	s. MPa, kgf/mm²	E. MPa, kgf/mm²	ε, %
0	84 ±5	136 ± 7	40
5	93 ± 5	140 ± 7	36
10	105 ± 5	150 ± 8	25
20	111 ± 5	170 ± 8	20

is provided at the expense of deagglomeration of agglomerated nanoparticles [12] (nanotubes) under the influence of thermal copolymerization waves and fixation of this status in polymer binder.

Behavior of the dynamic module (E′) and tangent of angle of mechanical loss (tgδ) for copolymer nanocomposites which content different amount of SWCNT is illustrated in Figures 7.5a and 7.5b.

As should be expected (Figure 7.5a), the values E′ upon the same filling degrees keep constant and only at the temperatures ~220°C their decrease occurs.

Obviously, this change of the dynamic module is associated with the increase of mobility of macromolecules and individual fragments of copolymer binder at the devitrification temperatures. Growth of the values E′ (curves 1–4, Figure 7.5a) is observed when amounts of nanotubes increase. Most probably, observed growth of the value E′ caused by the amounts of nanotubes occurs due to the intercalation of

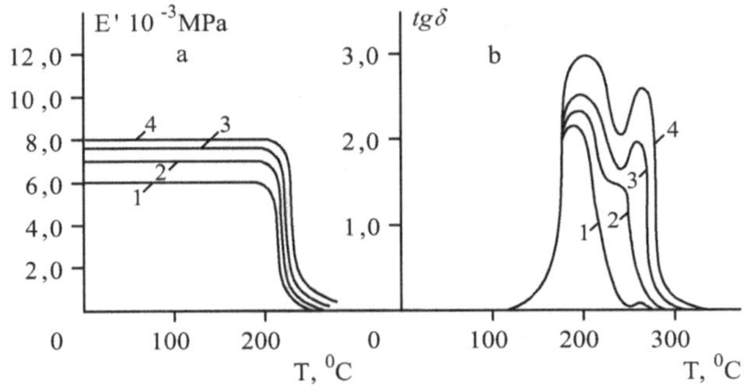

FIGURE 7.5 Change of the dynamic module (a) and tangent of angle of mechanical loss (b) caused by the temperature with different filling degrees. Filling degree corresponds to the data in Figure 7.1.

copolymer macromolecules or their fragments into the inner surface of nanotubes.

Behavior of tangent of mechanical loss angle (curves 1–4, Figure 7.5b) caused by the amounts of nanotubes has quite uncommon shape. Here two transitions are observed in the range of temperatures higher than 200°C which correspond to the devitrification of nanocomposite copolymer samples. In the beginning quite intensive primary (devitrification) transition is observed and then the secondary transition occurs. Herewith, the intensity of the secondary transition grows (Figure 7.5b, curves 2–4) when the amounts of SWCNT increase. This fact, the growth of the secondary transition intensity with the increase of amounts of nanotubes, confirms the assumption that the intercalation of individual elements or binder macromolecules into the inner surface of nanotubes takes place.

Curves of weight loss depending on the temperature with different amounts of SWCNT are given in Figure 7.6.

It is seen from the data in Figure 7.6 that weight loss for the pure copolymer of AAm with MMA starts at the temperature ~300°C (curve 1). Adding SWCNT causes some decrease of the initial temperature of thermal-oxidative degradation and quite tangible change of character of weight loss curves.

Indeed, as is seen from the curve 2 and 3 in Figure 7.6 in the temperature range ~450–500°C small plateau occurs (curve 2) and only after this occurrence the second stage of weight loss starts. When the amounts of SWCNT grow the plateau value increases (curves 2, 3).

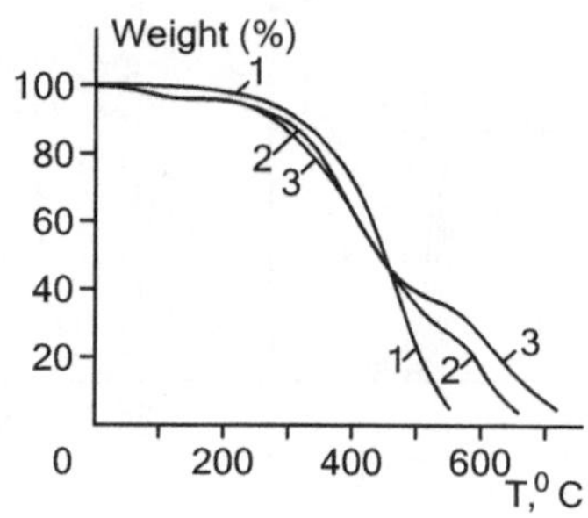

FIGURE 7.6 Weight loss of pure copolymer based on AAM with MMA and copolymer nanocomposites with different amounts of nanotubes (% wt. of total amount of comonomers): 1–0, 2–10, 3–20.

Most probably, observed two-stage character of nanocomposite weight loss curves is associated with the intercalation of AAM-MMA copolymers macromolecules into the inner surface of SWCNT which slows down the thermal-oxidative degradation process to some extent. The following fact remains incomprehensible: increase of SWCNT amounts causes some decrease of initial temperature of nanocomposites thermal-oxidative degradation. Results given in Figure 7.5 correspond to the paper conclusions [29] quite well and clarify the data [30] on the influence of SWCNT and MWCNT on the process of nanocomposites thermal-oxidative degradation.

Samples of copolymer nanocomposites which content 15–25% of SWCNT have practically zero electrical conductivity. In order to generate the electroconductive nanocomposites we added 5% (of comonomers weight) of aluminum nanopowder to the initial reaction mixture. As it turned out, for copolymer nanocomposites which contain 18–19% of single-wall nanotubes and 5% of aluminum nanoparticles the electrical conductivity reaches up to ~95,000 $Si \cdot m^{-1}$.

Thus, the results which were obtained in this paper show that in the process of the frontal copolymerization of AAM with MMA in the presence of single-wall nanotubes, there takes place intercalation of fragments or binder macromolecules into the inner surface of nanotubes which causes the increase of tensile strength, modulus of elasticity and decrease of deformability of nanocomposite samples. The frontal copolymerization of AAM with MMA in the presence of ~18% of single-wall nanotube additives and 5% of aluminum nanoparticles (according to the total weight of comonomers) leads to the formation of nanocomposites with insignificant current-carrying properties.

KEYWORDS

- **frontal polymerization**
- **heat transfer**
- **nanocomposites**
- **nanotubes**
- **nonlinearity**

REFERENCES

1. O'Connell, M. J., Boul, P., Ericson, L. M., Huffman, C., Wang, Y. H., Haroz, E. et al. Reversible water-solubilization of single-walled carbon nanotubes by polymer wrapping. Chem. Phys. Lett., vol. 342, 265–271, 2001.
2. Hill, D., Lin, Y., Qu, L., Kitaygorodskiy, A., Connel, J. W., Allard, Sun, Y. P. Functionalization of carbon nanotubes with derivatized polyimide. Macromolecules. 38, 7670–7675, 2005.
3. Neira-Velázquez María Guadalupe, Ramos-de Valle Luis Francisco, Hernández-Hernández Ernesto, Zapata-González Ivan. Toward greener chemistry methods for preparation of hybrid polymer materials based on carbon nanotubes. e-Polymers, 162, 1618–7229, 2008.
4. Hobbie, E. K., Bauer, B. J., Stephens, J., Becker, M. L., McGuiggan, P. Colloidal particles coated and stabilized by DNA-wrapped carbon nanotubes. Langmuir. 21, 10284, 2005.
5. Wagner, H. D., Vaia, R. A. Carbon nanotube-based polymer composites: Outstanding Issues at the interface for mechanics. Materials Today, 11(7), 38–42, 2004.
6. Barber, A. H., Cohen, S. R., Wagner, H. D. Measurement of carbon nanotube-polymer interfacial strength. Appl. Phys. Lett., 82, 4140, 2003.
7. Barber, A. H., Cohen, S. R., Kenig, S., Wagner, H. D. Interfacial fracture energy measurements for multi-walled carbon nanotubes pulled from a polymer matrix. Compos. Sci. Technol., 64(15), 2283, 2004.
8. Barber, A. H., Cohen, S. R., Wagner, H. D. Static and dynamic wetting measurements of single carbon nanotubes. Phys. Rev. Lett., 92(18), Art No.186103, 2004.
9. Buchachenko, A. L. New horizons of chemistry: single molecules. Uspekhi khimii, 75(1), 3–26, 2006.
10. Chi-Yuan Huang, Ching-Shan Tsai, Keng-Yu Tsao, Po-Chian Hu. Carbon black nano composites for over voltage resistance temperature coefficient. Proceedings of the World Polymer Congress – Macro, 41st International Symposium on Macromolecules, p. 41, 2006.
11. Wagner, H. D., Lourie, O., Feldman, Y., Tenne, R. Stress-Induced Fragmentation of Multiwall Carbon Nanotubes in a Polymer Matrix. Appl. Phys. Lett., 72(2), 188–190, 1998.
12. Davtyan, S. P., Berlin, A. A., Tonoyan, A. O., Rogovina, S. Z., Schik, C. Polymer nanocomposites with a uniform distribution of nanoparticles in a polymer matrix synthesized by the frontal polymerization. Rossiyskie Nanotekhnologii, 4(7–8), 489–498, 2009.
13. Arutyunyan Kh.A., Davtyan, S. P., Rozenberg, B. A., Enikolopyan, N. S. Curing of epoxy resins of bis-phenol A by amines under conditions of reaction front propagation. Dokl. AN SSSR, 223(3), 657–660, 1975.
14. Davtyan, S. P., Hambartsumyan, A. F., Davtyan, D. S., Tonoyan, A. O., Hayrapetyan, S. M., Bagyan, S. H., Manukyan, L. S. The structure, rate and stability of autowaves during polymerrization of Co metal-complexes with Acrylamide. Eur. Polym. J., 38, 2423–2431, 2002.
15. Davtyan, D. S., Tonoyan, A. O., Hayrapetyan, S. M., Manukyan, L. S., Davtyan, S. P. Peculiarities of frontal initiated polymerization of acrylamide. Izvestia NAN RA i GIUA, 52(1), 38, 2003.

16. Davtyan, S. P., Zakaryan, H. H., Tonoyan, A. O. Steady state frontal polymerization of vinyl monomers: the peculiarities of. Chemical Engineering Journal, 155(1–2), 292–297, 2009.
17. Begishev, I. P., Volpert Vit. A., Davtyan, S. P. Existence of the polymerization wave with crystallization of the initial substance. Dokl. AN SSSR, 273(5), 1155–1158, 1983.
18. Volpert Vit. A., Volpert Vl. A., Davtyan, S. P., Megrabova, I. N., Surkov, N. F. Two-dimensional combustion modes in condensed flow. SIAM. J. Appl. Math., .52(2), 368–383, 1992.
19. Davtyan, S. P., Tonoyan, A. O., Davtyan, D. S., Savchenko, V. I. Geometric shape and stability of frontal regimes during radical polymerization of methyl metacrilate in a cylindrical flow reactor. Polymer Sci. Ser. A, 41(2), 153–162, 1999.
20. Davtyan, D. S., Baghdasaryan, A. E., Tonoyan, A. O., Karapetyan, Z. A., Davtyan, S. P. The mechanism of convective mass transfer during the frontal (radical) polymerization of methyl metacrilate. Polymer Sci. Ser. A, 42(11), 1197–1216, 2000.
21. Davtyan, D. S., Baghdasaryan, A. E., Tonoyan, A. O., Davtyan, S. P. To the contribution of thermal convective mass transfer of the reactive mixture components to frontal curing of epoxy diane oligomers. Khimicheskaya fizika, vol.19, № 9, pp.100–109, 2000.
22. Davtyan, S. P., Shaginyan, A. A., Tonoyan, A. O., Ghazaryan, L. Oscillatory and spin regimes at frontal solidification of epoxy combinations in the flow. Compounds and Material with Specific Properties, Nova Science Publishers, Inc. ISBN 978-1-60456-343-6, Editor: Bob, A. Howell et al., p.88, 2008.
23. Davtyan, S. P., Berlin, A. A., Tonoyan, A. O. Advances and problems of frontal polymerization processes. Obzor. J. po khimii, 1(1), 56, 2011.
24. Tonoyan, A. O., Ketyan, A. G., Zakaryan, H. H., Sukiasyan Zh.K., Davtyan, S. P. Polyacrylamide/bentonite, polyacrylamide/diatomite nanocomposites obtained by frontal polymerrization. Khimicheski, J. Armenii, 63(2), 193, 2010.
25. Ivleva, T. P., Merzhanov, A. G. Structure and variability of spinning reaction waves in three-dimensional excitable media. Physical Review, E., 64(3), 036218, 2001.
26. Volpert, A. I., Volpert Vit.A., Volpert Vl.A. Traveling Wave Solution of Parabolic Systems. AMS Books Online, p. 455, 2003.
27. Ivleva, T. P., Merzhanov, A. G. Three dimensional modes of unsteady-solid flame combustion. Chaos, 13(1), 80, 2003.
28. Avetisyan, A. S., Tonoyan, A. O., Ghazanchyan, for example, Davtyan, S. P. The influence of high-temperature polymerization modes on monomer-polymer equilibrium. Vestnik GIUA, Issue 15(2), 32–39, 2012.
29. Chipara Mircea, Cruz Jessica, Vega Edgar, R., Alarcon Jorge, Mion Thomas, Chipara Dorina Magdalena, Ibrahim, A., Tidrow Steven, Hui David. Polyvinylchloride-Single Walled Carbon Nanotube Composites: Thermal and Spectroscopic Properties, Special Issue: Synthesis, Properties, and Applications of Polymeric Nanocomposites. Journal of Nanomaterials, Article ID 435412, 6 pp, 2012.
30. Wu, X. L., Liu, P. Poly(vinyl chloride)-grafted multi-walled carbon nanotubes via Friedel-Crafts alkylation. EXPRESS Polymer Letters, 4(11), 723, 2010.

CHAPTER 8

UPDATES ON POLYMERIZATION TECHNIQUES

G. E. ZAIKOV

Russian Academy of Sciences, Russia

CONTENTS

8.1 INTRODUCTION

In chemistry and in chemical technology often one uses such notions like: macromolecules, polymers, synthetic materials, plastics and elastomers. They have similar means but they are not synonyms.

Macromolecules are substances with a very large molecular mass, of linear structure or forming a special network. Sometimes it is difficult to give their chemical structures because the particular elements forming them exhibit an unrepeatable structure.

Polymers are macromolecules formed from repeatable units, called "mers." The number of mers in macromolecule determines the degree of polymerization.

Thermoplastic Polymers are utility materials, made from polymers, combined with different additives, such as fillers, plasticizers, stabilizers, dyes, pigments, modifiers and others. They can be processed in the melt.

Plastic Materials (Plastics) are synthetic materials made of polymers and additives.

Elastomers are synthetic materials, characterized by a large ability for deformations and elongation (up to 1200%), keeping at the same time intact the elastic properties. It means that after stopping the deformation force they recover their initial form.

A characteristic property of polymer structure is the presence of exceptionally large size constituent macromolecules, composed from a large number of atoms. Typical length of polymers oscillates between 100 nm (1 nm is a milionth part of millimeter) and 100,000 nm (100 μm). However, the diameter of individual linear polymer chain doesn't surpass the diameter of constituent molecules, i.e., about 1 nm. Therefore, the structure of polymers can be observed only by the very high-resolution microscopes (atomic force microscope (AFM), scanning tunneling microscope (STM)) and analyzed with help of the electron microscope (EM).

The length of particular macromolecules is determined by the number of constituent atoms, so their molecular mass. Because the particular macromolecules differ in the length of polymer chain for their description one introduces the notion of average molecular mass ($\overline{M}$). It appears that very often polymers prepared from the same monomer, characterized by the same average molecular mass $\overline{M}$ exhibit different properties. Therefore, the notion of polydispersity of polymers was introduced. It describes the molecular mass distribution.

There exist several experimental techniques allowing a practical determination of the polydispersity degree for a polymer. Between them the most important are:

- method of fractional precipitation and solubilization,
- chromatographic technique exploiting the adsorption ability of macromolecules,
- fractioning in a centrifuge,
- measure of the diffusion rate,
- light scattering measurements, and
- comparison of molecular masses determined by different techniques.

The polydispersity is usually expressed graphically using the molecular weight distribution curve. For this purpose, the experimentally found values for each fraction of polymer is spotted on a chart in the reference frame: degree of polymerization P – weight part of mass fraction dM/dP. Through the obtained in this way stepped chart an integration line is passed. This curve represents the weight parts of polymer fractions of with polymerization degree P (number of mers in the molecule) from P to (P + dP).

Taking the integral curve as the abciss one can plot in the coordinate system P, dM/dP the differential distribution curve. This curve exhibits a sharply marked maximum and allows a more illustrative observation of the degree of polydispersity of polymer than it is the case with the integral curve. Figure 8.1 shows, as examples, the differential distribution curves of molecular weights of cellulose nitrate and of polystyrene with different degrees of polydispersity.

The polydispersity of a polymer can be controlled, to a certain degree, by choosing in an adequate way the polymerization conditions. Polymers are obtained through:

- polymerization itself, that is, chemical binding of monomer molecules into one macromolecule without liberation of side products,

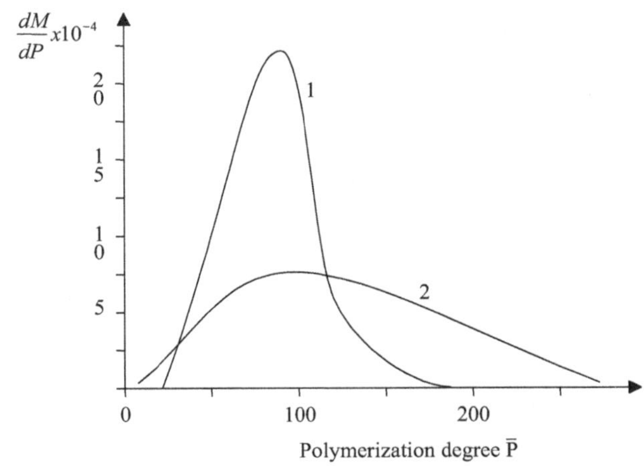

FIGURE 8.1 Differential distribution curves of molecular weight M for: 1 – cellulose nitrate with an average degree of polymerization 860; 2 – polystyrene with an average degree of polymerization 800.

- polycondensation (condensation polymerization), consisting on binding of a large number of molecules, containing reactive groups, into one macromolecule with liberation of small molecule side products (e.g., H_2O, HCl, NH_3, etc.),
- synthesis of macromolecules by living species (e.g., bacteria), and
- chemical modification of natural or synthetic polymers.

The type and the properties of obtained polymer in a large measure depend on the monomer used. As a result of polymerization of bifunctional monomers, with one double bond or two reactive groups in molecule, one obtains linear polymers (chain like), which can melt and are soluble in solvents of similar polarity.

In the case of polymerization of multifunctional monomers (with two double bonds or at least three functional groups) one obtains in the first reaction step linear polymers, containing reactive groups, able to interact with neighboring chains and form cross linkages between polymer chains. As result one obtains a spatially cross-linked polymer, forming a very large 3D macromolecule (a block polymer can be just one macromolecule). Such polymers are usually insoluble and do not melt.

Polymers play a very important role in a lot of vital processes. They find also a large and increasing number of applications in, practically, each branch of industry as well as in everyday life. At present the polymer fabrication in the world is so important that its size for a given country may be used as an indicator of the degree of development and modernity of its chemical industry.

Macromolecules, composed of thousands of atoms, are the basic components of the living world. Polymer macromolecules are part of a large number of materials used by man since the prehistory, such as leather, natural fibers (linen, cotton, wool, silk), wood and rubber. Their number has increased significantly in recent years. A large number of new polymers and composites were obtained. The common characteristic of chemical structure of polymers is the macromolecule, which usually is a long chain, made of hundreds or thousands of mers, connected together by chemical bonds. The polymer chain ability to take different geometrical forms provides properties unattainable in the case of substances made of smaller molecules.

The long polymer chains may form regular series, forming crystalline phases, characterized by a high hardness and mechanical strength. Due to the perfect ordering of molecules materials with strength over passing that of steel were obtained.

The development of polymer chemistry was particularly important in the last, XX centaury. New types of polymers, with different properties, were obtained. They found application as construction materials, foils, synthetic fibers, synthetic rubbers, adhesives, ion exchangers, products to change the soil structure, coagulating agents, in medicines, electric conductors and semiconductors, photoconductors, etc. The diversity and versatility of polymers, provided by the possible infinite modification of their molecular and macroscopic structure guarantees a further development and increase of applications of these materials.

The twentieth centaury has known an exceptional development of polymer chemistry. New types of polymers, characterized by new properties, were synthesized. They found numerous applications, such as construction materials, foils, synthetic fibers, rubbers, adhesives, paints, ionic exchangers, agents amending the soil structure, coagulants, medicine and medicine components, packaging materials, semiconductors and others. The diversity of polymer properties, possible due to an infinite variability of the molecule and macromolecule structures, is a best guarantee of further growth of their applications.

Nowhere a large diversity of polymers is observed as in biological materials. Among them one distinguishes proteins and polycarbohydrates, which are part of all living organisms: vegetal and animal. Another important biopolymers are enzymes (globular proteins), being catalysts of processes connected with the life, polynucleotides (DNA and RNA), responsible for genetic information during the whole development of living organisms. The biopolymers have to fulfill well-defined functions. The sequence of amino acids in the chains of proteins is well established and unchangeable, ideally repeated in all molecules of a given protein. Such repeatability was still not obtained by using the modern, available, polymer synthesis methods. The performances of nature in this field demand the polymer scientists to use more and more elaborated synthesis methods. The polymer science develops in an exceptionally dynamic way and the new discoveries will be of importance in different scientific and technical areas.

Polymers are basic elements of synthetic materials and nowadays they find applications in all areas of life and of industrial activity.

8.2 METHODS OF POLYMERIZATION OF VINYL COMPOUNDS

8.2.1 FREE RADICAL POLYMERIZATION OF VINYL COMPOUNDS

8.2.1.1 Effect of Monomers on the Course of Polymerization

Vinyl monomers are derivatives of ethylene, in which one or more hydrogen atoms have been replaced by other substituents. A characteristic feature of the construction of connections of this type is the electron pair-sharing, e, and thus the presence of a double carbon- carbon bond. The sp^2 orbitals present in monomer molecule make that each carbon atom is in the middle of a triangle, in which the vertices are: hydrogen atom of substituent and the second carbon atom.

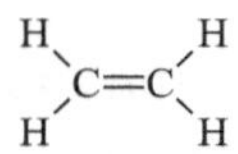

Each of the angles between the bonds in such a system is equal to 120°. By creating sp^2 orbitals, each carbon atom uses only two of its three p orbitals. The remaining p orbital is composed of two equal loops, one of which lies above and the second below the plane defined by the three sp^2 orbitals and is filled by one electron. By a combination of the two p orbitals of carbon a new π bond is created, consisting of an electron cloud located above the plane and defined by the atoms and another one lying below this plane. Due to the lateral orbital overlap, the binding energy of the double π bond between carbon atoms is greater than that a single δ bond.

The binding occurs only when the p-orbitals may overlap. It means that it is possible only when all six atoms responsible for the establishment of the bond (2 C atoms and 4 H atoms) are lying in the same plane. Therefore, an ethylene molecule is flat.

Thus the double C = C bond consists of a strong δ bond and a weak π bond. The total binding energy in this case amounts to 682 kJ/mol and is larger than the binding energy characteristic of a single carbon-carbon bond in the ethane molecule, which is of 368 kJ/mol. Therefore, the distance between carbon atoms in the molecule of an unsaturated compound is smaller than the distance between carbon atoms in the ethane molecule. It means that the double carbon-carbon bond is shorter than the corresponding single bond.

The free radical polymerization process consists on breaking the double bond between carbon atoms in the monomer molecule, followed by a chain reaction of its growing. From the above considerations it follows that the polymerization process is an exothermic reaction. The monomer molecules can polymerize only if the process is accompanied by a decrease of free energy ($\Delta G < 0$):

$$\Delta G = \Delta H - TDS$$

where: ΔG – change of the free energy of system, ΔH – change of the enthalpy of the system, equal to the heat of reaction with the opposite sign, and ΔS – change of the entropy of the system.

The polymerization process runs always with a decrease of the entropy of system. The value of $T\Delta S$ at 298 K is of 31.38–41.84 kJ/mol. For $\Delta G < 0$ the following condition must be fulfilled:

$$\Delta H < 31.38 \text{ kJ/mol.}$$

The double bond energy in vinyl monomers is of 609 kJ/mol. In the expense of that two single carbon-carbon bonds with energy of 2×351.5 kJ/mol = 703 kJ/mol are formed. The difference of these thermal effects is equal to the theoretical value of the polymerization heat Q:

$$Q = 703 - 609 = 94 \text{ kJ/mol}$$

In practice it turns out that there are significant differences between the various polymerization heats of different monomers. This is caused by both: the loss of coupling energy in the conversion of monomer molecules into the polymer molecule and the energy loss associated with the formation of stresses in the polymer chain as a result of mutual interaction of substituents (steric effect). The polymerization heats of selected monomers are listed in Table 8.1.

The steric effect in some monomers is so large that their polymerization is often nearly impossible. Examples of such monomers containing

TABLE 8.1 Heat of Polymerization for Unsaturated Monomers

No.	Monomer	Chemical formula	Heat of polymerization* (kJ/mol)
1	Ethylene	$H_2C=CH_2$	93
2	Propylene	$H_2C=CHCH_3$	85.7
3	Isobutylene	$H_2C=CH(CH_3)_2$	51.4
4	1,3-butadiene	$H_2C=CH-CH=CH_2$	72.3
5	Styrene	$H_2C=CHC_6H_5$	68.97
6	Vinyl chloride	$H_2C=CHCl$	95.7
7	Vinylydene chloride	$H_2C=CCl_2$	75.2
8	Vinyl acetate	$H_2C=CHCOOCH_3$	89
9	Acrylonitrile	$H_2C=CHCN$	72.3
10	Methyl acrylate	$H_2C=CHCOOCH_3$	78.5
11	Methyl methacrylate	$H_2C=CCH_3COOCH_3$	56.4
12	Acid methacrylate	$H_2C=CCH_3COOH$	66
13	Tetrafluoroethylene	$F_2C=CF_2$	155.4
14	Isoprene	$H_2C=CCH3-CH= CH_2$	74.4

*Heat of polymerization is defined as the energy necessary to change the enthalpy of a liquid monomer into an amorphous or a partially crystalline polymer.

exceptionally large substituents are the 1,2-dibenzil (1,2-diphenylethane) and the maleic acid.

The mechanism of polymerization is influenced also by the size and the type of the polarization of double bond in the monomer molecule, which depends on its structure. Depending on the type of functional groups adjacent to the double bond, monomers can be divided into three types:

- polymerizing by free radical mechanism (containing substituents affecting slightly π electrons);
- polymerizing according to the cationic mechanism (containing substituents dislocating π electrons); and
- polymerizing according to the anionic mechanism (having substituents attracting π electrons in double bond and facilitating formation of a carbanion).

Vinyl monomers polymerize readily through the free radical mechanism. An important factor determining their reactivity is the possibility of the resonance stabilization of formed free radicals. Larger is the value of energy corresponding to equilibrium between the resonance forms of free radicals coming from a given monomer, larger is the reactivity of this monomer in the polymerization reaction.

8.2.1.2 Course of Free Radical Polymerization

When testing the radical polymerization mechanism it was demonstrated that it takes place in three steps:

- initiation of polymerization,
- polymer chain growth (propagation), and
- ending of chain (termination).

8.2.1.2.1 Initiation of Polymerization

Initiation of free radical polymerization is a process in which the creation of a free radical on the carbon atom of monomer molecules takes place.

Factors enabling the initiation of the free radical polymerization are:

- free radicals, resulting from the degradation of peroxide or azo initiators, as well as during the redox reactions,
- heat energy (thermal polymerization),
- UV irradiation (photopolymerization),
- X-rays or gamma rays (radiation polymerization),
- ultrasounds (ultrasonic polymerization).

Most frequently the polymerization process is initiated by free radicals obtained through the decomposition of hydroperoxides, alkyl peroxides, dialkyl peroxides, acyl peroxides, carboxylic ester peracids, salts of (tetraoxo)sulphuric acid, hydrogen peroxide, aliphatic azo compounds and bifunctional azobenzoin initiators. The rate of decomposition of different initiators into free radicals depends on their structure and on temperature. A measure of the efficiency of the initiator in the polymerization process is the half-decomposition period.

Table 8.2 presents the characteristics and the half-decomposition periods of the most common free radical initiators.

TABLE 8.2 Characteristics of Selected Radical Polymerization Initiators

No.	Initiator	Chemical formula	Molecular mass.	Melting temp. [°C]	Boiling temp. [°C(kPa)]	Density [kg/m^3]	Refract. index $n^{20}D$	Half-decomp. period [H/°C]
1	dibenzoyl peroxide	C_6H_5COOO-COC_6H_5	242.22	100 with decomp.	-	-	1.4056	2/85
2	acetyl-t-butyl peroxide	CH_3COOO–$C(CH_3)_2CH_3$	135.15	-	42.5 (1.46)	9455	1.5007	14/100
3	benzoyl-t-butyl peroxide	C_6H_5COOO–$C(CH_3)_2CH_3$	194.22	8	76.2	1043	1.3872	20/100
4	di-t-butyl peroxide	$(CH_3)_3COOC(CH_3)_3$	146.22	-	12.5 (2.66)	793	1.4013	20/120
5	t-butyl hydroperoxide	$(CH_3)_3$-COOH	90.12	–4	5 (0.266)	896	1.524	20/115
6	k-cumene hydroperoxide	$(CH_3)_2C$-C_6H_5OOH	152.18	-	53 (0.311)	1062	-	25/113
7	azoisobutyric acid dinitrile (AIBN)	$(CH_3)_2C(CN)$-N=N-$(CN)C(CH_3)_2$	164.2	-	103 with decomp.	-	-	1.5/80
8	potassium persulfate	$K_2S_2O_8$	270.32	100 with decomp.	-	2477	-	1.25/80
9	ammonium persulfate	$(NH_4)_2S_2O_8$	228.21	120 with decomp.	-	1982	-	1.25/80
10	hydrogen peroxide	H_2O_2	34.01	–0.41	150.2	1422	-	-

A typical example of a initiator decomposition into free radicals is the chemical decomposition of dibenzoyl peroxide (called also benzoyl peroxide):

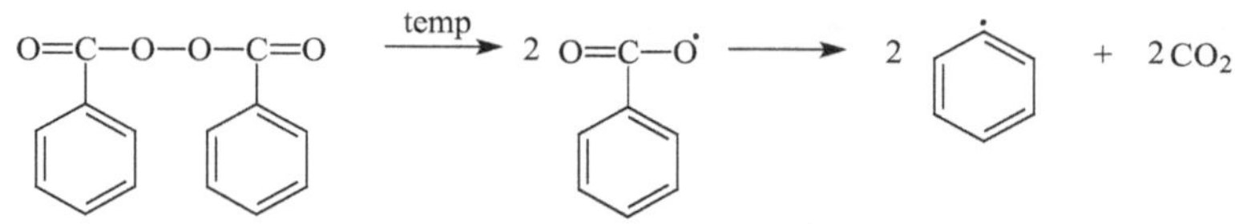

As shown by experimental studies in the polymerization process involving both benzoilowe radicals and phenyl radicals their ratio in the reaction mixture is 1:1.

Initiators may disintegrate into free radicals with a simultaneous liberation of nitrogen. The most widely used initiator from this group of compounds is the dinitryle of azoisobutyric acid, commonly called azoisobutyronitryle (AIBN):

The emerging cyanopropyl radicals not only can initiate the polymerization reactions but may also recombine with the formation of *dinitril tetramethyl succinate* or a corresponding ketenimine.

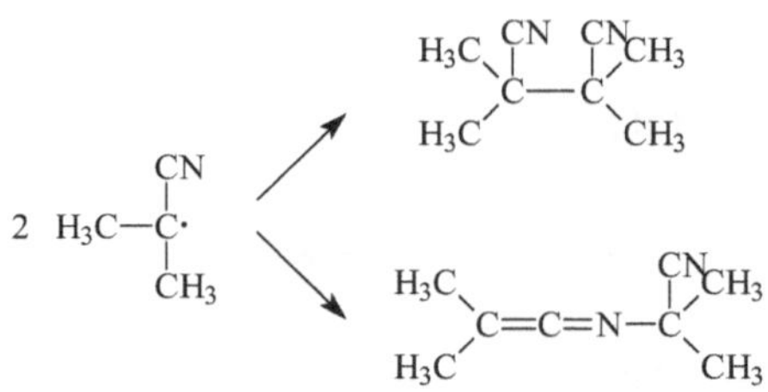

Free radicals are formed also in the redox reactions. A characteristic feature of these reactions is the possibility to obtain free radicals at low temperatures. They are particularly useful when initiating the low temperature and the emulsion polymerizations.

The most widely used redox systems, serving for the production of free radicals, are the mixtures of diluted in the reaction environment solutions of hydroperoxides or organic peroxides with transition metal ions:

$$HOOH + Fe^{2+} \longrightarrow Fe^{3+} + OH^{-} + {}^{\bullet}OH$$

$$ROOH + Co^{2+} \longrightarrow RO^{\bullet} + OH^{-} + Co^{3+}$$

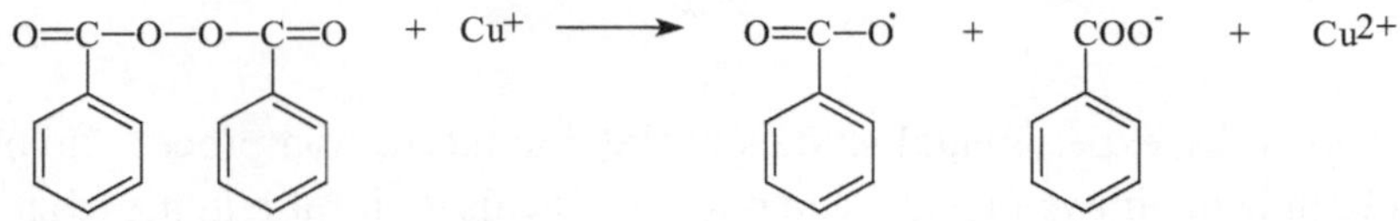

In the process of emulsion polymerization as initiating system one uses often a mixture of potassium persulphate and sodium thiosulfate:

$$S_2O_8^{2-} + S_2O_3^{2-} \longrightarrow SO_4^{-\bullet} + SO_4^{2-} + S_2O_3^{-}$$

$$SO_4^{-\bullet} + H_2O \longrightarrow HSO_4^{-} + {}^{\bullet}OH$$

Sometimes, as a reducing agent in the initiating system, the tertiary amines are also used:

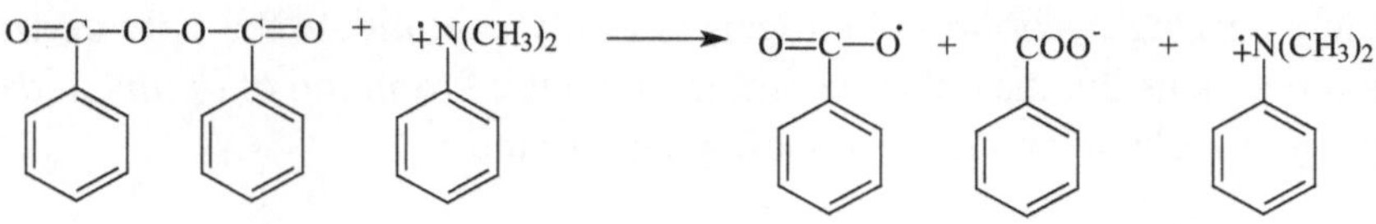

$$C_6H_5COO^{-} + C_6H_5N^{+\bullet}(CH_3)_2 \longrightarrow C_6H_5COOH + {}^{\bullet}C_6H_4N(CH_3)_2$$

$$2\,{}^{\bullet}C_6H_4N(CH_3)_2 \longrightarrow (CH_3)_2NC_6H_4{-}C_6H_4N(CH_3)_2$$

An interesting redox system serving to initiate polymerization reaction in the aquatic environment is also a solution of potassium permanganate and a reducing organic substance, such as, for example, oxalic acid, ascorbic acid, lactic acid, citric acid, thioglycollic acid, ethylene glycol or glycerol, which generates free radicals in acidic environment in the presence of monomer.

The mechanism of the polymerization reaction with such system is the following: manganese atom in potassium permanganate, which is in the +7 degree of oxidation is reduced as a result of the reaction with monomer molecule to the oxidation degree of +3, with a simultaneous formation of a by-product.

$$Mn^{+7} + \text{monomer} \longrightarrow Mn^{+3} + \text{oxidation by-product}$$

The manganese ions in +3 oxidation states may be subject to a reversible disproportionation reaction or react first with the water molecules, and then with the reducing agent (Red) and monomer (M) with the formation of free radicals (M*), initiating the polymerization.

$$2\ Mn^{+3} \rightleftharpoons Mn^{+4} + Mn^{+2}$$

$$Mn^{+3} + H_2O \rightleftharpoons (MnOH)^{+2} + H^{+}$$

$$(MnOH)^{+2} + Red \rightleftharpoons (Mn(OH)Red)^{+2}$$

$$(MnOH)^{+2} + M \rightleftharpoons (Mn(OH)M)^{+2}$$

$$(Mn(OH)Red)^{+2} + (Mn(OH)M)^{+2} \rightleftharpoons M^{\bullet} + (MnOH)^{+2} + Mn^{+2} + H_2O + Red$$

Sometimes the free radical polymerization can be initiated thermally without a use of initiators. This process generally takes place in an anaerobic system, in different ways, depending on the type of monomer used. An example of such a reaction may be the styrene polymerization:

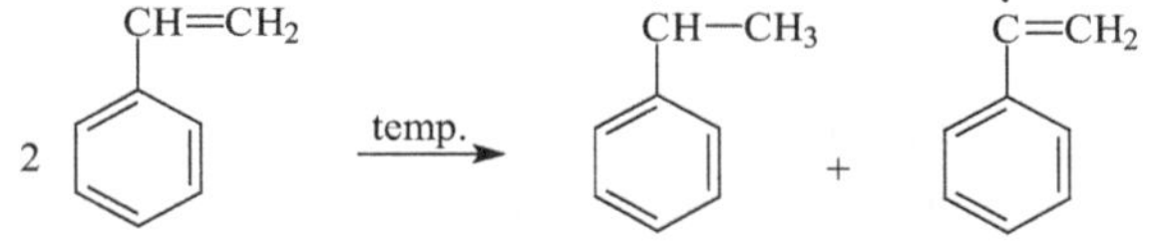

The free radical polymerization reaction may be initiated photochemically too. This process consists on the absorption by monomer molecule of a quantum of light energy resulting in its passage to an excited state. The photochemical initiation of polymerization takes place frequently in the presence of photosensitizers (optical sensitizers), which readily absorb the energy and play an important role in its transmission. The main advantage of the photochemical reaction is its total independence on temperature and the possibility of initiating the polymerization process at temperatures often lower than those used with other sources of free radicals generation.

The photochemical polymerization process can be easily controlled by using a light source with a narrow range of emission wavelengths and the possibility of introducing into the reaction environment other types of initiators, disintegrating into free radicals under the UV radiation.

Examples of such initiators are disulfides, benzoin and dibenzoil:

$$RSSR \xrightarrow{h\nu} 2\ RS\cdot$$

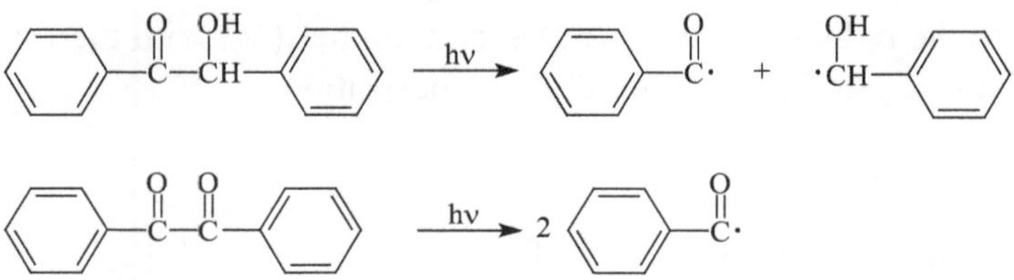

An important factor characterizing the stage of initiation of polymerization is the induction period. This is the time counted from the initiation of polymerization to the beginning of the growth of chains. It depends on the reaction temperature, structure and monomer concentration as well as on the presence of inhibitors.

Inhibitors are compounds, which inhibit the polymerization process during the initiation and growth of the chain. Typical inhibitors include hydroquinone, pyrocatechol and its derivatives, aromatic p-phenylenediamine amines, N-phenyl-2-naphthylamine, and also trinitrobenzene, picric acid, copper salts, and others. The reaction of free radical with a monomer molecule depends on the reactivity of double bonds of the latter.

For example, the free radical polymerization of ethylene runs extremely difficult and is necessary to be performed at high temperature (420–470 K) and at high pressure of the order of 100 MPa.

The presence of an electronegative substituent (halogen, nitrile or carboxyl group) or an electropositive (methyl or amine group) in the monomer molecule causes a polarization of double bonds, which manifests itself in the increase of the molecular dipole moment and the corresponding increase in the polymerization rate. From this rule, however, exceptions are because the butadiene molecules have no own dipole moment whereas they are easily polarized and thus readily polymerize.

8.2.1.2.2 Chain Growth Reaction

Second important step in the course of the polymerization process is the increase of the chain length, called propagation. This increase is associated with the attachment of new monomer molecules, initially to a free radical, created at the initial stage of the polymerization, and then to the continuously growing macroradical.

The chain growth rate is so high that practically it has no effect on the kinetics of polymerization. The polymer molecule composed from about 10,000 monomer molecules is formed in less than one second. One assumes that the activity of the growing polymer radical does not depend on the chain length. The activation energy of the polymer chain growth reaction is 16.7–41.9 kJ/mol.

While the speed of the polymerization initiation depends also on the monomer activity, the rate of increase of the chain length depends on both the monomer activity and the activity of the growing polymer radical.

It was shown that the polarized monomer molecule reacts easier with the free radical. However, free radicals, arising from the polarized monomer molecules, are always less active than the radicals of unpolarized monomer molecules. This is due to the fact that the activity of free radical depends on the ability of free electrons to act upon the π bonding of the monomer molecule. The shift of electron due to the coupling effect reduces the probability of the course of this reaction, and hence the radical activity. The coupling effect affects more the radical activity than that of monomer. Therefore, the chain growth reaction rate in the radical polymerization depends primarily on the activity of free radicals. The type of substituent in the monomer molecule has an impact not only on the rate of the macroradical growth, but also on the configuration of the resulting polymer macromolecule.

The addition of a monomer molecule, containing a substituent X, to the growing macroradical may occur in the "head to tail" (Scheme A) or "head to head" configuration (Scheme B).

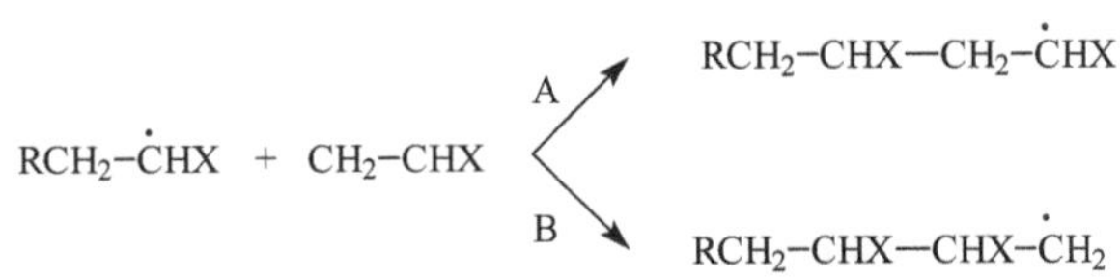

As a result of the reaction of two polarized groups: the growing radical and monomer molecule the addition takes place in agreement with the charge density distribution. The coupling effect provides the stability to the growing polymer radical and therefore more likely is obtaining the structure of "head to tail" type. In principle, more probable is also attachment of a monomer molecule in which the most stable radicals are formed. However, obtaining of polymers with a regular arrangement of mers by the free radical polymerization method is difficult. This is explained by small differences in the activation energy of monomer molecules addition in different positions.

A large amount of "head to head" connections was found in some polymers such as poly (vinylidene fluoride) [-CH_2CF_2-] and poly (vinyl fluoride) [-CH_2CHF-] by using the high-resolution nuclear magnetic resonance spectroscopy technique.

8.2.1.2.3 Ways to End the Chain Growth

The completion of the polymer chain growth during the free radical polymerization may occur in the following situations:

- through the reaction of the formed macroradical with another radical (the so-called recombination), resulting from the initiator disintegration or from the secondary reactions.

$$RM_n^{\bullet} + R_1^{\bullet} \longrightarrow RM_nR_1$$

$$RM_n^{\bullet} + H^{\bullet} \longrightarrow RM_nH$$

- through the recombination of two macroradicals;

$$RM_n^{\bullet} + RM_m^{\bullet} \longrightarrow R(M)_{n+m}R$$

where: M = CH_2CHX; RM^*_n and RM^*_m – growing macroradicals.

- through the disproportionation reaction by the transfer of hydrogen atom from one of the growing chain to another one.

$$R{-}(CH_2CHX)_n{-}CH_2\dot{C}HX + R{-}(CH_2CHX)_m{-}CH_2\dot{C}HX \longrightarrow$$

$$\longrightarrow R{-}(CH_2CHX)_n{-}CH{=}CHX + R{-}(CH_2CHX)_m{-}CH_2CH_2X$$

– through the transfer reaction of kinetic chain activity as a result of a collision with inactive molecule (monomer, solvent, polymer or otherwise), contained in the reaction environment. It is followed by the completion of the macroradical chain growth with a simultaneous emergence of a new radical.

$$R{-}(CH_2CHX)_n{-}CH_2\dot{C}HX + R_1H \longrightarrow R{-}(CH_2CHX)_n{-}CH_2CH_2X + \dot{R_1}$$

Completion of the chain growth as a result of recombination or disproportionation of reactants stops the growth of the polymer macromolecule and ends the kinetic chain free radical reaction. In that case no new free radicals, able to initiate the growth of a new macromolecule, are formed.

During the chain transfer reaction, new free radicals emerge, which, depending on their structure may show different activity.

The transfer of the chain to another monomer molecule is followed by the increase of next macroradical. In that case the polymerization reaction rate does not change. The transfer of the chain on a solvent molecule, able to create relatively persistent free radicals leads to the formation of oligomeric products terminated by large substituents, coming from the decomposition of solvent molecule. These substituents determine the properties of the resulting compounds, called telomeres. This issue will be discussed larger in the chapter devoted to the polymerization in solution.

The transfer of chain to the polymer molecule gives rise to the creation of a new macroradical in which the unpaired electron is not at the end of the chain, but in its middle. The formed in this way new chain is a branch of the already existing one. This type of reaction is called the grafting reaction.

$$R\dot{M_n} + RM_kMM_lH \longrightarrow RM_nH + RM_k\dot{M}M_lH$$

$$RM_k\dot{M}M_lH + mM \longrightarrow RM_k\underset{\dot{M}_m}{\underset{|}{M}}M_lH$$

Very often one introduces into the medium of polymerization reaction some compounds, which serve as a chain transfer agents. They are known in the polymer chemistry under the name of moderators or molecular weight regulators. As examples one can cite: chlorinated hydrocarbons, dodecyl mercaptan and thioglycollic acid. These compounds are characterized by the presence in molecule of a mobile atom, able to detach away on impact with the growing macroradical.

If we denote by AB the moderator molecule than its action can be demonstrated as follows:

$$RM_n^{\cdot} + AB \longrightarrow RM_nA + B\cdot$$

If the resulting radical B* has a similar activity to that of the completed macroradical, than the action of AB moderator is limited to reducing the molecular mass of the product and have no significant influence on the kinetics of polymerization.

However, if B* is less active than the completed macroradical, than the rate of polymerization is reduced. In that case the AB compound plays the role of retarder in this process.

A special case of delaying the polymerization process is its inhibition. The radical B* originating f from the inhibitive substance, that is from the inhibitor, is completely unreactive and causes a gradual termination (ending) of the kinetic chain growth reaction. The time needed to wear the total amount of inhibitor is called the period of inhibition.

8.2.1.2.4 Influence of Oxygen on the Course of the Polymerization Process

Oxygen, present in the reaction environment (usually air), plays an important role in the process of polymerization. In some cases, such as during the high temperature and high-pressure ethylene polymerization it plays the role of initiator. In other cases, oxygen, depending on the reaction conditions accelerates or delays the process of polymerization. In extreme cases oxygen acts as an inhibitor.

These phenomena are explained by the chemical reactions between oxygen and free radicals present in the reaction environment. In that case the peroxide radicals are formed, as shown below:

$$R\cdot + O_2 \longrightarrow R{-}O{-}O\cdot$$

The reactivity of these peroxide radicals determines the role oxygen plays in the given type of polymerization. Often, especially in the process of emulsion polymerization, one uses an anaerobic atmosphere by blowing the free space of reaction device with free of oxygen nitrogen.

8.2.1.3 Kinetics of Radical Polymerization

The rate of free radical polymerization depends on the speed of the initiation, growth and chain termination reactions.

The initialization process starts with the initiator decomposition into free radicals, which reacting with monomer molecules initiate the chain polymerization reaction:

$$I \xrightarrow{k_d} 2R^{\bullet}$$

$$R^{\bullet} + M \xrightarrow{k_1} RM^{\bullet}$$

where: I – initiator molecule, M – monomer molecule, R^*, M^* – free radicals, k_d – rate constant of the radicals formation, and k_i – initiation rate constant ($dm^3\ mol^{-1}\ s^{-1}$).

Not all free radicals initiate effectively the polymerization process as part of them is deactivated by the recombination process:

$$2R^{\bullet} \rightarrow R-R$$

The initiator decomposition reaction is of first order. The rate of formation of new free radicals V is given by the formula:

$$V_i = \frac{d[R^{\bullet}]}{dt} = -\frac{d[I]}{dt} = 2k_d[I]$$

where: $[R^*]$ – radical initiator concentration (mol/dm^3); [I] – initiator concentration (mol/dm^3).

Since not the all formed in this way free radicals initiate the polymerization process effectively a corrective factor f, called the capacity to initiate the reaction, was introduced. It defines the fractional part of free radicals reacting with monomer molecules and can be written in the following way:

$$V_i = 2fk_d[I] = k_i[I]$$

As mentioned previously, the chain reaction growth consists on the addition of consecutive monomer molecules to the nascent macroradical. Its speed V_w is given by the formula:

$$V_w = -\frac{d[M]}{dt} = k_w[M_n^\bullet][M]$$

where: k_w – the rate constant of chain growth ($dm^3\ mol^{-1}\ s^{-1}$), [M] – monomer concentration (mol/dm^3), $[M^*]$ – concentration of macroradicals (mol/dm^3).

In considering this it is assumed that:

- the reactivity of radicals does not depend on their chain length,
- the average length of the macroradical of the resulting polymer is large, (it allows to neglect the minor consumption of monomer in the initiation reaction and assume that its total amount is involved in the chain growth reaction)
- the number of radicals present in the system during the polymerization process is constant (it means that the rate of formation of radicals is equal to V_i and the decay rate to V_z).

The rate of the termination reaction V_z, due to the recombination of two macroradicals, is given by the following equation:

$$V_z = -\frac{d[M_n^\bullet]}{dt} = k_z[M_n^\bullet]^2$$

Given that:

$$V_i = V_z$$

one gets

$$k_i[I] = k_z[M_n^\bullet]^2 \qquad [M_n^\bullet] = \sqrt{\frac{k_i}{k_z} \cdot [I]}$$

It is assumed that the overall polymerization speed V is equal to the speed of chain growth V_w. By substitution of [Mn] in formula for V_w one obtains the following equation for V:

$$V = V_w = k_w \sqrt{\frac{k_i}{k_z} \cdot [I]} \cdot [M]$$

where: k_w – the rate constant for the polymerization reaction.

This means that the speed for free radical polymerization is directly proportional to the square root of the initiator concentration and the reaction is of first order with respect to the monomer concentration.

In the initial period of polymerization, with a low degree of conversion, the monomer concentration is constant (with its large excess) and therefore the polymerization rate depends solely on the square root of the initiator concentration

$$V = const \cdot \sqrt{[I]}$$

The constants and the activation energy values a for radical polymerization of selected monomers are given in Table 8.3.

In a completely different way run the polymerization at its final stage, when the degree of conversion of substrates is large. In this case a self-acceleration takes place and a departure from first order law is observed. This phenomenon is called the gel or Norrish–Tromsdorff effect. It is mainly due to the increased viscosity of the polymerization medium and the diminished probability of meeting macroradicals and their recombination. In that case one observes also a decrease of thermal conductivity of the system. It results in the temperature increase, thus in the acceleration of reaction. The gel effect leads also to the increase of molecular mass of polymer, because the inhibition of the termination process does not affect the growth rate of the chain.

The up to now known kinetic equations enable us to determine the average degree of polymerization P, which is expressed by the ratio of chain growth rate to the completion rate:

$$\overline{P} = \frac{V_w}{V_z} = \frac{k_w[M^\bullet][M]}{k_z[M^\bullet]^2} = \frac{k_w[M]}{k_z[M^\bullet]} = \frac{k_w[M]}{\sqrt{k_z k_i} \cdot \sqrt{[I]}}$$

At the initial stage of polymerization, assuming existence of an excess and constant monomer concentration, the average degree of polymerization,

In the case of photochemically and radiation initiated polymerization $E_i = 0$. Therefore, the polymerization energy $E < 0$ and the degree of polymerization increases with increasing temperature.

The activation energy of side branching reaction is larger than the activation energy for their growth. Therefore, an increase of temperature during the polymerization favors the formation of branched compounds.

Temperature affects also the structure of the resulting polymer. It contributes to the reduction of its stereoregularity (spatial arrangement of substituents in the polymer molecule) and causes an increase of polymer polydispersity (individual macromolecules differ significantly from each other in molecular mass). At higher temperatures another side reactions are possible, such as, for example, the degradation (breakdown of the polymer chain into smaller fragments) and crosslinking (formation of lateral bonds between the polymer chains).

Another important parameter influencing the course of polymerization is the pressure. A moderate increase of pressure shows no apparent effect on the polymerization reaction rate. Pressures above 100 MPa increase the rate constant of chain growth and by the same the speed of polymerization reaction. The increase of pressure results also in the increase of average molecular weight of the formed polymer and improves the regularity of its spatial structure.

During the polymerization in solution an important role plays also the monomer concentration. With the increasing monomer concentration the rate of polymerization and the polymer molecular mass increase. The influence of interactions between molecules of monomer and solvent are discussed in the chapter devoted to the polymerization in solution.

8.2.1.5 Study of the Polymerization Kinetics

The most widely used way to study the kinetics of polymerization is the dilatometric method. It allows to carry out the measurements on a continuous basis and provides precise and reproducible results.

Its principle is based on the measuring the difference between the specific volume of monomer and specific volume polymer. As a consequence of the running polymerization process is always the volume contraction of the studied system. This contraction is proportional to the changes in concentration of monomer or polymer in the reaction mixture. Measuring such changes and expressing them in function of time allows to determine the rate of polymerization. The densities and the contraction coefficients of selected monomers at different temperatures are listed in Table 8.4.

TABLE 8.4 Densities and Contraction Coefficients for Selected Monomers

No.	Monomer	Temperature [K]	Density [kg/m^3]	Contraction at 1% conversion [%]
1	Styrene	283	915	0.1327
2	Styrene	293	906	0.1405
3	Styrene	303	897	0.147
4	Styrene	323	878	0.163
5	Styrene	333	869	0.1704
6	Styrene	343	860	0.1778
7	Methyl methacrylate	293	941	0.2282
8	Methyl methacrylate	303	929	0.233
9	Methyl methacrylate	313	918	0.2379
10	Methyl methacrylate	323	907	0.2427
11	Methyl methacrylate	333	899	0.2468
12	Methyl methacrylate	343	887	0.2511
13	Methyl methacrylate	353	876	0.2552
14	Vinyl chloride	303	892	0.372
15	Vinyl chloride	313	872	0.4
16	Vinyl chloride	323	852	0.433
17	Vinyl chloride	333	831	0.47
18	Acrylonitrile	293	806	0.323
19	Acrylonitrile	303	795	0.332
20	Acrylonitrile	313	784	0.341
21	Acrylonitrile	323	773	0.35
22	Acrylonitrile	333	762	0.36
23	Acrylonitrile	343	751	0.369
24	Acrylonitrile	353	740	0.378
25	Methyl acrylate	293	960,7	0.2189
26	Methyl acrylate	303	936,2	0.2286

TABLE 8.4 Continued

No.	Monomer	Temperature [K]	Density [kg/m³]	Contraction at 1% conversion [%]
27	Methyl acrylate	323	912	0.239
28	Vinyl acetate	298	925	0.2206
29	Vinyl acetate	303	892	0.235
30	Vinyl acetate	355	850	0.2355

After S. Połowiński: Techniques of measurement and research in physical chemistry of polymers (Techniki pomiarowo-badawcze w chemii fizycznej polimerów, in Polish), WPŁ, Łódź, Poland 1975.

The implementation of the comprehensive studies of the kinetics of polymerization or copolymerization (polymerization of two different monomers) is a difficult task. According to the earlier provided information the kinetics of polymerization consists of initiation rates, growth, completion and a series of side reactions. It requires often an additional research using other methods.

The application of dilatometric method for kinetic studies of polymerization or copolymerization processes is usually based on the determination of the dependences of the speed of the process on the following parameters:

- concentration of initiator [I],
- concentration of monomer or comonomers (monomers in the copolymerization process) [M],
- value of absolute temperature (T),
- concentration of inhibitor,
- composition of comonomer mixture.

The quantities measured in the dilatometric method are the changes in the height of the liquid column (h) during the reaction time (t). The degree of transformation (W), which is a measure of the decrease of meniscus of the reaction mixture in the capillary of dilatometer, is obtained from the following formula:

$$W = \frac{\Delta V}{V_m L^T} x100$$

where: W – degree of conversion) [%]; ΔV – volume change of the system; V_m – initial volume of monomer; L^T – contraction coefficient of the system at temperature T.

$$L^T = \frac{V_m - V_p}{V_m} = \frac{\frac{1}{d_m} - \frac{1}{d_p}}{\frac{1}{d_m}} = 1 - \frac{d_m}{d_p}$$

where: V_m, V_p – specific volume of the appropriate monomer (m) and polymer (p) at temperature T; d_m, d_p – density of monomer (m) and of polymer (p) at the measurement temperature T.

The initial speed of polymerization V is calculated from the formula

$$V = \frac{dM}{dt} = \frac{\Delta V}{V_{zb}\Delta t} \cdot \frac{1}{MX} [mol / dm^3 s]$$

where: [M] – monomer concentration in mol/dm^3; ΔV – change of the volume of monomer (solution) at time Δτ; V_{zb} – volume of the dilatometer reservoir, equal to the initial volume; of monomer (block polymerization) or of solution (polymerization in a solvent); M – molecular mass of monomer; X – absolute change of the specific volume of the system, resulting from the full (100%) monomer conversion at temperature T, and given by

$$X = V_m - V_p = \frac{1}{d_m} - \frac{1}{d_p} = \frac{L}{d_m}$$

8.2.1.6 Stereochemistry of Radical Polymerization

As a result of the polymerization of unsaturated monomers containing one or two substituents at the same carbon atom one obtains polymers with different spatial structure:

- isotactic, which is characterized by one type of basic configuration units (with chiral or pseudochiral atoms in the main chain) with the same type of sequence,

- syndiotactic, in which the molecules can be described as an alteration of the enantiomeric basic configuration units,
- atactic, in which the particles have a random distribution of equal number of possible basic configuration units.

The stereochemistry of monomer molecules addition to the macroradical depends on:

- mutual interaction between the end carbon atoms of the growing macroradical and the approaching monomer molecule,
- configuration of the penultimate unit (mer) repeatable in the chain.

The terminal carbon atom with unpaired electron in the growing macroradical molecule exhibits probably a flat sp^2 hybridization. For the monomer CH=CXY (Y = H, X or R) polymerization to occur the molecule must approach the end carbon atom of the macroradical and set in the mirror position at which the same substituents are on the same side (scheme "a") or in nonmirror position at which the same substituents are on opposite side (scheme "b") as shown below:

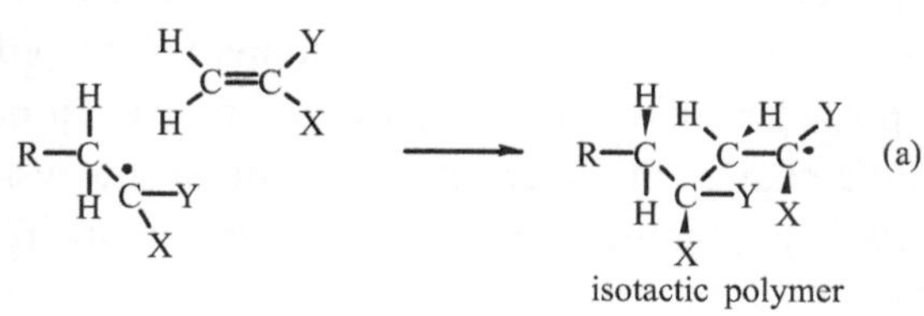

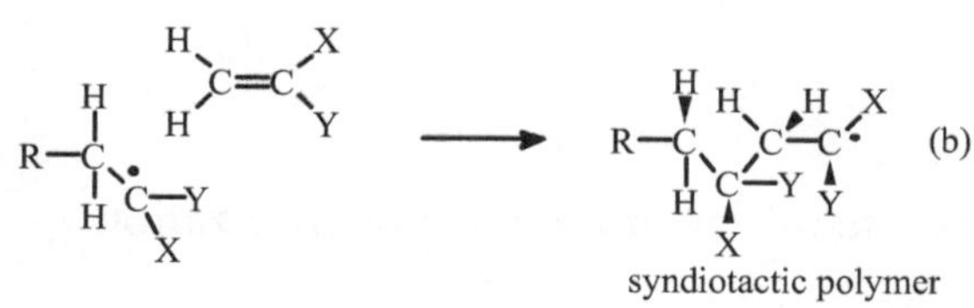

The stereochemistry of polymerization reaction is maintained when before the attachment of the next monomer molecule no free rotation around the macroradical takes place.

When the polymerization stereochemistry is preserved then the mirror approach (reaction "a") will always lead to the formation the isotactic polymer and the nonmirror approach to the formation of the syndiotactic polymer (reaction "b"), respectively.

If the mutual interaction between the reacting monomer molecule and the mer substituents of the penultimate carbon atom of macroradical is significant, the also the conformational factors may influence the way how the attachment is realized. It will depend on which of the two interactions: the steric or the electrostatic one will be the weakest one.

In the free radical polymerization process, conducted usually at elevated temperatures, these effects are insignificant and the reaction usually leads to the formation of atactic polymer only. However, in some cases, like, for example, in free radical polymerization of methyl methacrylate at temperature below 0°C one obtains a crystalline polymer with syndiotactic structure, as it was proven by the high resolution nuclear magnetic resonance spectroscopy. These results confirm the rule that according to which the degree of stereoregularity decreases with increasing temperature.

The stereochemistry of free radical polymerization depends on a set of steric and polar effects. A significantly larger stereoregularity is obtained in ionic polymerization, and in particular the coordination one.

The tactility of obtained polymer is characterized by the ratio k_{ws}/k_{wi} of reaction rate constants of chain growth with the formation of syndiotactic (k_{ws}) and isotactic (k_{wi}) structures. Namely when:

(a)

$$\frac{k_{ws}}{k_{wi}} = 1$$

an atactic polymer is formed,
(b)

$$\frac{k_{ws}}{k_{wi}} = \infty$$

a syndiotactic polymer is formed,
(c)

$$\frac{k_{ws}}{k_{wi}} = 0$$

an isotactic polymer is formed,

8.2.1.7 Polymerization of Dienes

Polymerization of dienes is a special case of the polymerization of unsaturated compounds and its course depends on the structure of diene used.

Dienes with isolated double bonds can polymerize with the formation of cross-linked polymers, because the two double bonds present in the molecule can react independently of each other. However, in some cases in addition reactions both double bonds of monomer molecules can take part, leading to the formation of cyclic polymers. This reaction occurs particularly when it leads to the formation of rings with five or six-members. During the cyclization reaction side vinyl groups are formed, which polymerize with formation of cyclic polymers.

In the case of cyclopolymerization the most favorable configuration is *cis*, while in the normal polymerization of vinyl monomers the dominating configuration is the *trans* one.

Dienes with π electron conjugated double bonds, such as 1,3- butadiene and its derivatives undergo an addition process in both positions: 1.2 and 1.4. As a result of the polymerization process of 1.2 type a polymer containing side vinyl groups is formed, while 1.4 addition reaction leads to a polymer with double bond in the chain. In that case both configurations *cis* and *trans* are equally possible.

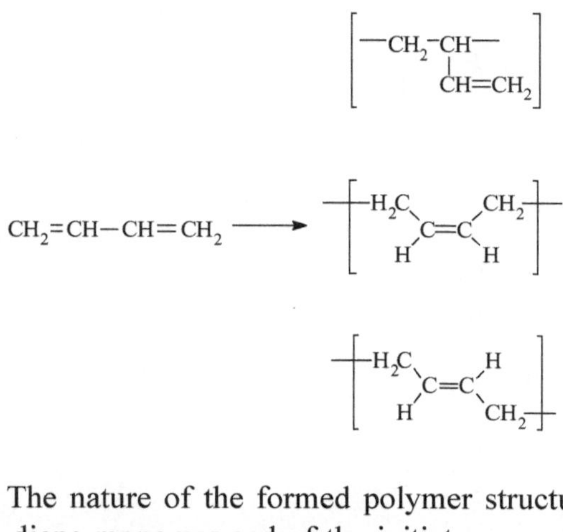

Polymerization of type 1,2

Polymerization of type 1,4 cis

Polymerization of type 1,4 trans

The nature of the formed polymer structure depends on the type of the diene monomer and of the initiator as well as on the polymerization reaction conditions. In general, lowering the reaction temperature leads preferentially to type 1.4, which is more desirable than 1.2 because of a greater flexibility of these polymers.

In the case of polymerization of substituted dienes such as 2-methyl-butadiene (isoprene), the structure of the resulting product may be more complicated because it is possible to create polymers with structure of type 1.2 and 3.4, as well as 1, 4 *cis* and 1, 4 *trans.* Moreover, structures "head to tail" and "head to head" type may be formed. The results of the analysis of synthetic polyisoprene have shown that, in general dominating is the "head to tail" arrangement and the 1, 4 *trans* configuration. The natural rubber is 1,4 *cis*-polyisoprene, whereas gutta-percha and balata have 1, 4 *trans* structure.

The parts of the structural elements of three most important diene monomers: butadiene, isoprene and chloroprene (2-chlorobutadiene) in the structure of different polymers obtained by free radical polymerization at different temperatures are given in Table 8.5.

In the case of chloroprene the content of structural groups of type 1.4 is much higher compared to the others. This may be due to the large difference in the electron density of two double bonds present in the monomer molecule.

TABLE 8.5 Structure of Diene Polymers Obtained by the Free Radical Polymerization Method

No.	Monomer	Polymerization temperature [K]	Content of the structure type [%]			
			1,4 *cis*	1,4 *trans*	1,2	3,4
1	Butadiene	253	6	77	17	-
		263	9	74	17	-
		278	15	68	17	-
		293	22	58	20	-
		373	28	51	21	-
		448	37	43	20	-
		506	43	39	18	-
2	Isoprene	253	1	90	5	4
		268	7	82	5	5
		283	11	79	5	5
		323	18	72	5	5

TABLE 8.5 Continued

No.	Monomer	Polymerization temperature [K]	Content of the structure type [%]			
			1,4 *cis*	1,4 *trans*	1,2	3,4
		373	23	66	5	6
		423	17	72	5	6
		476	19	69	3	9
		530	12	77	2	9
3	Chloroprene	227	5	94	1	0.3
		283	9	84	1	1
		319	10	85	2	1
		373	13	71	2.4	2.4

8.2.2 CATIONIC POLYMERIZATION

The cationic polymerization is the process of a monomer or a mixture of monomers conversion into the polymer by a cationic mechanism in the presence of catalysts.

8.2.2.1 Characteristic of Cationic Polymerization

Cationic polymerization of unsaturated compounds proceeds through the stage of carbanion cations, called also carbocations. Typical catalysts for this reaction are strong protic acids such as sulfuric acid, perchloric and trifluoroctane or the Lewis acids, which include halides of elements III, IV and V groups of the periodic table (Friedel-Crafts catalysts), such as: boron trifluoride, aluminum trichloride, tin tetrachloride and titanium tetrachloride. The activity of Friedel-Crafts catalysts increases significantly the presence of small quantities of cocatalysts, that is, the compounds which most often are the source of protons.

As cocatalysts one uses water, hydrochloric acids and chlorinated aliphatic hydrocarbons. The quantities of cocatalysts introduced into the reaction environment are very small, since the increase of their concentration facilitates the completion of the chain reaction, what leads to the formation of polymers of lower molecular mass. The use of water in excess leads to the catalyst deactivation as a result of its hydrolysis.

Monomers used in the cationic polymerization processes are characterized by a high electron density around the unsaturated carbon atom and show a tendency to form a relatively permanent carbanion cation. Among the others they include isobutylene, vinyl alkyl ethers, styrene, α-methylstyrene, isoprene, coumarone and indene.

As a result of the reaction of a Friedel-Crafts catalyst with a cocatalyst a complex compound is formed which dissociates into ions according to the scheme:

$$BF_3 + H_2O \longrightarrow [BF_3{*}H_2O] \rightleftharpoons H^+ + [BF_3OH]^-$$

$$SnCl_4 + HCl \longrightarrow [SnCl_4{*}HCl] \rightleftharpoons H^+ + [SnCl_5]^-$$

The resulting protons H^+ joins the monomer molecule, giving it the structure of the carbanion cation. In the case of substituted alkyl monomers the proton attachment follows the Markovnikov rule. It means that it is attached to the unsubstituted methylene group:

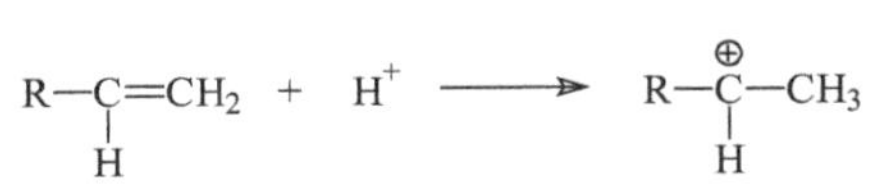

The resulting carbocation, similarly as the free radical, is very reactive because of the tendency of the endowed with a positive charge carbon atom to complete the octet of electrons.

In carbocation the carbon atom with a deficit of electrons binds three other atoms by means of its orbitals and therefore the system exhibits trigonal bonds (bonds are directed towards the vertices of an equilateral triangle). For this reason the center of active carbocation is flat. The carbon atom with a deficit of electrons and three atoms connected to it lie in the same plane. In accordance with the laws of electrostatics, the stability of a charged system increases with the increasing charge delocalization. The electron donating substituents located in the molecule (methyl groups) help to reduce the positive charge of the carbon atom with a deficit of electrons. The charge diffusion stabilizes the carbocation. It follows from this theorem that with the increasing number of alkyl groups in the molecule increases also the sustainability of the emerging from it carbocation. This can also explain the special flexibility of isobutylene for cationic polymerization.

During the cationic polymerization important is the kind of solvent used. A particularly positive role-play polar solvents such as, for example, nitrobenzene. In their presence the polymerization reaction proceeds many times faster than when using non-polar solvents. In the solvolysis process the solvent may facilitate the formation of carbocations through the so-called nucleophilic power.

The formation of tertiary cations occurs relatively easily and their reactivity to a small extent depends on the nucleophilic solvent. It depends mainly on its polarity. The formation of secondary cations requires a large nucleophilic assistance and their reactivity depends both on the nucleophilic power and on the polarity of solvent used. The nucleophilic assistance differs from S_{N2}-type attack that it does not lead to a product but only to the formation of carbocation, or a molecule with a carbocation character.

If the solvent used is characterized by a moderate polarity only than the bond between the cation and the anion can be mainly of electrostatic character. In this case the formed system is called the ion pair. The active center of the growing chain in the cationic polymerization process may occur as:

- free carbocation,
- carbocation molecule solvated by the solvent molecules,
- component of the ion pair of carbocation with the counterion derived from the catalyst,
- component of the solvated ion pair,
- pseudoion (covalent) active center.

The general mechanism of the cationic polymerization consists of three stages: initiation, chain growth and completion (termination).

As mentioned above, the most common way of initiating the cationic polymerization is the attachment of proton to the double carbon bond. There is a number of other ways to initiate this reaction. They include polymerization in the presence of diazonium salts, charge transfer complexes and catalyzed by metal salts in a heterogeneous system.

Very interesting is the recently developed cationic polymerization method, catalyzed with diazonium compounds. It runs at room temperature and the formed polymers exhibit a high molecular weight. An example of such a reaction is the polymerization of 4-methoxystyrene under the influence of 4-nitrobenzenediazonium hexafluorophosphate, in which the poly (4-methoxy-styrene) with a molecular weight of more than 100,000 is obtained:

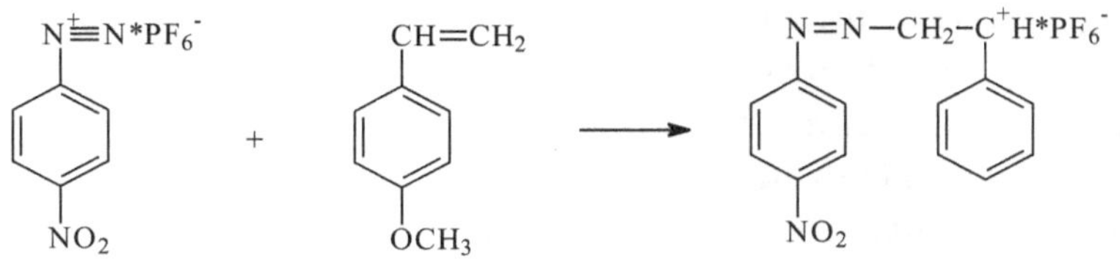

It should be noted that the diazonium compounds are unstable and decompose under heating or light illumination with the emission of nitrogen.

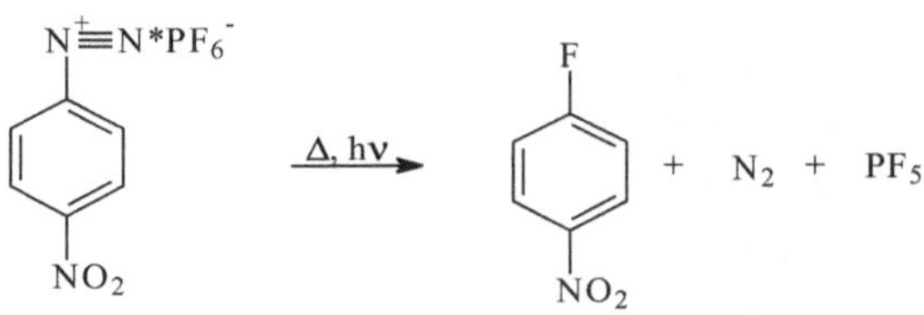

An example of the cationic polymerization initiated by a charge transfer complex is the reaction of Yamada, which can be illustrated by the polymerization of isobutylene with vanadium oxychloride complex with naphthalene in the heptane environment. The resulting complex decays with the formation of radical ions, which then react with isobutylene according to the following reactions:

$$VOCl_3 + C_{10}H_8 \longrightarrow \text{complex CT} \rightleftharpoons VOCl_3^{\cdot -} + C_{10}H_8^{\cdot +}$$

$$C_{10}H_8^{\cdot +} + H_2C{=}C(CH_3)_2 \longrightarrow C_{10}H_8 + \dot{C}H_2{-}C^{+}(CH_3)_2$$

$$2\dot{C}H_2{-}C^{+}(CH_3)_2 \longrightarrow (CH_3)_2C^{+}{-}CH_2{-}CH_2{-}C^{+}(CH_3)_2$$

To the new heterogeneous systems used to initiate the cationic polymerization belong perchlorates and trifluoromethyl sulphonates of magnesium, aluminum, cobalt, nickel and gallium. In contrast the lithium and silver salts do not show any catalytic activity. Initiating mechanism of this reaction relies on creation of an appropriate ion pair:

$$2\dot{C}H_2{-}C^{+}(CH_3)_2 \longrightarrow (CH_3)_2C^{+}{-}CH_2{-}CH_2{-}C^{+}(CH_3)_2 + nCH_2{=}C(CH_3)_2 \longrightarrow \text{polymer}$$

The formed carbocation reacts with the consecutive monomer molecules to create polymers. The cationic polymerization process requires high purity monomers and solvents, as impurities such as water, alcohols and acids play role of the chain transfer agent deactivating carbocations and decreasing the molecular weight of polymer.

The durability of the formed carbanion cation has a decisive influence on the growth rate of polymer chain. More lasting are the carbocations located at the chain ends larger is the reaction growth rate. So, this is an inverse relationship to that which takes place in the previously discussed process of free radical polymerization.

Presently this phenomenon is explained by the possibility of forming a π electron complex between carbocation located at the end of the chain and the new monomer molecule, rather than a covalent bond. It takes place during the attachment of a free radical to the monomer molecule. A large impact on the growth rate may have also the used solvent because of the differences in solvation of the growing ion. The use of a solvent of relatively low dielectric constant in the polymerization process causes that the ends of the chain appear mainly as a pair of ion with the counterion. Anion located near carbocation may also affect the growth rate, what makes the whole process more complex.

The formed during the polymerization carbanion macrocations may be subject of termination deactivity. The chain transfer can take place:

- in reaction with the monomer molecule:

- as a result of reaction with another polymer molecule with the formation of a more stable ion (grafting):

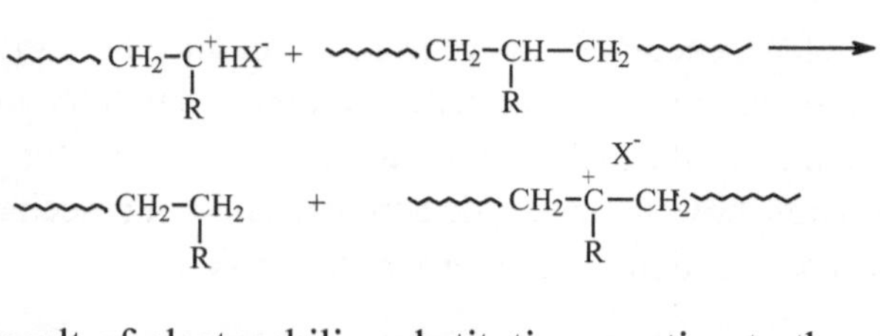

- as a result of electrophilic substitution reaction to the solvent molecule, for example, benzene:

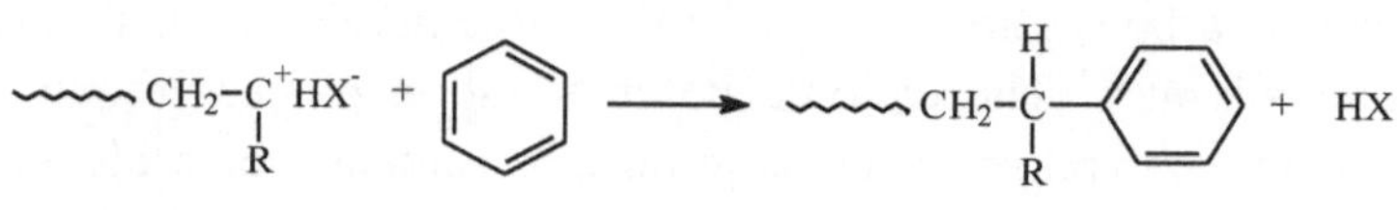

– as a result of the ring alkylation:

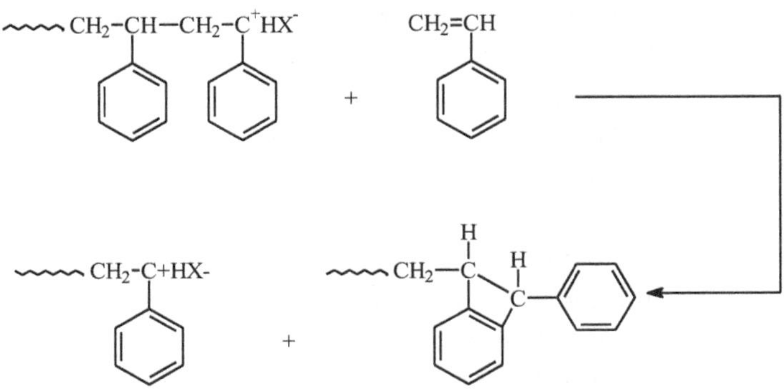

In the reaction consisting on the carbocation charge transfer to the carbon atom located in the main chain of another polymer molecule (grafting) a chain ramification takes place.

The completion of cationic polymerization occurs only when the catalyst is somehow deactivated during the process. Otherwise, the cationic polymerization shows several features of the so-called living polymerization process, what is of a great practical importance. In the living polymerization process the carbocation is not terminated and when adding next portion of the same or another polymer it is initiating further the polymerization reaction.

In several cases it is also possible to eliminate the transfer reaction. This is attributed to the covalent, pseudo ionic propagation reaction, running without involvement of reactive carbanion ions.

One of the examples illustrating the living cationic polymerization process is the polymerization of alkyl-vinyl ethers initiated by the mixture of hydrogen iodide with iodine. In this process the system is stabilized by a suitably strong interaction of carbocation with counterion:

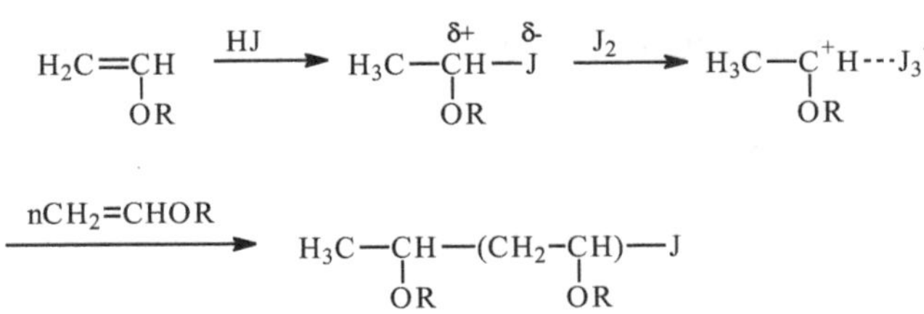

The concept of living polymerization will be detailed in Chapter 8.2.3.3. According to the mechanism of polymerization proposed by Higashimura its initiation consists on formation of an equimolar adduct of vinylether and hydrogen iodide. Then the binding of carbon – iodine in iodine adduct is activated with iodine, what in turn means a relaxation and polarization of the bond allowing attachment of further monomer molecules, up to their extinction in the reaction environment. The reaction, conducted in the presence of zinc iodide or other metal iodides as activators, proceeds with a conservation of the proportionality of molecular mass of the formed polymer to the degree of monomer conversion, even above the room temperature. The resulting polymer is characterized by a small dispersion of molecular masses (all resulting polymer molecules exhibit similar length chains).

The process of formation of living polymers allows the synthesis of block copolymers (in which the chain of one polymer is attached to the second polymer), because after all molecules of monomer A have reacted in the system there is no chain termination. Following introduction of monomer B to the system, capable of cationic polymerization, the reaction starts again with formation of a new block copolymer.

$$A\text{-}(A)_n\text{-}A\text{-}B\text{-}(B)_m\text{-}B$$

The formed block copolymer molecules are composed of linearly connected blocks either directly or by a constituent not being part of the blocks.

8.2.2.2 Kinetics of Cationic Polymerization

The kinetics of cationic polymerization determination was carried out in the presence of Friedel-Crafts catalysts and was specified for the cases in which the chain termination takes place. In the consideration the transfer reaction of chain activity to the monomer is ignored, because it runs with a low efficiency.

The rate of the catalyzing reaction is proportional to the concentrations of catalyst and monomer:

$$A + HB \rightleftharpoons AB^-H^+$$

$$AB^-H^+ + M \xrightarrow{k_i} HM^+AB^-$$

$$V_i = k_i[A][M]$$

where: A – cationic catalyst, such as, for example, BF_3, $AlCl_3$, $SnCl_4$, $SbCl_5$, etc.; HB – cocatalyst, for example, H_2O, CCl_3COOH and other; M – monomer; k_i – rate constant of initiation reaction; [A] – catalyst concentration; [M] – monomer concentration.

The reactions of the growth and of the chain termination can be described by the following equations:

$$HM_n^+AB^- + M \xrightarrow{k_w} HM_{n+1}^+AB^-$$

$$HM_n^+AB^- \xrightarrow{k_z} M_n + HAB$$

In the steady-state of catalysis the reaction rate (V_i) is equal to that of the chain termination (V_z):

$$V_i = V_z$$

$$V_z = k_z[HM^+]$$

$$k_i[A][M] = k_z[HM_n^+]$$

$$[HM_n^+] = \frac{k_i}{k_z}[A][M]$$

The overall polymerization rate (V_p) is equal to that of the chain growth (V_w):

$$V_p = V_w = k_w[HM_n^+][M]$$

where: k_w and k_z – rate constants for chain growth and termination, respectively.

Substituting the previously calculated value of the macrocation concentration $[HM^+_n]$ to the above formula one obtains

$$V_p = \frac{k_w k_i}{k_z}[A][M]^2 = k[A][M]^2$$

The activation energy of cationic polymerization is always smaller than that of the free radical polymerization, and is of 50.2–62.7 kJ/mol.

A characteristic feature of the cationic polymerization is its high rate at low temperatures. A classical example of such a process, which was applied on the industrial scale, is the polymerization of isobutylene in the presence of boron trifluoride, which runs at temperatures below 200 K. Polymers with very high molecular mass are then formed within a few seconds. At temperatures above 290 K formed are only oligomers of isobutylene with a low molecular mass.

The high speed of cationic polymerization process makes difficult the study of the kinetics of this reaction.

8.2.3 ANIONIC POLYMERIZATION

8.2.3.1 Characteristics of Anionic Polymerization Process

The reaction of the attachment of nucleophilic agent to the unsaturated double bond of monomer leads to the formation of carbanion monomer capable of reacting with the next monomer molecules with the creation of polymer macromolecule. This type of reaction, proceeding through the stage with carbanion anion is called by the anionic polymerization.

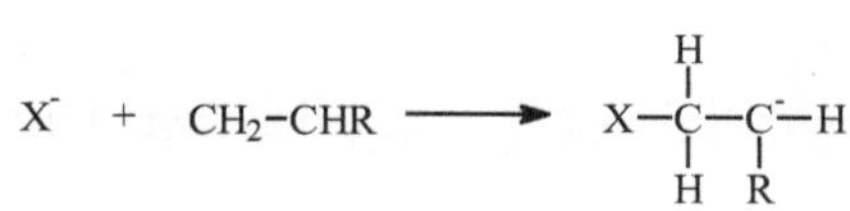

Using this method one can polymerize only the monomers, which contain substituents such as nitro group, nitrile, carboxyl, ester, vinyl and phenyl, capable of stabilizing the carbanion anion through the resonance or inductive effect.

Typical examples of monomers, which can polymerize in anionic way are: 2-nitropropene, acrylonitrile, methacrylonitrile, esters of acrylic and methacrylic acid, butadiene, styrene and izopropene.

A decisive influence on the effectiveness of the anionic polymerization has the chemical structure of used monomer. In the extreme case, very reactive monomers can be polymerized using very weak bases as nucleophilic agents.

Examples of such highly reactive monomers are: vinylidene cyanides, already polimerizing under influence of so weak nucleophilic agent as water and 2-nitropropene, whose polymerization is catalyzed by potassium bicarbonate:

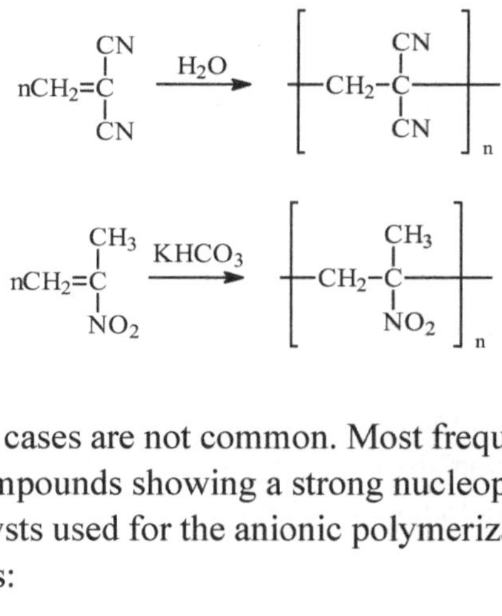

Unfortunately such cases are not common. Most frequently one has to use for this purpose compounds showing a strong nucleophilic effect.

Generally catalysts used for the anionic polymerization can be divided into two basic types:

- reacting through the connection of a negative ion, and
- reacting by an electron transfer.

A few examples of catalysts reacting through attachment of a negative ion are: amides, litoorganic compounds, Grignard compounds and other metalloorganics.

The second group of anionic polymerization catalysts include alkali metals and their additive complexes with some aromatic compounds.

The mechanism of anionic polymerization may change depending on the type of nucleophilic catalyst used, on reaction environment and on temperature.

The active site at the end of the growing chain can occur as:
– free anion

$$\sim\!\sim\!\sim\!\sim M^-$$

– anion solvated by the solvent molecules S

$$\sim\!\sim\!\sim\!\sim M^- \star\ yS$$

– component of the ion pair

$$\sim\!\sim\!\sim\!\sim M^-A^+$$

– component of the solvated ion pair

$$\sim\!\sim\!\sim\!\sim M^-A^+ \star\ yS$$

Between different forms of the active carbanion center an equilibrium state is established:

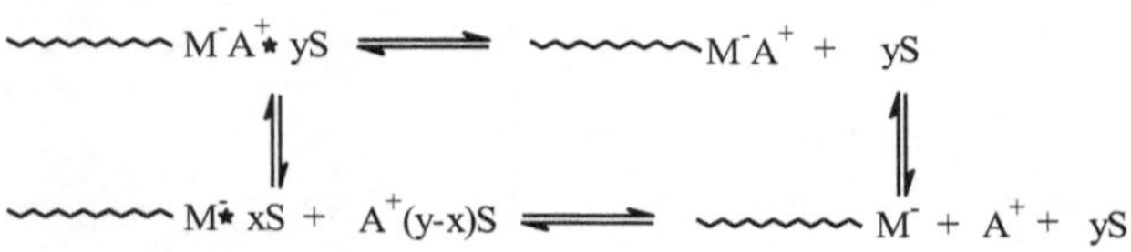

The active center, occurring in the form of a free ion, is more reactive in anionic polymerization than the active center being a component of an ionic pair. This is due to the polarization of monomer molecule approaching the active site.

Free ion produces an electric field stronger than a pair of ions and thus facilitates the polarization.

8.2.3.2 Mechanisms of Anionic Polymerization

The anionic polymerization, similarly as the cationic polymerization, takes place in three stages:

– initiation:

$$AB \rightleftharpoons A^- + B^+$$

$$A^- + M \xrightarrow{k_i} AM^-$$

– chain growth (propagation)

$$AM^- + nM \xrightarrow{k_w} AM_n^-$$

– chain termination:

$$AM_n^- + H^+ \xrightarrow{k_z} AM_nH$$

where: A, B – catalyst; M – monomer, and k_i, k_w, k_z – rate constants of initiation, growth and chain termination, respectively.

A typical example of anionic polymerization, running with participation of free carbanions, is the polymerization of styrene initiated by potassium amide in liquid ammonia. The mechanism of this reaction presents as follows:

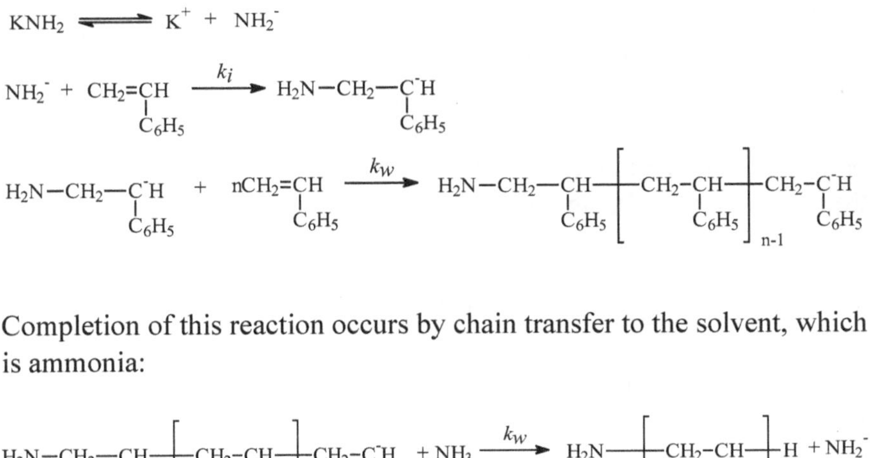

Completion of this reaction occurs by chain transfer to the solvent, which is ammonia:

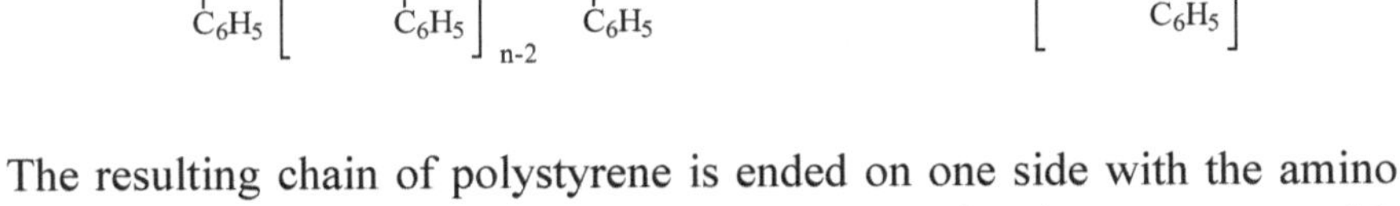

The resulting chain of polystyrene is ended on one side with the amino group. The anion NH_2, formed from the solvent molecule, reconstruct with the K^+ cation the potassium amide.

Basing on the above equations one can write the following formulas for the rates of: initiation (V_i), growth (V_w) and chain termination (V_z) reactions:

$$V_i = k_i[NH_2^-][M]$$

$$V = V_w = k_w[M_n^-][M]$$

Taking account of stationary conditions, according to which $V_i = V_z$, one gets

$$k_i[NH_2^-][M] = k_z[M_n^-][NH_3]$$

$$[M_n^-] = \frac{k_i}{k_z}\frac{[NH_2^-][M]}{[NH_3]}$$

By substituting the calculated quantities to the equation defining the reaction rate V one gets the final result:

$$V = V_w = \frac{k_w k_i}{k_z}\frac{[NH_2^-][M]^2}{[NH_3]}$$

It follows from the above considerations that the rate of the anionic polymerization of styrene in the presence of potassium amide is proportional to the concentration of catalyst and the square of monomer concentration.

The degree of polymerization of substrates in this reaction is directly proportional to the monomer concentration and inversely proportional to the concentration of ammonia:

$$\bar{P} = \frac{V_w}{V_z} = \frac{k_w[M_n^-][M]}{k_z[M_n^-][NH_3]} = \frac{k_w[M]}{k_z[NH_3]}$$

An example of the anionic polymerization with a participation of ion pairs is the polymerization of styrene in the presence of butyl lithium:

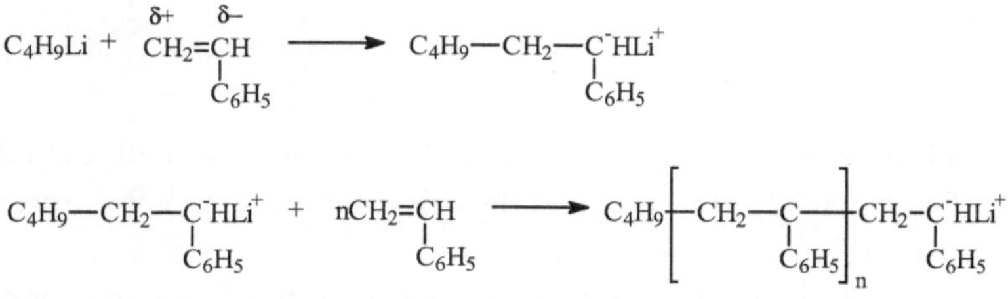

When such a reaction is carried out at room temperature, in a mixture of benzene with tetrahydrofuran, then it involves the participation of both unsolvated and solvated tetrahydrofuran ionic pairs. For this reason the kinetic scheme of such a process is more complicated, since one has to take into account three different rate constants of chain growth (depending on solvation) as well as the equilibrium constants between the various forms of active centers.

In a quite different way proceeds polymerization with an electron transfer. In that case, during the reaction of monomer with a metal from the first group of the periodic table of elements, an anionic radical is formed, which dimerizes immediately with formation of a bianion:

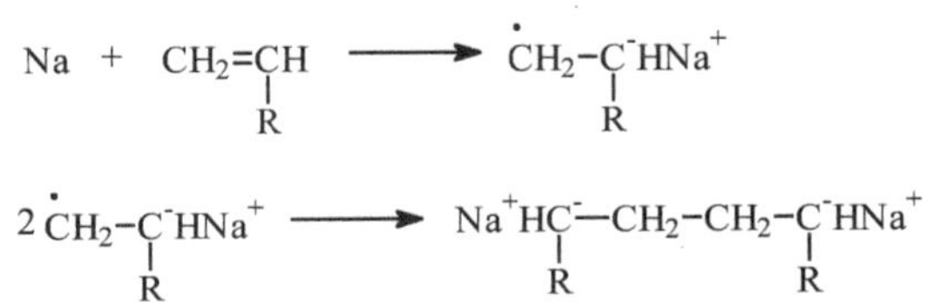

The evidence demonstrating such course of the reaction is the formation of corresponding dicarboxylic acids after an addition of carbon dioxide to the equimolar mixture of catalyst and monomer.

In some cases a mixed free-radical – anionic polymerization process is observed.

In a similar way proceeds the process of styrene polymerization, assisted by sodium naphthalene, formed easily in the reaction of sodium with naphthalene:

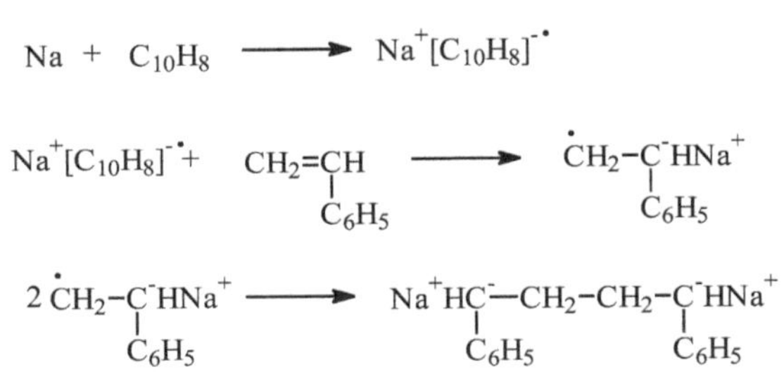

There exist also another methods of anionic polymerization, catalyzed by pyridine derivatives or by phosphines. An example of the latter reaction is the polymerization of maleic anhydride in the presence of triphenylphosphine, which is not homopolymerizing in the presence of radical initiators. The present reaction proceeds through the stage of ylide by using a free π electron pair.

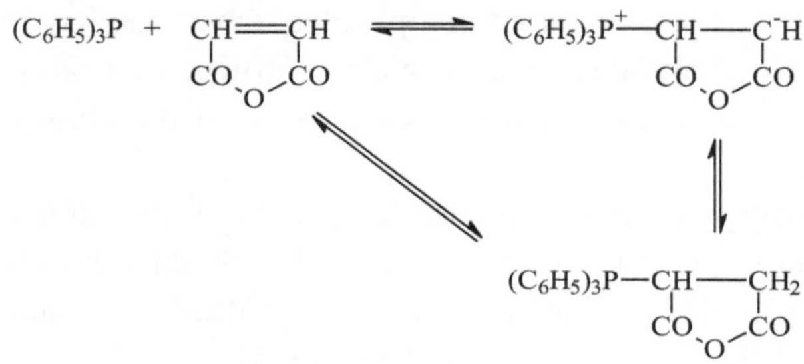

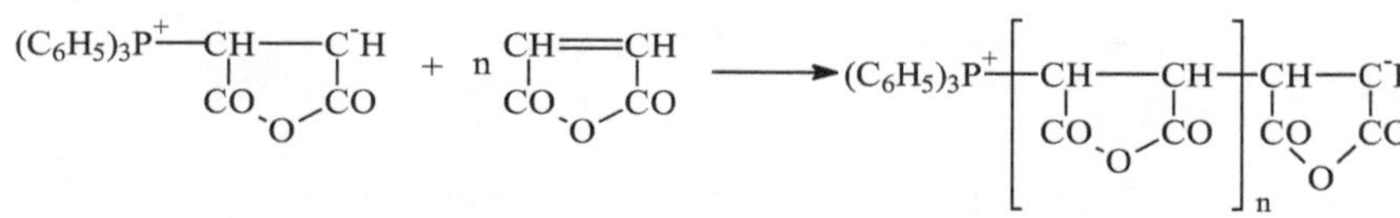

The activation energy of this reaction depends on the type of solvent used and amounts to 26.4 kJ/mol in acetic anhydride and to 39.8 kJ/mol in dimethylformamide.

The termination or the chain transfer reactions in anionic polymerization runs through:

- elimination of proton with formation of an unsaturated bond at the chain end,
- transfer of proton from the solvent molecule to monomer or to polymer,
- isomerization to an inactive ion,
- an irreversible reaction of the active carbanion center at the end of the chain with a monomer or a solvent molecule.

In a clean reaction environment the process of the termination or the chain transfer is rare. The formed polymer contains at the end of the chain carbanions, which can be, even after a prolonged storage, an active center, starting the polymerization process after introduction of a new portion of monomer.

8.2.3.3 "Living" Polymerization

The concept of "living" polymerization was introduced for the first time by M. Szwarc in 1956. According to his definition the "living" polymerization takes place if:

$$k_i > k_w \text{ and } k_z = 0$$

where: k_i, k_w, k_z – rate constants of initiation, growth and termination of polymer chain, respectively.

If these conditions are satisfied, the concentration of active centers is constant and equal to the initial concentration of nucleophilic agent (catalyst), the polymerization rate is constant. The average molecular weight (M) increases linearly with the increasing degree of conversion. The distribution of molecular weights is narrow and close to the Poisson distribution

$$\frac{\bar{M}_w}{\bar{M}_n} < 1.1$$

Introduction of the concept of living polymerization and its practical use was possible owing to the research done on the course of anionic polymerization of styrene and dienes. During the styrene polymerization, using sodium naphthalene as catalyst; its green color changes to red of styrene anion.

The red color is maintained even at 100 percent of conversion, and the addition of another portion of styrene or other polymerizing anionic monomer induces a further polymerization. In the living polymerization all macromolecules present in the system are terminated by active centers, which retain their activity during the polymerization, even after depletion of monomer. This creates a practical possibility of obtaining block copolymers by a successive introduction of different monomers into the system:

$$-[M_1]_{n-1} - \bar{M} + (m+1)M_2 \rightarrow -(M_1)_n - (M_2)_m - \bar{M}_2$$

The possibilities of obtaining block copolymers from "living" polymers are presented in Table 8.6.

TABLE 8.6 Block Copolymers Obtained from "Living" Polymers

No.	"Living" polymer	Second monomer
1	Polystyrene	isoprene
2	Polystyrene	α-methylstyrene
3	Polystyrene	1-vinyl naphthalene

TABLE 8.6 Continued

No.	"Living" polymer	Second monomer
4	Polystyrene	acrylonitrile
5	Polystyrene	methacrylonitryle
6	Polystyrene	methyl methacrylate
7	Polystyrene	methyl methacrylate
8	Polystyrene	2-vinylopyridine
9	Polystyrene	4-vinylopyridine
10	Polystyrene	ethylene oxide
11	Polystyrene	dimethylosiloxane
12	Polyisoprene	styrene
13	Polyisoprene	ethylene oxide
14	Polı(α – methylstyrene)	styrene
15	Poly(4-vinylopyridine)	methyl methacrylate
16	Poly(2-vinylopyridine)	styrene
17	Poly(4-vinylopyridine)	4-bromostyrene
18	Poly(methyl methacrylate)	butyl methacrylate
19	Poly(methyl methacrylate)	isopropyl acrylate
20	Poly(methyl methacrylate)	acrylonitrile

Some of them are used on industrial scale. The systems, which accomplish all the strict requirements of the "living" polymerization, include only the anionic polymerization of vinyl and cyclic monomers.

Carbanions derived from such monomers like styrene and butadiene, are not subject to any other reactions outside the propagation. During the anionic polymerization of methacrylates a chain transfer on the Esther-COOR group of monomer or polymer can take place. Therefore, a formation of living polymers in this system is difficult.

It turned out that in many cases one uses the term of "living" polymerization for processes that only partially fall within the definition given by Szwarc. Therefore, introduce the concept of the "pseudo-living" or "quasi-living" polymerization. These concepts apply when the termination and transfer of chain rate constants are equal to zero and the condition: $ki > kw$ is not satisfied, or the propagation reaction is reversible, or a reversible chain transfer to polymer takes place.

Living polymers can also react with the appropriate chemical reagents (XY) to give polymers containing terminal functional groups, such as macromeres:

$$—[M]_n—M^- + XY \rightarrow —(M)_n—MY + X^-$$

8.2.3.4 Star-Shaped Polymers

In the last few years, the anionic polymerization obtained by living polymers was used for the synthesis of a new class of polymers called star-shaped polymers.

The synthesis of such polymers is represented schematically in the following examples:

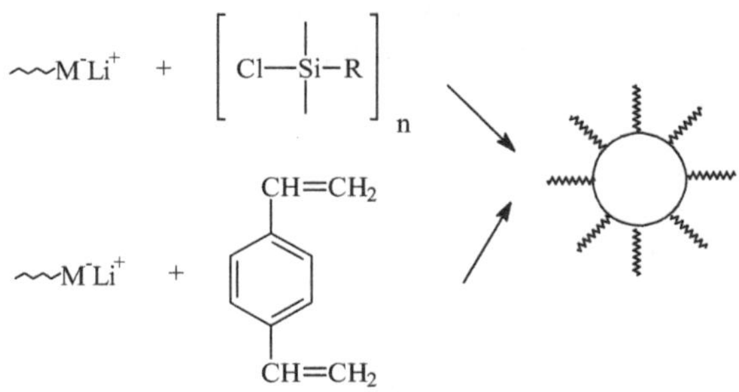

The first type synthesis allows a better control of the arm lengths, which exhibit similar average molecular weight.

In the case of the application of p-divinylbenzene for this purpose the core macromolecule is a cross-linked polymer. Despite that the presence of long, linear arms, derived from the "living" polymer makes the macromolecule soluble. It turned out that not all active centers in the cross-linked core of poly divinylbenzene are accessible in the same way. For this reason the initiation of the reaction proceeds slowly and leads to the star-shaped polymer with arms of different lengths. An important advantage of this method is the possibility to obtain the star block copolymers. An example of such a compound is a copolymer containing up to 30 arms. Each of them is a block copolymer of styrene (M = 2700–30,000) and ethylene oxide (M = 3000–6000). In addition to this information it should be added that recently the Japanese researchers elaborated a method of synthesis of star polymers based on the Grignard compounds:

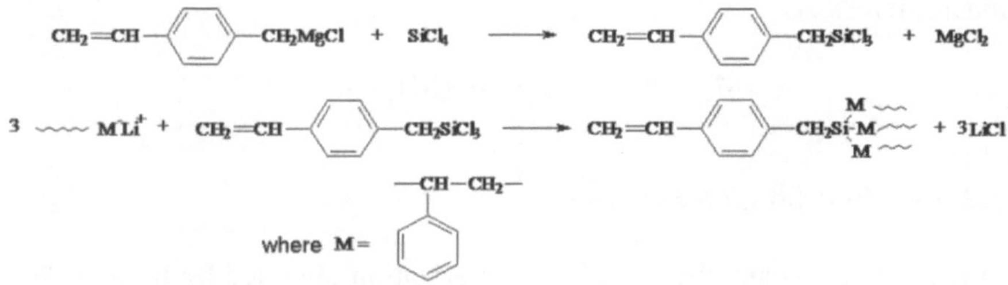

The presence of free vinyl groups in these products was used to obtain grafted copolymers. Grafted copolymer is a grafted polymer formed of more than one type of monomer.

In 1988 J. Roovers with coworkers have synthesized star-shaped polymers with a high packing. For that they have used the hydrosilylated poly (1,2-butadiene) with a low molecular weight:

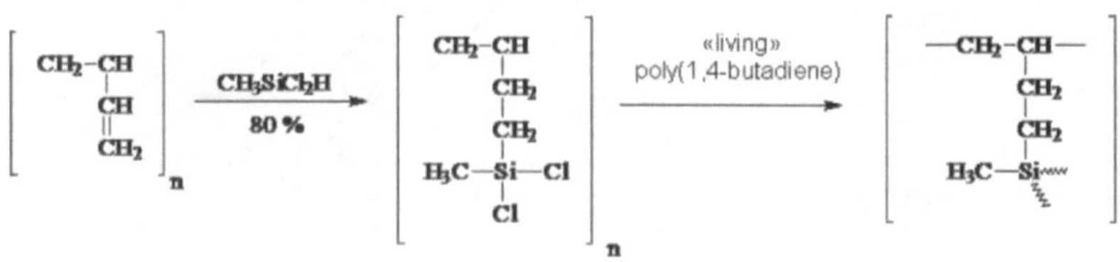

The obtained in this reaction polymer had about 270 arms and the average molecular weight of arms was of about 40,000. The average molecular weight of the entire macromolecule was of 11,000,000.

Strictly speaking the obtained reaction product is a grafted copolymer. However, due to the small size of the core (average molecular weight of used poly (1,2-butadiene is of about 9000) in comparison with the whole macromolecule (M = 11,000,000) and the large number of chains the synthesized macromolecule s a model of a star-shaped molecule.

The above-described star-shaped polymers are soluble and can serve as models to verify the theory describing conformations of macromolecules in solutions with a high degree of branching.

As mentioned previously, the anionic polymerization of methacrylates is accompanied by the chain transfer reaction. A more accurate understanding of this process has enabled the development of anionic polymerization with a group transfer group (Group Transfer Polymerization, GTP), which can be illustrated by the following example:

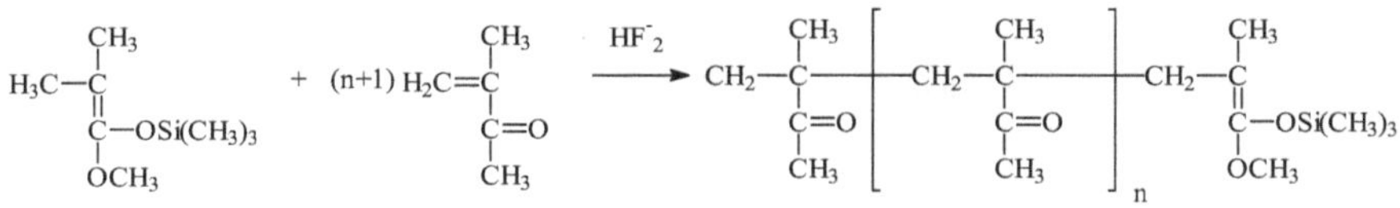

Basing on the modeling studies, which have shown that in the process of propagation the initial dissociation of O-Si bonds and the formation of an ion active site are not required, the following mechanism of this reaction is assumed:

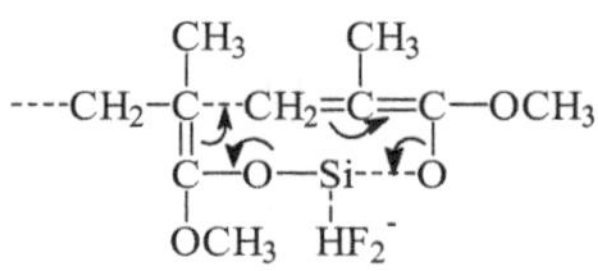

The polymerization process with the group transfer was used for the synthesis of ladder polymers:

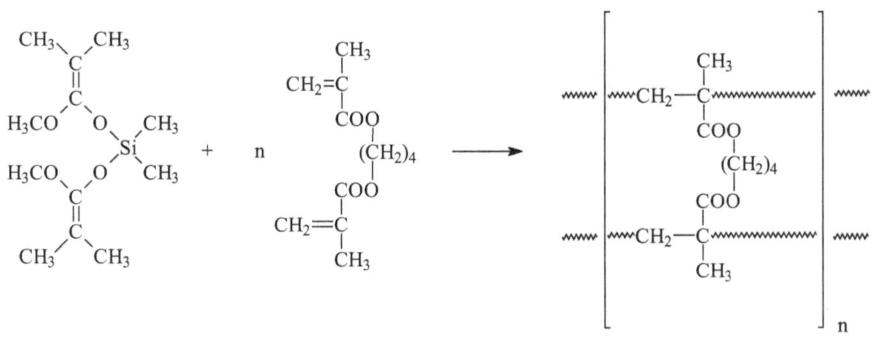

Introduction of ethylene or propylene to the dimethacrylate system gives rise to formation of a cross-linked core (P_n). The resulting product has structure of a star-shaped ladder polymer:

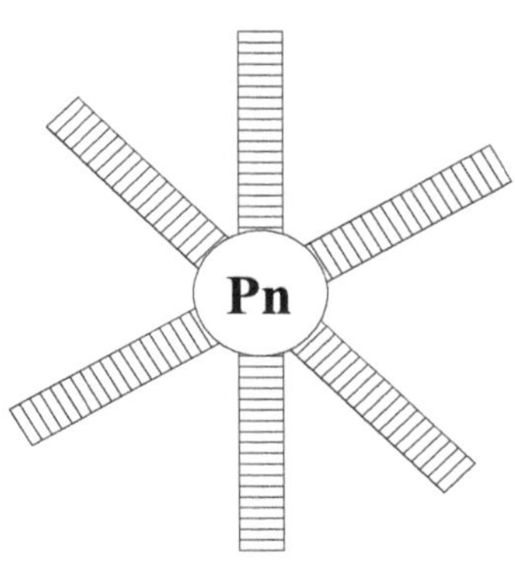

From the above considerations it follows that the possibilities of application of anionic polymerization are large with a real perspective of its use on the industrial scale.

8.2.3.5 Stereochemistry of Anionic Polymerization

The spatial structure of the products of anionic polymerization depends on the type of catalyst used and to a large extent on the degree of counterion association. An important role plays here the type of solvent used and the temperature of process conduction.

When using anionic catalysts in a homogenous environment at low temperatures the polar solvents favor the formation of syndiotactic structure, while the non-polar solvents privilege the formation of isotactic polymer. It is believed that the mechanism of polymerization in this case is connected with the formation of ion pairs.

Anion located at the end of the chain has a flat structure and is stabilized by resonance. The monomer molecule can come close to the active center from the front, giving the syndiotactic structure of polymer, or from behind, creating an isotactic structure. It depends on the degree of associative links of carbanion with the counterion and on the size of substituents in the polymer molecule.

In particular, of importance is the use of anionic polymerization to obtain polydienes. Depending on the type of used catalyst and on operating conditions one can obtain in this way polydienes of different structures, including polymers with overwhelming "*cis*-1, 4" structure, similar to that of natural rubber. The influence of the used catalyst on the course of anionic polymerization of butadiene and isoprene and the structure of formed polymers are given in Table 8.7.

It is worthy to note that the first synthetic polybutadiene, produced on a large scale, was obtained by the anionic polymerization method in the presence of metallic sodium. From the names of butadiene and natrium (sodium) it took the trade name "Buna."

8.2.3.6 Effect of Solvent on the Course of Anionic Polymerization

The course of anionic polymerization in solution depends on the type of solvent used and particularly on its polarity. Polar solvents solvate easier

TABLE 8.7 Effect of Catalyst in Anionic Polymerization of Butadiene and Isoprene on the Structure of the Resulting Polymer Chain

No.	Catalyst	Monomer	Solvent	Chain microstructure [% mol]			
				cis-1,4	*trans*-1,4	1,2	3,4
1	C_2H_5Li	butadiene	n-hexane	43	50	7	-
2	C_2H_5Li	butadiene	tetrahydrofuran	0	9	91	-
3	Na	butadiene	-	10	25	65	-
4	$C_{10}H_8Na$	butadiene	tetrahydrofuran	0	9	91	-
5	K	butadiene	-	15	40	45	-
6	$C_{10}H_8K$	butadiene	tetrahydrofuran	0	17	83	-
7	Rb	butadiene	-	7	31	62	-
8	Rb	butadiene	tetrahydrofuran	0	25	75	-
9	Cs	butadiene	-	6	35	59	-
10	n-C_4H_9Li	isoprene	n-heptane	93	0	0	7
11	n-C_4H_9Li	isoprene	tetrahydrofuran	0	30	16	54
12	$C_{10}H_8Na$	isoprene	n-heptane	0	47	8	45
13	$C_{10}H_8Na$	isoprene	tetrahydrofuran	0	38	13	49
14	$C_{10}H_8CH_2K$	isoprene	n-heptane	0	52	10	38
15	$C_{10}H_8CH_2K$	isoprene	tetrahydrofuran	0	43	17	40
16	Rb	isoprene	-	5	47	10	38
17	Cs	isoprene	-	4	51	8	37

the catalyst. It was checked carefully in the case of anionic polymerization initiated by lithium or by litorganic compounds. In addition, the larger value of dielectric constant of polar solvents facilitates dissociation of ionic bonds of carbon-lithium anion, present in the catalyst. The polarity of the used solvent affects the chain termination rate as well as the structure of the resulting polymer. Using of nonpolar solvents causes formation of oligomeric side products during the anionic polymerization. A very useful solvent to carry out the anionic polymerization is tetrahydrofuran. It was also demonstrated that the polydispersity of resulting polymer decreases with the increasing size of counterion.

The solvent polarity affects significantly the structure of polymer. Isoprene polymerized by anionic polymerization method in hydrocarbons forms 70–90% of *cis*-1,4-variety of polyisoprene, whereas in tetrahydrofuran it yields mainly the poly (3,4-isoprene). A similar effect was observed during the polymerization of 1,3-butadiene. During the polymerization in hydrocarbons mainly poly (1,4-butadiene) is obtained, whereas in tetrahydrofuran the main product is poly (1,2-butadiene). These drastic changes occur even when a small amount of polar solvent such as tetrahydrofuran is added to the hydrocarbon reaction environment in which the anionic polymerization runs. Other, ether or anionic co-solvents do not show such large influence on the reaction like does tetrahydrofuran. The ratio of the variety *cis* to the variety of *trans* polybutadiene obtained in hydrocarbon environment is usually of 2–3 and is maintained even at high content of vinyl groups originating from poly (1,2-butadiene).

8.2.4 COORDINATION POLYMERIZATION

8.2.4.1 Characteristics of Coordination Polymerization Process

The coordination polymerization is a new type of synthesis of macromolecular compounds to obtain polymers with a regular spatial structure. It was preceded by the basic researchs presented in the field of synthesis and application of organometallic compounds. The studies have led in 1953 to the discovery of a new complexing compound consisting of triethylaluminium and titanium trichloride.

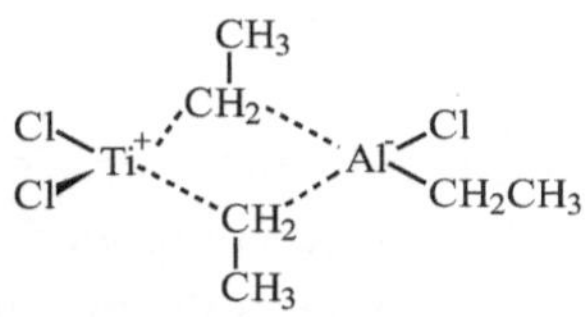

It turned out that Ziegler obtained a complex catalyzing polymerization of ethylene at normal pressure.

At the same time an Italian scholar G. Natta received for the first time the isotactic polypropylene by polymerization of propylene on the catalyst complex composed of titanium trichloride and triethylaluminium.

These discoveries have revolutionized the research on the polymerization process. The catalysts based on organometallic complexes are called Ziegler–Natta catalysts.

The importance of discovery is the best demonstrated by the launching in Ferrara (Italy) of industrial production of isotactic polypropylene already in 1957 by an Italian company Montecatini.

But it turned out that the various organometallic complexes vary in effectiveness and stereospecificity in catalytic activities in the polymerization process and in different behavior for different monomers. An example is the already discussed complex: titanium trichloride with aluminum triethyl. It is a very good catalyst for the synthesis of polyethylene, but is not suitable for polymerization of propylene due to the small stereospecificity. The quantity of produced isotactic polypropylene is then twice less than the one obtained using the Natta complex of titanium trichloride with aluminum triethyl.

In further studies it appeared that the activity of Ziegler–Natta catalysts depends not only on the chemical structure of the complex molecule but also on its crystal structure, and therefore on the way of its preparation.

For example, the titanium trichloride can be prepared by reducing titanium tetrachloride with hydrogen, metallic titanium or with aluminum. It can be activated additionally by using physical or chemical methods. The greatest activity, without additional treatments, shows δ trichloride of titanium. The characteristics of titanium trichloride as a catalyst for propylene polymerization are presented in Table 8.8.

Currently a large number of Ziegler-Natta catalysts is known. They are produced by the reaction of transition metal halides derivatives (Tr, Cr, Th, Zr, V, Nb, Ta, Mo, Co) with organic derivatives of metals from groups I, II

TABLE 8.8 Characteristics of Titanium Trichloride as a Catalyst for Propylene Polymerization

No.	Designation of the form	Preparation	Stereospecificity (amount of isotactic polypropylene) [%]
1	α (alfa)	Reduction of TiC14 by hydrogen or titanium at 673 K	80–90
2	β (beta)	Reduction of TiC14 with hydrogen in an electric arc or decomposition of CH_3TiCl_3 in hydrocarbons	40–50
3	γ (gamma)	Heating of beta form for 2–3 hours or decomposition of CH_3TiCl_3 without solvent	80–90
4	δ (delta)	Reduction of $TiCl_4$ with metallic aluminum (the product contains still $AlCl_3$)	>90

After F. Andreas, K. Grobe, Propylene chemistry (Chemia propylenu, in Polish), WNT, Warszawa 1974, p. 372.

and III (Li, Mg, Al), playing the role of reducing agents. The products of these reactions have the structure of complex compounds. The influence of the used metal halide on the catalytic activity of the complex compound with aluminum triethyl during the propylene polymerization is presented in Table 8.9.

The strong interest of industry for Ziegler – Natta catalysts led to the discovery of new types of bonds of diverse, often much more active and applicable for different purposes forms. Currently catalytic systems operating in a homogenous as well as a heterogeneous environment system are known. They allow to make the coordination polymerization in solution or in a fluidized phase.

Several recent studies have shown that the activity of the classical Ziegler-Natta catalyst can be increased by introducing into the system additional chemical compounds, most frequently the magnesium chloride. Sometimes similar effect is obtained by the addition of ethyl benzoate. By the fact it was shown that the time of course of the polymerization

TABLE 8.9 Influence of Metal Halides on the Catalytic Activity of the Complex with Aluminum Triethyl

No.	Metal halide	Chemical formula	Yield of isotactic poly- propylene [%]
1	Titanium dichloride	$TiCl_2$	80–90
2	α-titanium trichloride	$TiCl_3$	85–90
3	β-titanium trichloride	$TiCl_3$	40–50
4	Titanium tribromide	$TiBr_3$	44
5	Titanium triiodide	TiJ_3	10
6	Zirconium dichloride	$ZrCl_2$	55
7	Vanadium trichloride	VCl_3	70–75
8	Chromium chloride	$CrCl_3$	36
9	Titanium tetrachloride	$TiCl_4$	45–50
10	Titanium tetrabromide	$TiBr_4$	40–42
11	Titanium tetraiodide	TiJ_4	46
12	Titanium tetraalkoxy	$Ti(OR)_4$	traces
13	Zirconium tetrachloride	$ZrCl_4$	52
14	Vanadium tetrachloride	VCl_4	48
15	Vanadium chloride	$VOCl_3$	32

reaction has no apparent effect on the average rate of isotacticity and on the molecular weight.

The influence of the catalyst, operating in a heterogeneous system, on the rate of isotacticity and efficiency of the reaction is given in Table 8.10.

Some new types of Ziegler-Natta catalysts such as the complex of tetrabenzyltitan with methyl aluminosiloxane allow obtaining the syndiotactic polymers with a good yield.

A characteristic feature of the coordination polymerization, which distinguishes it from other processes of this type, is the fact that each attached molecule is at the beginning coordinated by a metal atom of the catalyst, for example, the atom of titanium. In this way it gets a well-determined orientation with respect to the growing chain. It facilitates also the formation of polymers with stereoregular structure. The addition of the monomer molecules to the growing chain of polymer takes place at its base and not at the final fragment of the chain, as it was the case in the classical free radical or ionic polymerization. The growth of chain of such a polymer can be compared to the hair growth.

TABLE 8.10 Effect of Catalyst on the Course of the Coordination Polymerization of Styrene and Propylene at Temperature of 313 K

No.	Catalyst	Cocatalyst	Styrene polymerization		Propylene polymerization	
			Activity [mol/ (Ti*h)]	Isotatic rate [%]	Activity [mol/ (Ti*h)]	Isotatic rate [%]
1	$Ti(OC_4H_9)_4$/ Mg(OH)Cl	$Al(CH_3)_3$	0.4[a]	~100	100[b]	12
2	$TiCl_3/MgCl_2$	$Al(C_2H_5)_3$	25[c]	96,7	14,000[d]	0.9
3	$TiCl_3$ (Solvay type)	$Cp_2Ti(CH_3)_2$	30[c]	99,2	70[d]	99.2
4	$TiCl_4$/ $COOC_2H_5$/ $MgCl_2$	$Al(C_2H_5)_3$	190[c]	93,4	8360[d]	59.7

Explanations: C_p – cyklopentadiene;

[a] 40 cm³ of heptane and 10 cm³ of styrene, 24 H;

[b] 30 cm³ of heptane and propylene at a pressure of 2 MPa, 24 H;

[c] 40 cm³ of heptane and 10 cm³ of styrene, 0.5 H;

[d] 100 cm³ of heptane and propylene at a pressure of 0.1 MPa, 2 H;

After K. Soga et al., Macromol. Chem. Rapid Commun. 11, 229 (1990).

8.2.4.2 The Mechanism of Coordination Polymerization

The mechanism of the coordination polymerization may vary depending on the used coordinate catalyst. If for the synthesis of a catalytic complex a metal halide is used, in which the metal atom is at a higher degree of oxidation, then the coordinate – radical polymerization takes place. This is possible due to the possibility of free radical formation by the reaction of reduction.

$$TiCl_4 + AlR_3 \longrightarrow RTiCl_3 + AlR_2Cl$$

$$RTiCl_3 \longrightarrow R^{\bullet} + TiCl_3$$

$$R^{\bullet} + nM \longrightarrow \text{polymer}$$

The formed free radicals "R •" initiate the process of polymerization. Its efficiency depends on the type of monomer used. Polymers obtained by the coordinate – radical polymerization exhibit a less regular structure than those obtained by the coordinate – anionic polymerization. This is due to the presence of free radicals in the system initiating the polymerization of monomers without participation of active centers of the coordinate catalyst.

If in the formation of the Ziegler-Natta catalyst complex is involved a component, which is an electron acceptor than the acidity of such a system is sufficient to initiate the coordinate – cationic polymerization. In the reaction of such kind involved are monomers capable of resonance stabilization of carbonion cation, for example, vinyl ethers. The reaction is probably initiated by the proton produced by reaction of an acid fragment of catalyst with water, which in turn joins the monomer molecule, coordinated by the metal atom.

Due to the nature of the initiation the resulting polymer does not contain alkyl groups, originating from the catalyst.

The polymerization on Ziegler-Natta catalysts most frequently proceeds along the coordinate – anionic mechanism in active sites located on the surface of catalyst. It is believed that the organometallic component of catalyst activates a specific place of the polymer chain, alkylating the transition metal atom located on the surface. The literature describes a number of possible courses of such a reaction. Currently as the most possible for this reaction are considered two mechanisms:

- bimetallic mechanism, and
- monometallic mechanism.

According to the bimetallic mechanism of monomer polymerization on the catalytic titanium-aluminum system, it is assumed that at the first stage a complex between the π bond of monomer and the titanium atom is formed, which is connected with the aluminum atom by alkyl groups. It is followed by polarization of the transition metal-alkyl group bond and the formation in the transition state of a hexacomponent ring, incorporating the monomer molecule:

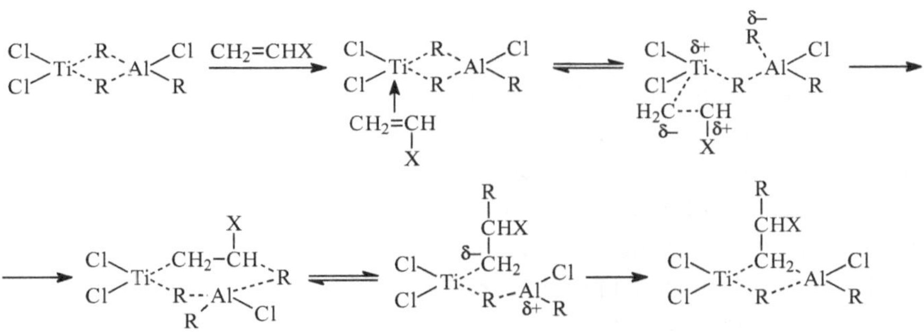

The reaction proceeds further in analogic way with the next monomer molecules, up to the formation of polymer.

The monometallic mechanism, developed by P. Cossee, presupposes the existence of an active center, which is the transition metal atom with a coordination number of 5. This center is activated by replacement of the alkyl group by a metaloorganic cocatalyst (e.g., trialkylaluminium).

The titanium or other transition metal atom coordinates the monomer molecule, which then joins. At the same time the location of vacancy in octahedron moves. The polymer chain moves back, so that the free space (vacancy) takes its starting position. This last step is necessary to preserve the stereoregularity of polymer.

The reaction of coordination polymerization by the monometallic mechanism can be expressed as follows:

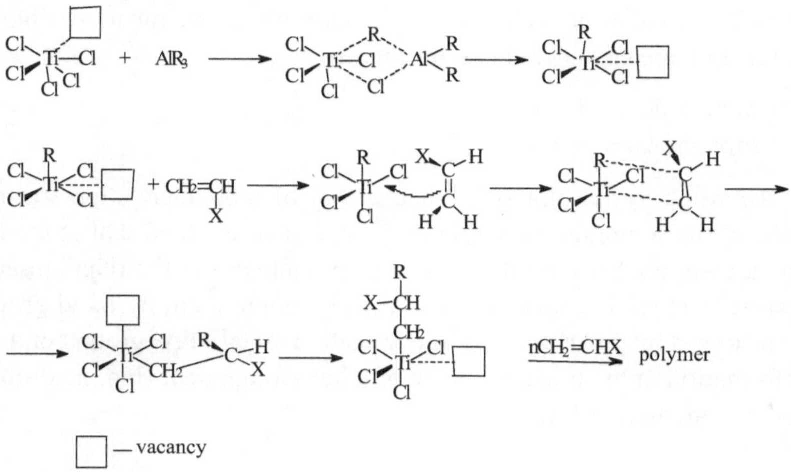

For both mechanisms of the coordination polymerization the polymer chain grows away from the surface of the catalyst as a result of attaching the next, pre-coordinated monomer molecules. The alkyl group, coming from trialkylaluminium or another metaloorganic cocatalyst becomes to be the final group of the polymer chain.

In the Ziegler – Natta polymerization takes also place the transfer reactions of the chain activity as demonstrated by the fact that the molecular weight reaches a relatively constant value in a short time after the initiation of reaction. These reactions include:

– transfer to the monomer molecule;

$$Cat{-}CH_2{-}\underset{}{\overset{X}{CH}}{-}\left[CH_2{-}\overset{X}{CH}\right]_n{-}R + CH_2{=}CHX \longrightarrow Cat{-}CH_2CH_2X + CH_2{=}\overset{X}{C}{-}\left[CH_2{-}\overset{X}{CH}\right]_n{-}R$$

– transfer to the alkyl group of organometallic compound;

$$Cat{-}CH_2{-}\overset{X}{CH}{-}\left[CH_2{-}\overset{X}{CH}\right]_n{-}R + AlR_3 \longrightarrow Cat{-}R + R_2Al{-}CH_2{-}\overset{X}{CH}{-}\left[CH_2{-}\overset{X}{CH}\right]_n{-}R$$

– transfer of a proton, associated with the termination of the chain growth;

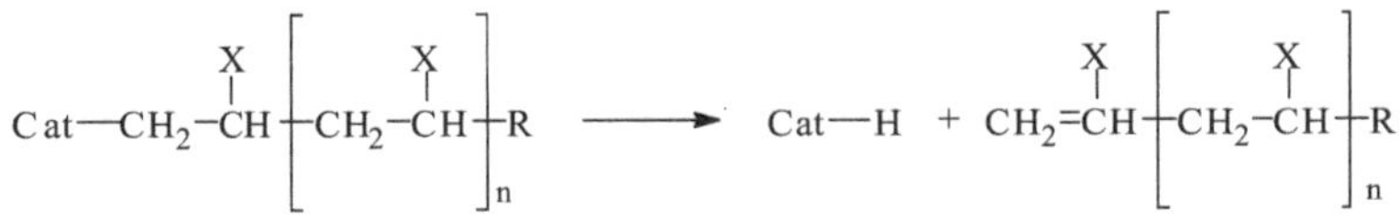

The presence of methyl groups and unsaturated bonds in polymer, confirmed by the infrared spectroscopy demonstrates the ability of such a course of reaction.

Introduction to the system of compounds active hydrogen atoms causes the termination of the chain growth. It proves the coordinate – anionic mechanism of this reaction.

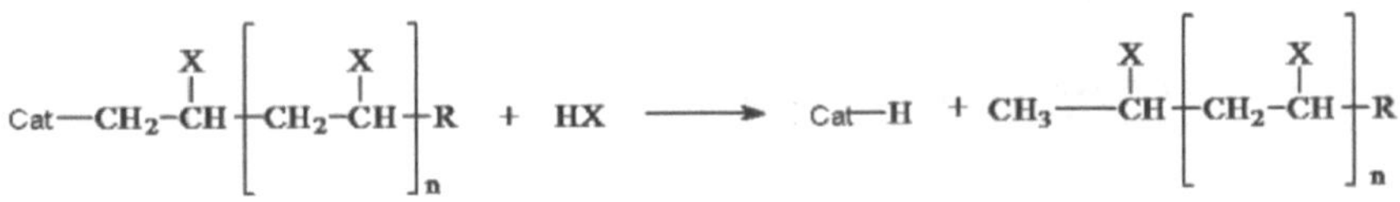

Another evidence confirming the coordinate – anionic mechanism of reaction is the relatively long life of growing chains. It makes possible obtaining of block copolymers by adding a second monomer after the reaction of the first. The process of polymerization in the presence of Ziegler-Natta catalysts is not affected by the typical radical transfer reactions. It means that the process is not a free radical one.

8.2.4.3 Coordination Polymerization of Dienes

As a result of polymerization of 1,3-butadiene and its derivatives one obtains synthetic rubbers (elastomers) whose properties depend on the structure of the formed products. The stereospecific Ziegler-Natta catalysts offer new opportunities for the synthesis of rubbers with a defined structure. It turned out that by choosing an appropriate catalyst one can be obtain the following polymer structures: "cis-1.4," "trans-1.4," isotactic "1.2," and syndiotactic "1.2," all in a relatively pure form. The influence of the catalyst structure on the stereospecific polymerization of butadiene is shown in Table 8.11.

TABLE 8.11 Effect of Catalyst on the Stereospecific Polymerization of Butadiene

No.	Catalyst	Yield [%]	Polymer structure	References
1	R_3Al-VCl_4	97–98	trans-1,4	J. Polym. Sci. 48, 219 (1960)
2	R_3Al-VCl_3	99	trans-1,4	–
3	R_3Al-$VOCl_3$	97–98	trans-1,4	–
4	di(π-cyclooctadiene) nickel–HJ	~100	trans-1,4	J. Polym. Sci. B. 5, 785 (1967)
5	R_3Al-TiJ_4	93–94	cis-1,4	J. Polym. Sci. 48, 219 (1960)
6	R_3Al-$CoCl_2$•pyridine	90–97	cis-1,4	Dokł. Akad. SSSR. 135,847 (1960)
7	R_2AlCl-$CoCl_2$	96–97	cis-1,4	J. Polym. Sci. 48, 219 (1960)
8	R_3Al-$Ti(OC_6H_9)_4$	90–100	1,2	J. Polym. Sci. 38, 45 (1959)
9	$Al(C_2H_5)_3$-$V(OCOCH_3)_3$	~90	syndiotactic 1,2	J. Polym. Sci. 48, 219 (1960)
10	$Al(C_2H_5)_3$-$Cr(C_6H_5CN)_3$ Al/Cr = 2	~100	syndiotactic 1,2	Macromol. Chem. 77, 114 (1964)
11	$Al(C_2H_5)_3$-$Cr(C_6H_5CN)_3$ Al/Cr = 10	~100	isotactic –1,2	Macromol. Chem. 77, 126 (1964)

The structure of the emerging polyene depends on the structure and the crystal form of the catalyst, the valence of transition metal and the configuration of the monomer molecule (*cis* and *trans*).

Depending on the used catalyst one or both diene double bonds are coordinated. Coordination of one of two bonds only leads to the polymerization such as "1.2." The coordination of both bonds provides polymer with "1.4" structure. According to another theory the coordination of π-allyl structure takes place and the polymer structure is determined by the movement direction of the approaching monomer molecule. If the monomer molecule is approaching the center of an active catalyst from the CH_2-titanium bond side the formed polymer has the "1.4" structure type. When the monomer molecule is approaching from the CH_2-titanium bond then the polymerization is of "1.2" type.

A simplified diagram is as follows:

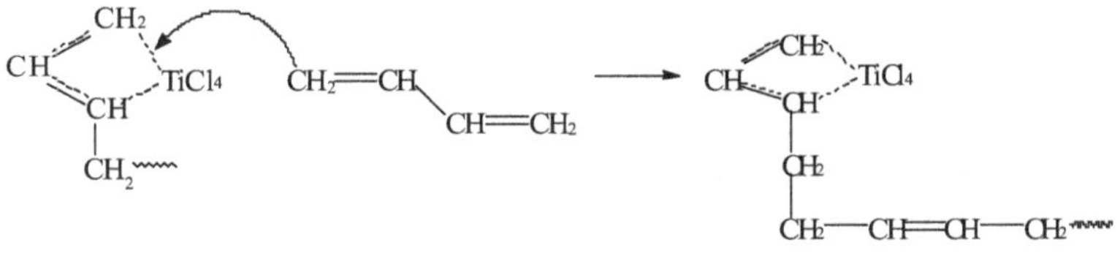

structure 1,4

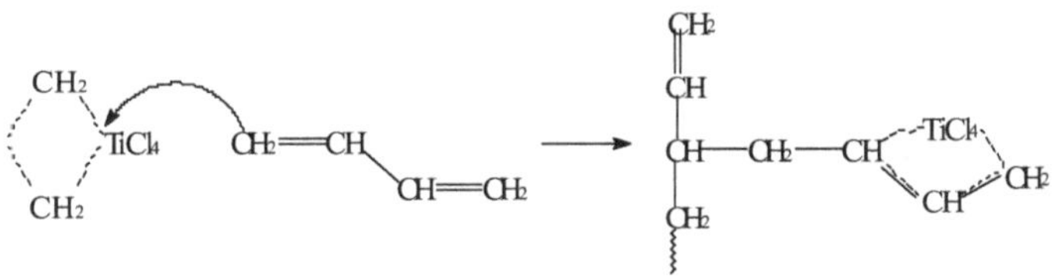

The coordinate catalysts are also capable of coordination polymerization of dienes with isolated bonds. A typical example of this reaction is the polymerization of 1.5-hexadiene in the presence of $TiCl_4$-Al (C_2H_5) complex. One obtains in this case polymer with repetitive cyclopentanone rings and interconnected via methyl bridges of *cis* form in positions 1 and 3:

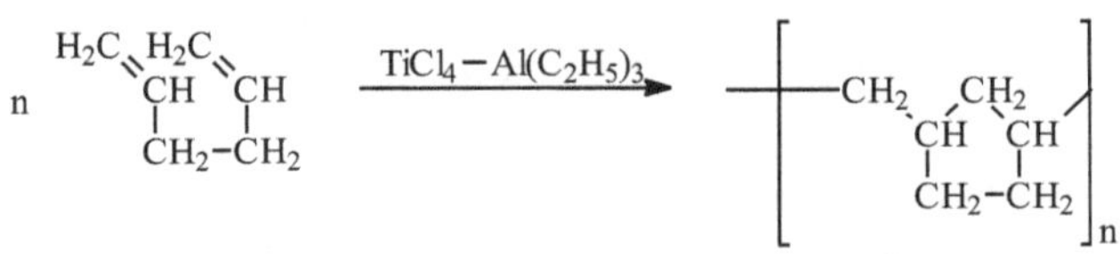

If soluble catalysts are used then the "1.2" polymerization can be carried out. Similarly, from the 2,6-dimethyl-1,6-heptadiene is formed a polymer with cyclohexane rings:

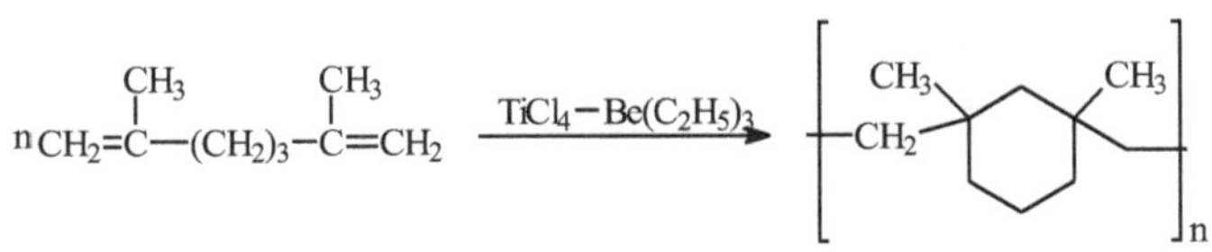

Another example is the polymerization of 1,5-cyclooctadiene leading to the formation of a polymer with a bicyclic repeat unit:

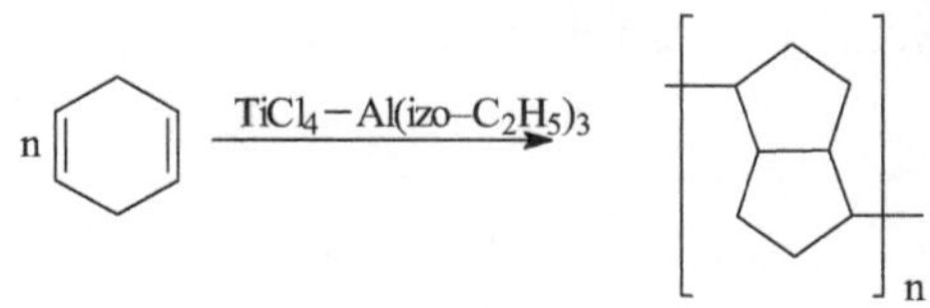

The coordinate catalysts enable the synthesis on an industrial scale of isoprene rubbers, which in terms of composition, structure and physico-chemical properties are similar to natural rubber. Also with these catalysts one can obtain the *cis*-butadiene rubbers of a regular structure, known for valuable utilization performances.

The difference in the structure explains several properties of these polymers. Atactic polymers are formed of poorly packed molecules and do not form crystallites. Thus they exhibit a lower static strength. The syndiotactic and isotactic polymers crystallize relatively easily and have a greater static strength and density than the atactic. The formation of varieties of syndiotactic polymers is more likely than of the isotactic ones. Syndiotactic molecule is characterized by a spiral conformation, because it provides the regularity of the recurrence of the mers, containing the tertiary carbon atoms.

The data presented in Table 8.12. confirm these conclusions.

A series of catalytic systems was also developed, in which the mechanism of action is similar to Ziegler-Natta catalysts. These are the oxides of nickel, cobalt, vanadium and molybdenum deposited on the surface of aluminum and of chromium oxide on silica gel. Such catalysts contain different promoters in the form of metal alkyl. They can be also used for the synthesis of polydienes.

TABLE 8.12 Crystallizable Stereoisomers of Polybutadiene-1,4-Diene

Stereoizomer	Melting temperature [K]	Glass transition temperature [K]	Crystal repeat unit [nm]	Density [g/cm^3]
1,2-isotactic	393	263	65.0	0.96
1,2-syndio isotactic	427	-	51.4	0.96
trans-1,4	408	190	49.0	1.01
cis-1,4	263–274	163	-	-

After B.A. Dogadkin: Elastomer Chemistry (Chemia elastomerów, in Polish), WNT, Warszawa 1976, p.88.

8.2.4.4 Metallocene Catalysts

The name of metallocenes is derived from ferrocene, which is a complex compound of iron with dicyclopentadiene. The metallocene complexes are new types of Ziegler-Natta catalysts. The general formula of these catalysts is identified as Cp_2MeCl_2, where Cp denotes the cyclopentadiene ring or its derivatives, Me is metal atom, which the most often is zirconium (Zr), hafnium (Hf), titanium (Ti), scandium (Sc), thorium (Th) or another rare earth element.

Some examples of chemical structures of metallocene catalysts are shown below:

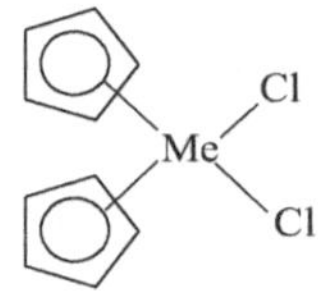

General formula

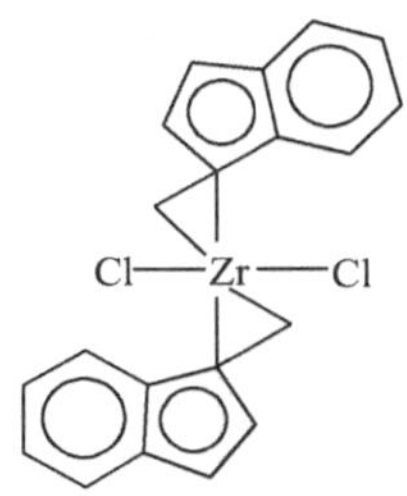

diindeno dichloro zirconium.

The activity of these catalysts increases after the addition of a cocatalyst, which is an oligomeric methylaluminoxane:

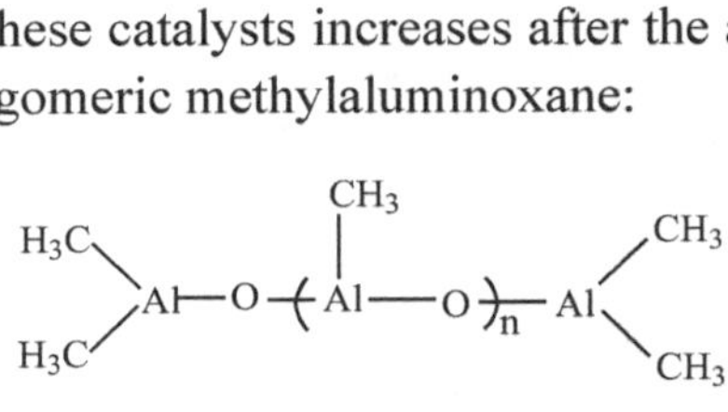

where n = 2 – 20.

Often this compound is presented in a simpler form:

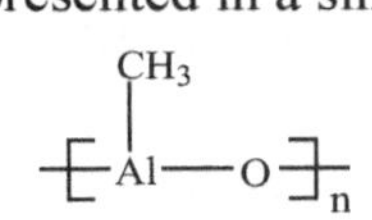

During the reaction the metallocene catalyst molecule with methylaluminoxane a displacement of the chlorine atom to the molecule of cocatalyst and of methyl group to the metal atom takes place with the formation of an active center $[Cp_2Me^{+} CH_3]^{\cdot}$.

The reaction proceeds according to the scheme:

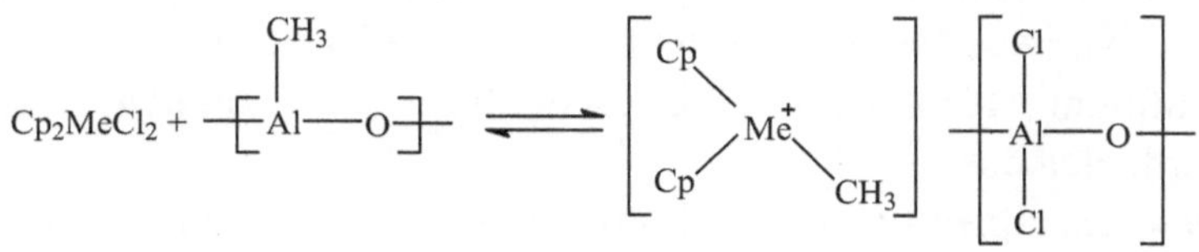

To the formed catalytic center is attached the monomer molecule through π electrons of double C=C bond, being properly oriented at the same time. As a result of successive attachments of monomer molecules one gets an increase of the chain, resulting in the formation of an isotactic polymer.

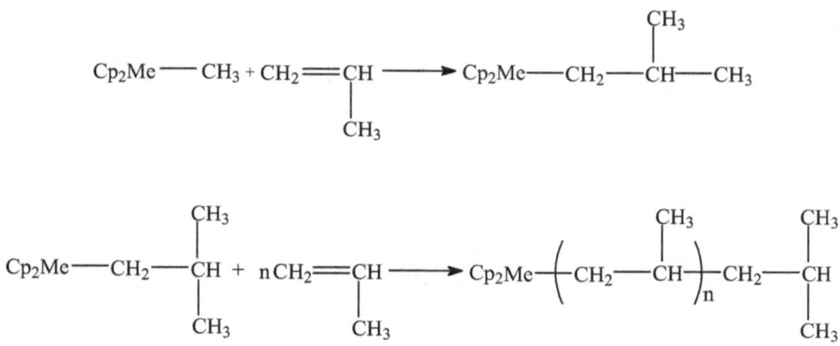

The chain termination reaction proceeds through β-elimination with formation of a double bond at the chain end or by the reaction with a hydrogen molecule.

8.2.4.5 Coordination Catalysts Based on Metal Oxides

Oxides of some transition metals, deposited on well – shredded material such as diatomaceous earth, clay, alumina, charcoal or silicon dioxide cause polymerization of ethylene and of some other vinyl monomers. The mechanism of this process is probably very similar to the mechanism of Ziegler-Natta catalysis. The initiation of the polymerization reaction takes place in the active centers of catalyst in which are placed metal atoms such as chromium, molybdenum, vanadium, nickel, cobalt, niobium, tantalum, tungsten and titanium.

Catalysts of this type are obtained by two methods. In the first, the medium is saturated with metal ions, and then is heated in the presence of air to a high temperature in order to obtain a metal oxide. The second method consists on the deposition on support of a metal oxide with a higher degree of oxidation, such as, for example, anhydride chromic acid (CrO_3) and its subsequent reduction by hydrogen or by carbon monoxide.

The type of the reductive agent used and the conditions under which this process is realized influence significantly the subsequent activity of the catalyst. Through the reduction of deposited chromium trioxide by hydrogen one obtains a green colored catalyst (the color comes from the formed Cr_2O_3), whereas in the case of reduction with the carbon monoxide the catalyst takes a blue color. In the case of transmission of excess carbon monoxide over the catalyst it changes its color to violet and becomes inactive. This is due to the absorption of carbon monoxide on the catalyst surface and the blocking of active centers.

The catalysts based on metal oxides are susceptible to poisoning by compounds such as water, oxygen and acetylene.

The catalyst activation affects not only its activity but influences also the molecular weight of the emerging later polymer. For example, the complex CrO_3 with diatomaceous earth, activated at the temperature of 500°C (773 K), causes formation of polyethylene with an average molecular weight 3.5 times greater than the product obtained on the catalyst from the same starting material activated at the temperature of 850°C (1123 K). The average molecular weight of the produced polymer depends also on the content of metal oxide such as, for example, chromium oxide on the support.

When the chromium concentrations is reduced from 0.75% to 0.001% the average molecular weight of formed polymer increases by 40%. This is due to the increase in the number of ethylene molecules per one active center of the catalyst.

The monomer polymerization on catalysts always runs in three stages:

- adsorption of monomer (initiation),
- reaction on the surface (chain growth), and
- desorption (termination of the chain).

The course of polymerization of ethylene on chromium catalyst (Cat–Cr) can be presented schematically as, for example,

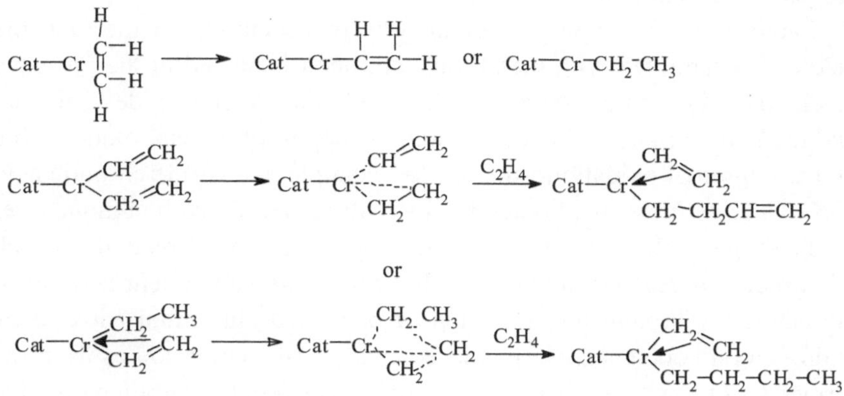

The chain termination reaction is presented as follows:

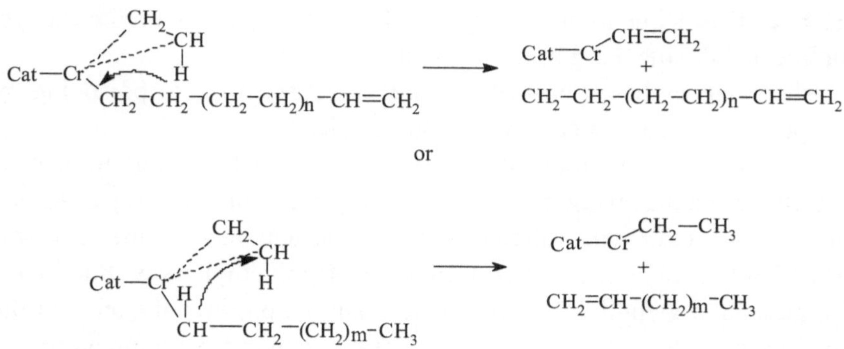

From the above considerations it follows that the center of active catalyst, based on chromium oxde, not only participates in the growth of the polymer chain, but affects also the orientation of monomer molecules. The catalyst operates in a stereospecific way and therefore facilitates the formation of tactic polymers.

One of possible ways for manufacturing polyethylene on an industrial scale is the Philips method, introduced in 1957. The catalyst used for the reaction is the chromium trioxide (CrO_3) disposed on a support composed of a mixture of silica and alumina. In this method a 5–7% solution of ethylene in selected hydrocarbons is prepared. Then it is heated to 420 K under a pressure of 3.5 MPa. Under these conditions ethylene is polymerized virtually in a single cycle. The solvents used, which may be aliphatic or aromatic hydrocarbons, allow to control the molecular mass of produced polyethylene as well as the reaction rate. It was found at the same time that the rate of polymerization increases with increasing molecular weight solvent.

The structure and the properties of polyethylene obtained with the catalyst based on chromium trioxide are similar to those of polyethylene synthesized using the Ziegler-Natta catalyst.

8.2.5 ELECTROCHEMICAL POLYMERIZATION

The electrochemical initiation of polymerization involves the formation of radicals, ions, or ion radicals as a result of the reduction or oxidation reactions on a monomer, or a specially introduced electrochemical initiator. These reactions occur directly on the cathode or anode surface.

In organic chemistry well known is the electrosynthesis reaction of Kolbe, in which the electrolysis of sodium salts of carboxylic acids, carried out under appropriate conditions, leads to the formation of free radicals at the anode. These free radicals undergo decarboxylation and linking with the formation of saturated hydrocarbons:

$$R{-}COONa \rightleftharpoons R{-}COO^- + Na^+$$

on anode:

$$R{-}COO^- - e \longrightarrow R{-}COO^{\bullet} \longrightarrow R^{\bullet} + CO2$$

When conducting the reaction in the presence of monomer, the formed free radicals can initiate the polymerization reaction. The method is used for polymerization of styrene, butadiene, isoprene, vinyl acetate and methyl methacrylate. Similar effects can be achieved by the use of other active radicals such as trifluoromethyl $\bullet CF_3$, or fluorine $\bullet F$. These radicals are formed at the anode as a result of the electrolysis of trifluoroacetic or hydrofluoric acid. Application of these systems allows performing the free radical polymerization of tetrafluoroethylene and trifluorochloroethylene at the anode, at normal pressure and at a temperature below 273 K. The polymerization of these compounds, when initiated by peroxides, takes place at high pressures and at high temperatures. The effect of temperature and of current density on the process of electrochemical polymerization is illustrated by the results obtained during the polymerization of N-vinylcarbazole in the solution of methylene chloride at a concentration of around 0.1 mol/dm^3 and in the presence of tetrabutylammonium tetrafluoroborate (32 mmol/dm^3). The data presented in Table 8.13, shows clearly that lowering the reaction temperature affects in a decisive extent the molecular weight of the formed polymer.

The type of electrode material used influences strongly the molecular weight of polymer formed during the electrochemical polymerization. This is well illustrated by the example of polymerization of methyl methacrylate at the presence of nitric acid. The obtained results are presented in Table 8.14.

8.2.6 RADIATION-INDUCED POLYMERIZATION

The radiation-induced polymerization is a process initiated by the ionizing radiation. There are several known methods of its initiation by α, β, γ rays as well as with neutron beams or with Roentgen radiation. However, the most frequently it is initiated by γ rays or by a beam of fast electrons. The mechanism of such a radiation polymerization may differ depending on the type of monomer used and the reaction temperature. It may be also of free radical, cationic or anionic character.

TABLE 8.13 Electroinitiated Polymerization of N-Vinylcarbazole.

No.	Current density [mA/cm^2]	Reaction temperature [K]	Conversion degree [%]	Molecular mass*
1	1	298	98	18,500
2	2	298	92	26,400
3	5	298	90	40,300
4	7	298	88	37,000
5	10	298	88	37,000
6	10	273	90	401,000
7	10	223	96	602,000
8	10	203	94	757,000

After E.B. Mano, B.A.L. Calafate: J. Polym. Sci. Polym. Chem. Ed. 21, 829 (1983).

* Molecular mass determined by the Viscosimetric method in benzene and at 298 K.

TABLE 8.14 Effect of the Anode Material on the Molecular Weight of Poly (Methyl Methacrylate) Formed During the Electrochemical Polymerization Methyl Methacrylate Performed with the Current Intensity of 5 mA and in the Presence of Nitric Acid During 80 Minutes

No.	Material of anode	Efficiency [%]	Molecular mass [M_v]
1	Graphite	81.8	275,000
2	Platinizated graphite	42.5	295,000
3	Gold	11.1	420,000
4	Platinum	4.24	680,000
5	Aluminum	0.86	290,000

After G. Pistoria, O. Bagnarelii, M. Maiocco: J. Appl. Electrochem. 9, 343 (1979).

Studies on the radiation-induced polymerization were carried out mainly on the example of vinyl monomers, which polymerize by free radical mechanism, leading both reactions in the mass of monomer as well as in solution.

The rate of free radicals formation in the polymerization process induced by radiation is directly proportional to the intensity of radiation (I) and depends on the effectiveness of the radiation polymerization.

The quantity indicating the effectiveness of the radiation polymerization is the coefficient G, called also the radiation efficiency. It is equal to the amount of monomer molecules participating in the polymerization reaction after absorption of 1.602×10^{-17} J (100 eV) of radiation energy:

$$G = \frac{w \cdot 6.023 \cdot 10^{23} \cdot 100}{m \cdot D \cdot M \cdot 6.24 \cdot 10^{13} \cdot 10^{6}} = \frac{w}{m \cdot D \cdot M} \cdot 0.97 \cdot 10^{26}$$

where: w – yield of polymer, kg; m – mass of monomer subjected to radiation, kg; D – amount of absorbed dose, M rad (1 M rad = 10^4 J/kg); M – molecular weight of monomer; 6.24×10^{13} – energy equivalent of 1 rad, eV/g (1eV/g = 1.602×10^{-22} J/kg); 6.02×10^{23} – Avogadro number.

The value of G varies in a wide range, depending on polymerization conditions and the type of monomer. Table 8.15 lists the G values for radiation polymerization of styrene or methyl methacrylate by β and γ radiation at several different temperatures, irradiation power and polymerization rates for both initiation and polymerization processes.

TABLE 8.15 G Values for Polymerization of Styrene and Methyl Methacrylate

No.	Monomer	Radiation	Power [rad/min]	Temperature [K]	Rate of polymerization [m/s×10⁵]	Rate of initiation [m/s×10⁸]	G
1	Styrene	γ	4100	345	5.84	4.4	0.78
		γ	3200	298	1.01	2.08	0.46
		γ	4400	255	0.24	2.74	0.56
		β		303.5	-	-	0.22
2	Methyl methacrylate	γ	4100	343	49.9	28.4	4.88
		γ	4200	298	21.1	36	5.74
		γ	4400	255	7.25	53.8	7.68
		β		303.5	-	-	3.14

After A. Charlesby: Chemia radiacyjna polimerów, WNT, Warszawa 1962, p. 360.

The total activation energy for radiation polymerization varies within the limits of 0–41.8 kJ/mol. A small value of this energy causes that the radiation polymerization reaction rate is temperature independent within a few tens of Kelvins. A characteristic feature of the radiation polymerization is the presence of the retrospective effect (post-effect). It consists on the fact that after the cessation of irradiation the polymerization continues over many hours. This phenomenon concerns primarily the polymerization in solid phase or in solution precipitation. The reason for this phenomenon is the reduced mobility of macroradicals in solid phase and difficulties ending the chains by recombination.

The mechanism of radiation polymerization reaction in the liquid phase can be determined by examination its kinetics. Namely, if the polymerization rate is directly proportional to the square root of the absorbed radiation dose by monomer and the resulting polymer molecular weight is inversely proportional to this value it shows the free radical polymerization mechanism:

$$V = k\sqrt{P_D}$$

$$M = \frac{K}{\sqrt{P_D}}$$

where P_D denotes the radiation dose.

Additional proofs on the presence of free radical mechanism in the polymerization process are:

- inhibition of the reaction with the introduction of conventional radical inhibitors,
- increase in reaction rate with increasing temperature, and
- always positive values of the activation energy (25–41.8 kJ/mol).

If the rate of polymerization is directly proportional to the radiation dose then the mechanism of reaction is of anionic or cationic type:

$$V = kP_D$$

In addition, the course of radiation polymerization running according to the ionic mechanism is confirmed by the lack of effect of radical inhibitors (free radical scavengers) on the process of polymerization and low values of the activation energy.

There are also some known examples of polymerization running along the mixed radical-ionic radiation polymerization mechanism.

An interesting example of the application of radiation polymerization is the polymerization of clathrate (channel complexes) monomer with urea or its derivatives. The channel walls are formed by twisted into a spiral urea molecules, linked together by hydrogen bonds. Inside the channels are arranged in an orderly manner the immobilized monomer molecules associated in a complex manner with urea. Under the influence of radiation, they undergo polymerization with the formation of an ordered polymer structure. At the end of the process urea is dissolved in water and the resulting polymer is isolated by filtration or centrifugation.

Another major application of the radiation polymerization is the use of radiation to initiate the emulsion polymerization. In the case of styrene the radiation polymerization in emulsion runs many times faster than the radiation polymerization in bulk and gives a product with significantly higher molecular weight. The increased efficiency of the polymerization process in emulsion can be also caused, partly, by the formation of additional free radicals in water, which can initiate polymerization of styrene, and partly, by the higher G value for the formation of radicals in this medium. In addition, reducing the rate of termination increases the molecular weight because more monomer molecule per initiating free radical undergoes the polymerization. The activation energy of emulsion polymerization is of 15.5 kJ/mol. Comparative results of the radiation polymerization of styrene in bulk and in emulsion is given in Table 8.16.

The radiation polymerization, induced by an intense Co^{60} radiation, was used for emulsion polymerization of vinyl acetate in experiments and on industrial scale.

8.2.7 POLYMERIZATION IN PLASMA

8.2.7.1 Characteristics of Plasma

Plasma is a gas, ionized by any method, in which the inertial motion of electrons prevails over the directed one. The degree of ionization of this gas is always large, reaching in extreme cases 100% of all molecules. In plasma the quantities of molecules charged positively and negatively are equal. It gives almost zero-th spatial charge. The high density of charged particles affects significantly the electrical conductivity of plasma.

Plasma cannot be seen as a normal, electrically neutral mixture of gas like the electron gas. Between the charged and the uncharged particles one observes interaction of forces associated with the electric polarization. For this reason all the neighboring molecules, apart from each other, interact mutually. The plasma properties are not only the sum of the properties of individual molecules, but equally are not the sum of the properties of gases, which compose it.

TABLE 8.16 Radiation Induced Polymerization of Styrene in Bulk and in Emulsion

No.	Polymerization	Dose [rad/min]	Temperature [K]	Reaction rate [%/h]	Molecular weight of polymer ($\times 10^3$)
1	In mass	4400	255	0.1	17–26
		3200	303	0.42	38–100
		4100	345	2.53	165–348
2	In emulsion	1000	298	36	800–1500
		1000	308	45	885–1726
		1000	318	54	1200–2060

After A. Charlesby, Radioactive Chemistry of Polymers (Chemia radiacyjna polimerów, in Polish), WNT, Warszawa 1962, str. 365.

In order to characterize the energy state of molecules present in the plasma an energetic scale of temperature was introduced. Its basic unit is electronvolt eV (1eV = 1.602x 10 $^{-19}$ J). The model chosen for such a scale is a gas whose molecules have only two degrees of freedom. For such a gas molecule the energy is given by the following equation:

$$E = k\mathrm{T}$$

where E = energy; k = Boltzmann's constant (1.381x10^{-23} J/K); and T = temperature (Kelvin).

From this equation we get:

$$T = \frac{E}{k} = \frac{1eV}{k} = \frac{1{,}602 \cdot 10^{-19} J}{1{,}381 \cdot 10^{-23} J / K} \approx 1{,}16$$

The above calculations show that the plasma energy of 1 eV on the energy scale corresponds to the temperature of 11,600 K.

It turned out that the properties of plasma are different at different energy ranges and therefore the concept of high temperature plasma (particles with energy of up to hundreds of electronvolts) and low-temperature plasma (the energy of particles not exceeding a few electron-volts) was introduced.

For the synthesis of polymers only the low temperature plasma, in which the energy of molecules does not exceed a few electronvolts is used The temperature of electrons in plasma is always significantly larger than the temperature of ionized gas. Therefore, it is possible to run the reaction at a temperature close to ambient temperature where the free electrons have enough energy to break the covalent bonds. For this reason this type of plasma is also called the non-equilibrium plasma.

The most important way of plasma generation, as a medium to conduct the polymerization, is the electric glow discharge. For that a direct (DC) or alternate (AC) current with frequencies ranging from 50 Hz to 5 GHz can be used. A typical plasma polymerization is conducted at low gas pressure of the order of $10^{-1} - 10$ Pa. The course of process in plasma may vary depending on the method of conducting the reaction. For this reason a distinction between the following is made.

– plasma polymerization, and
– plasma-initiated polymerization.

8.2.7.2 Plasma Polymerization

The plasma polymerization is a way to obtain specific polymeric materials which cannot be obtained by other methods. The plasma polymer is significantly different from the polymer formed from the same monomer by using other methods. In the case of a plasma polymer it is impossible to determine the mer unit. For this reason, in order to distinguish this type of compounds from the conventional polymers, the prefix pp (plasma polymerized) before the polymer name is introduced, for example, pp-styrene (polystyrene equivalent).

Almost all organic compounds undergo polymerization in plasma (not only the traditional monomers) as well as a significant number of inorganic compounds.

The chemical reactions occurring in plasma are very complicated and the mechanism of plasma polymerization has not yet been fully understood. The competing reactions running side by side depend always on the level of energy in plasma. These include, in the first order, the molecules decay processes and their re-synthesis. The resulting plasma polymers are often built not from the monomer molecules but from their degradation products, combined with other molecules contained in plasma gas such as nitrogen, carbon monoxide or water.

It is assumed that the primary mechanism for the chain growth in plasma polymerization is a fast gradual polymerization in which dominating are reactions between the active molecules. Such reactions can be also called as the polyrecombination.

In order to obtain polymeric materials with reproducible properties it is necessary to control precisely the reaction parameters such as: concentration of monomer, pressure of reaction mixture, discharge power, type and temperature of substrate, polymerization time and reactor geometry.

In many cases, the plasma polymer deposition process can be described using a complex power parameter:

$$W/FM$$

where: W –value of the delivered power; F – flow rate (mol/min); and M – molecular weight of monomer.

The plasma polymerization runs at large values of W/FM. The yield of polymerization grows always up to a critical value of parameter $(W/FM)_C$.

There exists also a dependence between the critical value of the power parameter $(W/FM)_C$ and the sum of the binding energies of all bonds present in the monomer molecule:

$$(W / FM)c = \alpha\phi$$

where α is a constant characteristic for the used reactor,

$$\phi = -\frac{\sum(binding\ energies)}{\bar{M}}$$

During the plasma polymerization the value:

$$\frac{W / FM}{\alpha\phi} > 1$$

Depending on the reaction parameters the resulting polymer is emitted in the form of powder, a continuous film, or an oily substance.

From the practical point of view the most important is the way to produce ultra-thin plasma polymer films, which the most frequently are highly branched and cross-linked. Therefore, they are insoluble and forming a good coating material.

8.2.7.3 Polymerization Initiated by Plasma

In this method, the plasma state is only used in the initial stage, to initiate the polymerization. The formed active molecules are transferred from the plasma phase to the monomer phase (liquid or solid), in which the chain growth reaction runs already without the plasma participation. As a result of this process one gets a conventional polymer, built of mers, which often exhibits a very high molecular weight. This method is particularly useful for the polymerization of water-soluble monomers such as acrylamide and its derivatives.

8.2.8 TECHNIQUES OF POLYMERIZATION CONTROL

8.2.8.1 Polymerization in Bulk

Polymerization in the bulk, often referred to as the block polymerization, runs in the monomer medium in which the resulting polymer is dissolved. With the increasing degree of conversion increases also the viscosity of the reaction mixture, and then, at a large degree of conversion the gelation occurs and the formed polymer takes shape of the reaction vessel. For this reason the block polymerization is carried out mostly in two steps. In the first step of the process, known as the prepolymerization, a solution of polymer in the monomer of a certain viscosity is obtained, which is then entered into the forms. Examples of such forms used to obtain a transparent organic glass are two plates, assembled from a glass mirror, sealed with a hose made from poly(vinyl chloride) and covered with a thick paper. The space between the plates is poured with the prepolymer. The final polymerization is carried out in the forms at elevated temperatures, the height of which depends on the type of monomer used and on the planned molecular weight of formed polymer.

By increasing the reaction temperature one increases the rate of polymerization, but decreases the molecular weight of the product. Polymerization in bulk has several serious drawbacks, resulting mainly from the high viscosity of the reaction mixture and the resulting poor thermal conductivity of the formed polymer. As a consequence of this fact, within the formed bloc emerge areas with a local overheating. This is due to the large exothermicity of the polymerization reaction. An inhomogeneous distribution of temperature in the block results also in locally different speeds of the polymerization process. Therefore, different degrees of polymerization in different parts of the block (lower inside and higher at its surface) are obtained. This leads to the formation of internal stresses, causing after a certain time, micro fractures of polymer (so-called silvering). A local overheating may also cause evaporation of monomer. The resulting steam cannot get out due to the high viscosity of the system. It causes formation of bubbles, which lowers the quality of the final product. It shows that the bulk polymerization, apparently very simple, is difficult to implement and requires good skills and experience of technical personnel involved in production. Despite its flaws the bulk polymerization in some cases is irreplaceable (e.g., flooding of organic preparations or minerals with a transparent polymer).

On the industrial scale known are also the semicontinuous or continuous bulk polymerization methods.

The continuous method allows to get a product which is practically free of monomer. Its molecular weight is higher than in the case of periodic polymerization in molds. It makes also possible the automated management of the process. It ensures the reproducibility of the polymer properties, improves the working conditions, increases productivity of the equipment and lowers the price of the final product.

The most common method of the continuous polymerization is the tower method. The principle of its operation lies in the fact that monomer polymerizes as it flows through the column and is continuously removed by the fresh monomer or solution of prepolymer. Eliminating the possibility of the product mixing at different stages of the reaction allows to obtain a homogeneous polymer with specific, fixed characteristics. This can be achieved, ensuring at the same time a hundred percent yield, by correlating the reaction times with the height of the column (tower) and an appropriate reaction rate.

8.2.8.2 Block-Precipitation Polymerization

The block-precipitation polymerization is a variation of the bulk polymerization during which the formed polymer is insoluble in monomer and precipitates in the form of a sediment.

The technical use of this method has encountered difficulties of the engineering nature. These difficulties are related mainly to the evacuation of the polymerization heat, mixing, and removal of polymer deposited on the inner walls of the reactor.

For polymerization one uses only monomers with the initiator dissolved in. As initiator usually an organic peroxide in an amount of about 0.1% is used.

The liquid monomer is in equilibrium with its vapors and the generated heat of reaction is evacuated, mainly by evaporation and by condensation on the walls of the reactor as well as in the condenser located directly above the reactor.

In the first stage of the reaction a large amount of heat is released. The resulting polymer begins to precipitate in the form of very fine particles of sediment, which already at the monomer conversion of 1% cause clouding of the reaction medium.

The concentration of these particles increases gradually giving rise to the formation of elementary primary grains with diameter of about 0.1 mm. At the higher conversion rate (7–8% of monomer) the primary grains merge into larger aggregates of polymer. Their size depends on the intensity of mixing. The increase of the mixing speed causes a reduction of the average diameter of grains. At this stage of the reaction its medium is still liquid and the grains suspended in polymer have already a shaped structure. During the further polymerization the number of grains is not increasing.

In the second stage of polymerization, with the increasing degree of monomer conversion (from 8% to 90%), the molecule le diameter increases to about 0.5–1.5 mm. The grains aggregate, creating porous clusters. At that time the initially liquid reaction environmental thickens gradually. At the end of reaction one obtains a polymer in powder form. After reaching the desired degree of conversion (70–90%) the unreacted monomers are driven off, condensated and returned to repolymerization.

8.2.8.3 Polymerization in Gas Phase

The gas phase polymerization is carried out in the case when using gaseous monomers, characterized by a low critical temperature. The best known such monomers are ethylene's and tetrafluoroethylenes. The polymerization is technically difficult to perform, as it requires application of high pressure, of the order of 10 MPa.

The polymerization reaction in a given case is carried out in adiabatic reactors of tube or autoclave type to which one introduces a gaseous monomer, after having pressured it via a two-stage compression to the desired pressure. For this purpose the dry monomer coming from a dry gasometer is mixed with the products coming from degassing liquid polymer. Then it is displaced with a compressor under 25 MPa pressure into a mixer, where it meets the circulating gas. The raw material should be of sufficiently low dew point to prevent clogging of pipes by precipitating crystalline hydrates of ethylene, formed during the cooling and compression of wet ethylene. Then the gas is subjected to the second step of compression to the working pressure, which depends on the kind of produced polymer. The reaction proceeds as follows: cold gas, after mixing it with initiator such as, for example, sprayed peroxides or oxygen, is gradually introduced into reactor on the suction side of the first compressor in such an amount that its content in the reaction zone, depending on the species of the produced polymer, amounts to 0.0045–0.0185 wt.%. The adiabatic reactor starts by heating its walls with hot air. After initiation of the polymerization process the heating is stopped and the cooling is started.

Immediately after leaving the reactor, the reaction mixture expands in a computer controlled reducing valve to 25 MPa. Then it is headed into the separator, operating at approximately 423 K (150°C). From the separator the gas escapes through the stepwise condenser jacket – tube and cyclone coolers to the fuel gas network or to other production, for example, manufacture of ethyl benzene, ethylene oxide, ethyl alcohol, etc. The liquid polymer flows through a pressure-reducing valve to the degassing tank. Then it is extruded in a wire form and cut into pellets with the appearance of rice grains, which are transferred by water jet onto a vibrating sieve. After an initial drying the compressed air the pellets are sent to the departments of averaging, homogenization, dyeing, and expedition. The flow sheet of the polyethylene production process in the gas phase is shown in Figure 8.2.

Another way of gas-phase polymerization is the anionic polymerization of 1,3- butadiene (divinyl). In this case the monomer in gas phase passes over a solid catalyst, which may be a sodium, potassium or lithium. The resultant polymer is deposited in the form of rubber type mass. Compared with the bulk polymerization the benefits in gas phase polymerization are: a better heat dissipation and the possibility of obtaining the polymer in large lumps in organic colloidal solution.

8.2.8.4 Suspension Polymerization

The suspensive polymerization, also called bubble polymerization or polymerization in suspension, runs under the influence of an initiator dissolved in dispersed in aqueous solution of organic colloids monomer or in

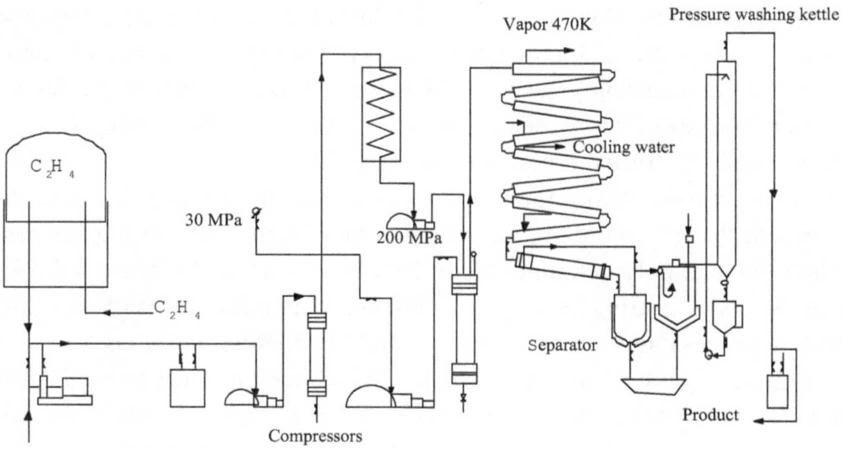

FIGURE 8.2 Scheme of ethylene polymerization in gas phase.

aqueous suspension of inorganic salts, being stabilizers, that is, the fixing agents of suspension.

The suspension polymerization process follows a typical bulk polymerization mechanism. Each drop of dispersed monomer behaves like a "microblock." In comparison with the block polymerization it allows to maintain easier a constant temperature and evacuate well the reaction heat.

A much higher rate of suspension polymerization as compared to the block polymerization indicates that there are also important reactions occurring in the surface layers at the interface monomer-emulsifier. For a strongly dispersed monomer the suspension polymerization is similar to the emulsion polymerization.

On the course of the suspension polymerization and the properties of the formed polymer a major impact have: protective colloids, that is, suspension stabilizers, initiators of polymerization and mixing of the system.

The dissolved in the aqueous phase protective colloids, being stabilizers of suspension, counteract the agglomeration of beads during the polymerization. They affect also their size. At the interface water-monomer a protective sheath is created, which prevents the merger (coalescence) of monomer droplets. An excessive reduction in viscosity of the aqueous phase can result in the formation of emulsion, while too high viscosity of this phase impedes the monomer dispersion during the mixing. It can lead to the formation of a coarse-grained polymer.

In the process of suspension polymerization one uses different additives modifying the action of protective colloid and supporting some shortcomings characterizing the given colloid. These additives, known also as auxiliary stabilizers of suspension, improve the hydrophilic-hydrophobic balance of colloid and adjust the surface tension at the interface monomer-water.

The type and the amount of used suspension stabilizer is chosen experimentally as a scientific theory allowing the resolution of this issue in a theoretical way is still lacking. Attempts to generalize in this area rely on introduction of (HLB) number as proposed by Griffin, determining the hydrophilic – hydrophobic or hydrophilic – lipophilic balance of given suspension stabilizer. It takes a value between 0 and 100. The main suspension stabilizer is primarily characterized by solubility or wetting in water, while the supporting stabilizer dissolves better in monomer. The HLB number takes an optimum value when at the maximum concentration of suspension stabilizer the interfacial surface tension on the border of the monomer-water reaches a minimum. As the suspension stabilizers one uses water-soluble organic macromolecular compounds such as poly (vinyl alcohol), copolymer of styrene with maleic anhydride, poly (sodium methacrylate), methylcellulose, methylhydroxypropylocellulose, carboxymethylcellulose, copolymer of allyl alcohol and vinyl acetate, gelatin and starch. One can use also inorganic suspension stabilizers such as calcium phosphate, aluminum and magnesium hydroxides, talc and silicates.

Poly(vinyl alcohol), formed by hydrolysis of poly(vinyl acetate) and used as a protective colloid, should have 8–20% of unsaponified acetate groups. The product containing less than 3% and more than 20% of acetate groups does not have sufficiently good protective properties. The molecular weight of poly (vinyl alcohol) used does not affect in a decisive way on the course of polymerization. A large impact on the size of beads in this case exerts pH of the reaction medium. This stabilizer works more effectively in the acidic medium, where one gets small beads. In alkaline environment, with unchanged other parameters, the formed beads are of larger dimensions. Note that during the polymerization the pH of the aqueous phase is changing. It is a result of a progressive hydrolysis of acetate groups of poly (vinyl alcohol) used and the accumulation of acidic products from the decomposition of benzoyl peroxide and other initiators.

Other macromolecular protective colloids are salts of poly (acrylic acid) or poly (methacrylic acid).

A very good suspension stabilizer is the methylcellulose and its derivatives. The advantage of these compounds is their total non-toxicity, resistance to mold and nonionic nature. The first characteristic determines the possibility of using the polymer for medical purposes, while the latter provides good dielectric properties.

The stabilizing ability of methylcellulose as a protective colloid in the suspension polymerization process depends on its molecular weight. It was found that only the methylcellulose, forming an aqueous solution of low viscosity is able to counter effectively the phenomenon of agglomeration of monomer particles. With increasing concentration of methylcellulose increases the fragmentation of the monomer droplets. At the same time the product benefits of fine-grained structure without a capacity to absorb the softener. Therefore, in order to obtain a porous product one has to keep low concentrations of methylcellulose in the aqueous phase.

There exist a number of patented solutions on the use of kits with different main and subsidiary suspension stabilizers. Noteworthy is the use as a secondary stabilizer the copolymer of vinyl acetate with allyl alcohol, which is more efficient than the well-known secondary stabilizer – glycerol stearate.

The allyl alcohol copolymer with vinyl acetate, used as a self-contained stabilizer during the polymerization of vinyl acetate, forms a porous product with a relatively large dispersion of grain sizes.

Using this copolymer as an auxiliary stabilizer next to the product of oxyethylation and alkylphenol gives a polymer with a porous structure and a homogeneous grain.

The poly (ethylene glycol) used as an independent suspension stabilizer leads to the formation of porous grains with a large dispersion of grain sizes. This phenomenon can be explained by the fact that only the two compounds together create an appropriate hydrophilic-hydrophobic balance. This happens as a result of the simultaneous occurrence in these compounds of hydroxide and ether groups. Therefore, the simultaneous use of both stabilizers in an appropriate proportion allows getting a product with a homogeneous grain.

When applying organic origin stabilizers to the suspension polymerization to some extent the emulsion polymerization takes also place. The post-reaction solution has a white color, which is caused by the presence of the emulsion polymer. To prevent this undesirable phenomenon, causing product loss and contamination, some patents recommend using polymerization inhibitors, soluble only in aquatic medium, such as, for example, copper salts. An alternative solution is addition of electrolytes to water.

In last years more and more frequently as protective colloids the previously mentioned inorganic substances are used. The behavior of a solid at the border of two mutually not solving liquids depends on its wettability by the liquids. In turn the wetting conditions depend on the size of the surface tension at the interfaces solid – water-monomer ($\delta_{1,2}$), solid-monomer ($\delta_{1,3}$) and water – monomer ($\delta_{2,3}$).When the contact angle θ <90 ° then the powdered solid is adsorbed by water. In this case the surface tension at the interface solid-monomer is larger than that at the phase boundary solid – water, that is, $\delta_{1,3} > \delta_{1,2}$

Stabilizers of this kind allow getting suspension of a non-polar liquid in water (emulsion of "oil in water" type). If the wetting angle is greater than 90° ($\delta_{1,3} < \delta_{1,2}$) then the powder is adsorbed by the non-polar liquid and one obtains emulsion of "water-in oil" type. For the contact value equal to 90 degrees, depending on such factors as number of separate phases, the way how the phase is dispersed, etc., formation of both types of emulsions is possible.

The size of beads formed during the dispersion of emulsion depends on the type and the quantity of protective colloid used. More of colloid is introduced into the system greater surface will be able to protect. Thus smaller balls will be obtained. The persistence of the protective layers formed from a fixed emulsifier depends on the size of its molecules. Emulsifier of this type must be thoroughly crushed in order to cover completely the surface forming the beads. However, the excessive fragmentation of the emulsifier adversely affects the emulsion stability.

As inorganic stabilizers in the suspension polymerization processes one uses mostly the calcium phosphate in ratio $CaO:P_2O_5 = 1.35$. This is a specially prepared mixture of tribasic calcium phosphate and calcium hydroxide.

The other this type protective colloids include aluminum and magnesium hydroxides. They can be precipitated directly in the aquatic environment through an action on the water-soluble salts of these metals with sodium hydroxide. In this way the cumbersome process of filtration, fragmentation and sieving, is omitted. The advantage of inorganic stabilizers is the ease of their separation from the polymer by dissolving in hydrochloric acid after the polymerization process.

In order to obtain beads of equal size one adds to water 0.001 wt.% of potassium persulphate. In practice a mixture of inorganic and organic stabilizers is often used.

Another important factor influencing the course of polymerization and the quality of the formed polymer is the polymerization initiator. As initiator always substances soluble in monomer are used. Usually these are organic peroxides such as benzoyl and lauroyl peroxides. They are chosen depending on temperature and duration of their half-decay. For the polymerization of monomers with a low boiling point such as, for example, vinyl chloride a very effective initiator is the isopropyl dipercarbonate, whose half-life period of decay is 2 h at 323 K (50°C). The lauroyl peroxide at this temperature has the half-life period of decay 25 times longer. Increasing the initiator concentration in the polymerization system shortens the time of polymerization, but reduces also the molecular weight and the porosity of the product. An excessive shortening of the polymer chains is also caused by the increase of the reaction temperature. The polymer chain length can be adjusted by the application of appropriate molecular weight regulators such as chlorinated hydrocarbons, isobutylene, dienes and mercaptans.

Mixing of components plays an important role in the course of suspensive polymerization, because its intensity influences the size of beads, their shape and the polydispersity. A less rigorous stirring can promote too quickly the agglomeration and the fragmentation of the product. The problem of mixing is particularly important in the initial stage of polymerization, when the amount of reacted monomer is of 10–60%. Stopping the mixer, even for a very short period of time, causes an irreversible agglomeration. The conglutinated beads form a rubbery, extending mass, which cannot be re-dispersed, even by a very intensive mixing.

The advantages of suspension polymerization are:

- possibility of synthesis of a grinded polymer, easy to extract by filtration or centrifugation,
- obtaining of a product with reproducible properties and low degree of polydispersity, and
- ease of heat dissipation secreted during the polymerization reaction.

The method has found a wide application for obtaining both the linear polymers, which can be then treated by injection molding and extrusion, as well as for the synthesis of cross-linked polymers. The cross-linked products in the form of beads may be subjected to a further chemical modification, leading to the generation of polymeric ion exchangers or polymeric catalysts. The porous polymers can be obtained by adding the liquid hydrocarbons (e.g., decane) to the system.

The suspension method is not applicable for the synthesis of elastomers.

8.2.8.5 Emulsion Polymerization

The principle of emulsion polymerization consists on dispersion of monomer in water using an emulsifier, followed by polymerization at the presence of water-soluble initiators. A polymer dispersion system, with a high degree of fragmentation, is then created in water, often called latex. The advantage of carrying out the emulsion polymerization process is the facilitated reception of the heat generated by the exothermic reaction of the aquatic environment, as well as the opportunity to run the reaction in a continuous way, what allows a substantial automation of the production process and lowering its cost. Due to the low viscosity of the emulsion, even at large concentrations, one can produce by these method viscous and rubbery polymers.

The monomers used in the polymerization process must be of a high degree of purity and be completely free of inhibitor. Water used in the polymerization process has to be demineralized, free of organic compounds and its electrical conductivity should be below 1 μS. It is recommended to remove oxygen from the water. Oxygen inhibits the polymerization reaction and causes formation of unstable, low-molecular weight products. The weight ratio of water to monomer can vary within the limits: 1:1 to 1:2.

As processing aids in the process of emulsion polymerization one uses emulsifiers, emulsion stabilizers, initiators, and regulating substances.

Emulsifiers are the surface-active compounds capable of obtaining a permanent emulsion of monomer in water. They can be divided into four groups:

a) anionic-active,
b) cationic-active,

c) nonionic, and
d) permanent.

The anionic-active emulsifiers most frequently are the soluble salts of fatty acids, alkyl sulfonates, alkylaryl sulfonates and alkyl sulfates. These compounds are the most frequently used emulsifiers in the emulsion polymerization.

The nonionic emulsifiers are rarely used in the emulsion polymerization. This group includes: esters of glycerol and fatty acids, products of ethylene oxide addition to alcohols, phenols or fatty acids, poly (vinyl alcohol) and others.

To the fourth group of emulsifiers belong: calcium phosphate, talc and oxides of aluminum and magnesium.

Characteristic properties of emulsifiers are given in Table 8.17.

Most of the emulsifier molecules are composed of hydrophobic and hydrophilic parts. The hydrophobic part of an emulsifier is aliphatic radicals, which give the molecule the ability to dissolve the emulsifier

TABLE 8.17 Properties of Emulsifiers

No.	Emulsifier	Critical micellar concentration		Medium	Molecular weight of micells	Specific surface [Å2]
		[mol/dm3]	[%]			
1	Potassium laurate	0.0125	0.3	Water	11,900	32
2	Potassium stearate	0.0005	0.16	Water	-	-
3	Potassium oleate	0.0012	0.04	Water	-	28
4	Potassium abietate	0.012	0.39	Water	-	25–40
5	n-Dodecylosulfonate of sodium	0.0016	0.055	Water	8200	32–40
6	n-Dodecylosulfate of sodium	0.0087	0.25	Water	17,100	-
7	Cetyltrimetlyl-ammonium bromide	0.001	0.036	0,13 M KBr	61,700	-
8	Cetyl-pyridine chloride	0.006	0.21	0,0175 M NaCl	32,300	46
9	Dodecylamine hydrochloride	0.014	0.31	Water	12,300	26

TABLE 8.17 Continued

No.	Emulsifier	Critical micellar concentration		Medium	Molecular weight of micells	Specific surface [Å2]
		[mol/dm3]	[%]			
10	Oxyethylated oktyphenol n = 9	0.0002	0.012	Water	66,700	53
11	Oxyethylated octyphenol n=30	0.00025	0.026	Water	-	101
12	Sorbite monolaurate	0.002	0.067	Water	-	36
13	Block copolymer (80:20) of ethylene oxide with propylene oxide M = 8200	6.9×10^{-6}	0.006	Water	8400	920
14	Sodium di(2-ethylhexyl) sulfosuccinate	0.0007	0.03	Water	21,300	25

molecule in the organic phase (monomer), while the hydrophilic parts are able to dissolve in the aqueous phase. Thanks to that the emulsifier molecules can adsorb at the border surface water-monomer, orienting in such a way that the hydrophilic part heads towards the water and the hydrophobic part to the monomer, respectively.

As the reaction initiators one uses: hydrogen peroxide, cumene hydroperoxide, persulphates and redox systems. The effect of the initiator type on the course of emulsion polymerization is depicted in Figure 8.3.

During the test of the emulsion polymerization mechanism it was found that when the monomer droplets in the emulsion have a diameter of 0.5–10 nm then the grains of the manufactured polymer are ten times smaller. It follows from this fact that the number of original "spores" of grains, contained in 1 cm^3 of emulsion, must be much larger than the number of monomer droplets. The number of polymer grains (in 1 cm^3) increases significantly with the increasing concentration of emulsifier. Therefore, it is assumed that the original spores of grains are formed in micelles of emulsifier. The emulsifier molecules (e.g., sodium palmitate) particles are not separate molecules in aqueous solution, but they are combined into aggregates called micelles.

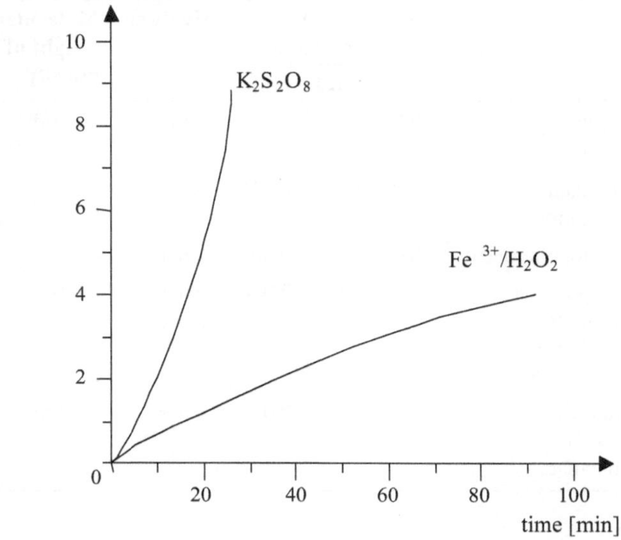

FIGURE 8.3 Effect of the initiator type on the course of emulsion polymerization [after A. S. Dunn, Surf. Inf 1974, 50 (1991)].

In this way one can explain the observation that the surface tension of the solution of a surfactant-active molecule is small in comparison with water. But it rises rapidly from the start of polymerization reaction. It is assumed that the individual emulsifier molecules are arranged in parallel, or directed to inside in such a way that the hydrophilic groups ($-COO^-$ or $-SO_3^-$) form the outer layer in contact with water. The course of emulsion polymerization is shown in Figure 8.4.

In inner part of the micelle, where are the hydrophobic groups, may be located also the monomer molecules. In this way one can explain why the small solubility of styrene in water (0.02% at 233 K) increases several times after an addition of 2% of emulsifier.

X-ray diffraction studies of micelles confirm their layered structure. Addition of a micelle to the hydrocarbon system makes diffraction pattern diffuse. This is interpreted as due to the penetration of hydrocarbon inside the core. The size of micells depends on the emulsifier concentration and increases with increasing penetration of water-insoluble monomers inside.

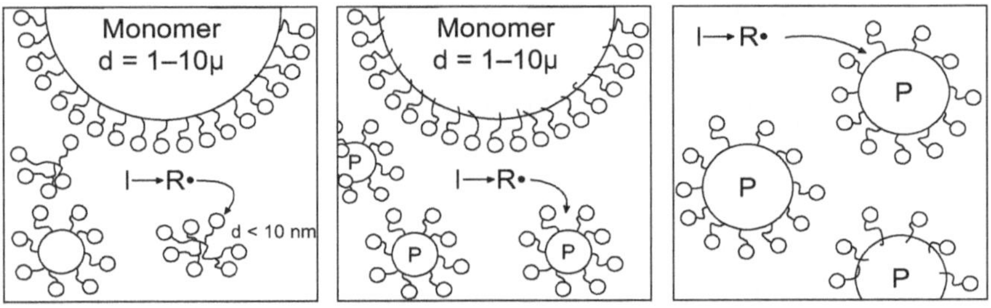

FIGURE 8.4 Chart showing he course of emulsion polymerization.

Probably the kinetic chain initiation takes place in micelle formed by absorption of a radical created in the aqueous phase. In micelle, in which a high concentration of monomer is observed, the chain grows proceeds until absorption of a next radical. Due to the very small size of the spore, in which the polymerization started, it seems likely that the second radical will react with the growing chain and will break its growth. Bu it will not start a new chain.

The growing chain takes the needed for its growth monomer molecules first from inside the micelles, and then pulls the monomers dissolved in the aqueous phase. The monomer concentration in the aqueous phase remains constant as long as there exist the droplets. The formed polymer molecules suck monomer and fills like a sponge. Small grains are of gelatinous form, but the pulled through the gel monomer does not dilute it, because it polymerizes immediately. Micelle is disrupted due to the running reaction of polymer growth. Its molecules surround the polymer molecules preventing them from coagulating and conglomerating. Further polymerization takes place in polymer grains. Monomer diffuses into the grains from suspension droplets through the aqueous phase. A characteristic property of the emulsion polymerization and its kinetics is that due to very small size of the polymer grains maximum only one active radical is in. If a second radical arrives during the chain growth it causes its disruption in the recombination reaction. Only after the cessation of the chain growth a new radical entering the grain initiates the growth of a new chain.

The emulsion is composed mainly of polymer grains formed in micelles at the initial stage of polymerization.

For a sufficiently large number of grains and in a given time one half of these molecules possess a radical, whereas the other half not. If we denote by N the number of polymer micelles in 1 cm^3 of emulsion the number of molecules in 1cm^3 possessing a radical is N/2.

The rate of attachment to the growing polymer radical is given by:

$$-\frac{d[M]}{dt} = k_w[M^\bullet][M]$$

To N/2 radicals is attached at one second (N/2)k_w [M] mole of monomers. As it was shown above, N/2 is the number of radicals in 1 cm^3 of emulsion. This value remains constant during the polymerization process. The rate of polymerization can be described by the formula:

$$V_p = \frac{N}{2}k_w[M]\sqrt{N_A}\left[molecules\right]xcm^{-3}xs^{-1}$$

Because the monomer concentration [M] in the aqueous phase is constant (saturated solution) and practically there is no influence on its concentration in micelles, the rate of emulsion polymerization depends only on the number of grains of produced polymer. However, the number of grains of polymer depends on emulsifier used and on initiator concentrations. Greater is the number of free radicals of initiator greater is the number of primary spores of polymerization.

The initiation of a chain can lead to self-manufacture of polymer grain only when there is a quantity of emulsifier sufficient to surround the seed with a monomolecular protective layer and prohibit connections with other grains of polymer, and thereby protect against coagulation. If there is more polymer grains than it corresponds to the critical amount of emulsifier than their shell is diminishing by combining two or more of them into larger grains. Sometimes, especially at the end of polymerization, coarse grains of coagulum are formed.

In technical conditions at 1% concentration of emulsifier and initiator concentration of about 0.1% in 1 cm^3 the number of molecules (grains) in polymer is of the order of 10^{14}.

A special feature of the emulsion polymerization is the large quantity (N) of polymer grains in 1 cm^3. Therefore, the degree of polymerization (P) can be controlled by the emulsifier concentration, regardless of the concentration of initiator. As result the harmful influence of initiator concentration on the chain length can be compensated. The growing chain radicals are isolated from each other by the action of emulsifier. Even at a large concentration of macroradicals (which takes place at the emulsion polymerization) there is no termination by recombination of growing polymer chains.

The emulsion polymers are characterized by a greater degree of polymerization and higher molecular weight than the block or suspensive polymers formed at the same temperature and at the same concentration of initiator. As an example one can quote the styrene polymerization. In the suspension polymerization of styrene at 80°C one gets a product with an average molecular weight of about 200,000, while the emulsion polymerization at the same temperature and at the same concentration of initiator (0.1%) gives a product with a five orders of magnitude greater molecular weight.

In contrary to the block polymerization where increasing the rate of polymerization by raising the temperature or increasing the initiator concentration leads to a lower degree of polymerization, in the emulsion polymerization it is possible to increase the reaction rate by increasing the concentration of emulsifier, without causing a reduction in the degree of polymerization.

For these reasons the emulsion polymerization was very widely used in the technology of polymers.

8.2.8.6 Polymerization in Microemulsion

In the recent years, more and more important theoretical and practical importance gains polymerizations carried out in mini-and in microemulsions. The latter are defined as the most complex quaternary systems composed of a monomer, a phase diffuser, a surface-active compound (surfactant) and a cosurfactant. Depending on the degree of hydrophobicity or hydrophilicity of monomer used water or an appropriate hydrocarbon can be used as a phase diffuser.

Microemulsions are characterized by their homogeneity They are optically isotropic, exhibit a low viscosity and show a much greater stability of emulsion. They are also thermodynamically stable and form spontaneously a continuous phase of the two not mixing ingredients.

There are three types of microemulsions:

- oil in water,
- water in oil, and
- bicontinous strip structure of microemulsion, formed in systems in which the quantities of monomer and water are comparable.

The type of formed microemulsion depends primarily on the type of surfactant and cosurfactant used.

The most frequently as cosurfactant one uses a small molecular mass alcohol, containing from three to ten carbon atoms in the chain, which reduces the packing density of surfactant and lowers the interfacial tension.

Bicontinous microemulsions of water in oil type require the use of a larger amount of cosurfactant than macro-emulsion of oil in water type. These microemulsions are distinct by the structure approaching that of liquid crystals, but with a smaller order.

When studying the course of polymerization of styrene it was shown that the formed microemulsions are characterized by transparency and are thermodynamically stable. The diameter of the molecules in micro-emulsions is between 40 and 100 nm.

There are also so-called mini-emulsions, which in contrast to the micro-emulsions, form opalescent milky-colored systems and are similar to the traditional emulsions. The diameter of the molecules in mini-emulsions is in 100–400 nm range.

A modification of the micro-emulsion polymerization is polymerization in reverse micelles, called inversed or micellar polymerization. This issue concerns the polymerization of hydrophilic monomers, from which one obtains the water-soluble polymers. Therefore, as the phase diffuser mostly hydrocarbons such as isooctane, heptane, benzene or toluene are used.

In a reverse micelle the aliphatic chains of surfactant are directed to the dispersal phase, while the hydrophilic groups point inside the micelle. The contained in the system water and the monomer molecules penetrate inside the micelle.

The amount of water contained inside the reverse micelles ("water pool") decides on the size and the properties of the formed aggregate. Water contained in the aqueous space of micelle differs from the volume water by physico-chemical properties. The degree of ordering, mobility, viscosity, polarizability, dielectric constant and the chemical activity of water depends on the size of the area occupied by water and change with the distance from the center of the aggregate.

Surfactants, containing a large polar group in molecule (ammonium salts), form small micelles with a diameter of 3–10 nm, characterized by a low water content (up to 3 molecules). Anionic surfactants such as sodium di (2-ethylhexyl) sulfosuccinate, containing a small polar group with two hydrocarbon chains in molecule, form without addition of a cosurfactant almost homodysperse, large aggregates of the size almost independent of the surfactant concentration. At high water content in the system the microemulsions of water in oil are formed.

During the polymerization in inverse micelles initiators soluble in the organic phase, such as dinitryl asoisobutyric acid or organic peroxides, are used.

Polymerization begins when the formed in organic phase free radical penetrates into the "swollen" micelle, in which are monomer molecules. Since in microemulsion only a small number of monomer molecules is present, the chain growth takes place by attachment of monomer molecules located in other micromicelles as result of a monomer collision or diffusion, as shown in Figure 8.5 (Option II).

During the polymerization of acrylamide in microemulsion, in toluene environment and in presence of sodium di(2-ethylhexyl) sulfosuccinate, initiated by UV radiation at a temperature of 10°C, the reaction initialization rate is directly proportional to the radiation intensity. In this

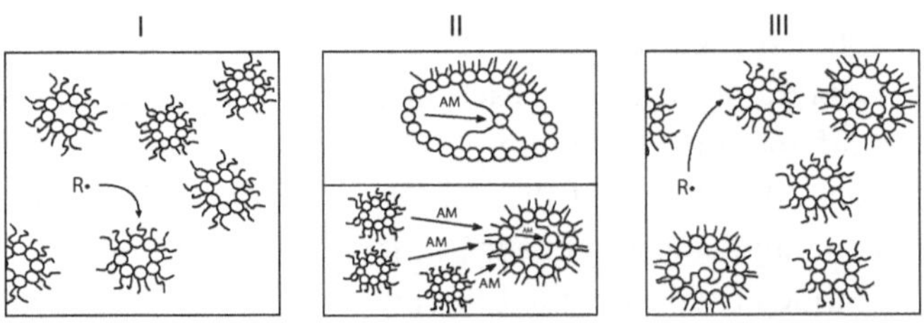

FIGURE 8.5 Flow diagram of micellar polymerization: I – initiation of polymerization, II – increase of the chain length, (a) by collision (fusion), (b) by monomer diffusion, III – complete polymerization (after KA Wilk, B. Burczyk, T. Wolf: 7th International Conference Surface and Colloid Science, Compiegne, 1991).

process the initial large number of micelles, which in the initial period of polymerization is of about $10^{21}/dm^3$ decreases to 10 latex molecules in 1 dm^3 of disperse system with dimension of 40 nm, as a result of collisions.

The completion of the polymer chain growth takes place as a result of deactivation by collision with another free radical.

8.2.8.7 Polymerization in Solution

The course of polymerization in solution depends to a large extent on both the type and the amount of solvent used. For some solvents, which dissolve both the monomer and the resulting polymer, the polymerization process proceeds in a homogenous environment. The course of this reaction is strongly influenced by the concentration parameter. With its decline decreases the probability of effective collisions of monomer molecules with the growing macroradical, and simultaneously increases the possibility of chain transfer. This causes the formation of polymer molecules with molecular weight less than obtained during the polymerization in solutions with higher concentration.

The major advantages of the polymerization process in a homogeneous solution is the possibility to obtain polymers with a relatively small dispersion of molecular mass and the possibility of direct use of polymer solutions for manufacturing synthetic fibers by the solution spinning methods. They can also be used for the preparation of solvent adhesives and lacquers.

The use of polar solvents to the polymerization process carried out by ion mechanism improves significantly the efficiency of these reactions. However, certain difficulties are encountered with the methods of polymer extraction from solutions. The reaction in these systems virtually never runs to the end and a certain amount of unreacted monomer rests always in the solution. Also a part of the solvent is absorbed in the mass of the formed polymer and causes a change in its final properties.

To the methods used to extract the polymer from an aqueous solution belong the steam distillation and the precipitation techniques with the use of diluents miscible with solvent and not dissolving the formed polymer.

During the polymerization reaction in solution possible are also side processes, which sometimes become to be the main processes. An example of such a process is the telomerization reaction. This reaction takes place when the solvent molecules tend to create permanent, stabilized by the resonance, free radicals or ions. Particularly useful for this purpose are halogenated organics, such as carbon tetrachloromethane (tetrachloride), chloroform and ethylene chloride. The solvent molecules can then take part in the chain transfer.

As result of such a reaction telomere molecules are formed, which differ from the oligomer molecules by their ends. These are large substituents coming from the decomposition of the solvent molecule (e.g., Cl and such groups as CCl_3, $CHCl_2$, CH_2CH_2Cl, etc.), which exert a decisive influence on the properties of the final product. For this reason, telomeres cannot be equated with oligomers, which do not have reactive terminal functional groups.

The telomerization reaction is often carried out deliberately to get low-molecular weight reactive substances, which serve as raw materials for the synthesis of other polymers or cyclic compounds.

Another way to conduct the polymerization in solution is the solution-precipitation method. It involves the use of a liquid dissolving only the monomer and not dissolving the polymer. Due to that the formed product precipitate from the reaction environment in the form of a very fine powder or suspension, which is separated by centrifugation or filtration. In this process the increasing macroradical, at a certain chain length, ceases to be soluble and precipitates from solution. As result, the resulting polymer exhibits a particularly low polydispersity of molecular weight. This method was successfully applied to the synthesis of solid polymers with a relatively low molecular weight. As examples of such a reaction are polymerizations of styrene in alcohol, methyl methacrylate in saturated hydrocarbons as well as acrylonitrile in water, respectively. The type of solvent used exerts also a great influence on the course of copolymerization in solution.

8.2.9 COPOLYMERIZATION

Copolymerization is a process of joint polymerization of at least two different monomers. Obtained in this way polymers are called copolymers, in contrast to homopolymers, which are produced by polymerization of a single monomer.

Depending on the chemical structure of the starting monomers, their mutual quantitative relationship and the method of conducting the reaction one can obtain copolymers of different composition and different properties. Physicochemical properties of copolymers depend on their composition, but in most cases the relationship is not additive. Often the introduction to macromolecule during the copolymerization of a relatively small amount of another monomer provides the polymer with completely new properties. As an example one can cite a copolymer of isobutylene with a small amount (0.6–3.0%) of isoprene. In this way the formed polymer gets properties of rubber and can be vulcanized with sulfur, while the polyisobutylene does not have such a capacity.

Generally the copolymers are classified in the following way:

- statistical copolymers,
- alternating copolymers,
- block copolymers, and
- grafted copolymers.

The statistical copolymers are products characterized by a disordered distribution in formed macromolecules of used in reaction mers A and B.

Alternating copolymers are products of a regular structure and their macromolecules are made up from mers of neighboring A and B mers:

$$\sim\sim ABABABABAB \sim\sim$$

Block copolymers have a chain structure and the macromolecule is composed of separate segments (blocks) made up exclusively of monomer A or monomer B.

$$A_m B_n \quad or \quad A_m B_n A_k$$

The block copolymers are the most frequently obtained during the formation of macroradicals in a mixture of polymers. For this purpose one can use the mechano-chemical method involving the rolling of the mixture of polymer or its grounding in a ball mill. However, the post-reaction mixture, in addition to block copolymers, contains in this case always a certain amount of homopolymers. Pure block copolymers can be obtained by using the previously discussed reaction of "living polymers," containing end-groups able to initiate polymerization of another monomer B.

The reaction proceeds according to the scheme:

$$Na^{+}A^{-}\text{—}(A)_{n}\text{–}A^{-}\text{—}Na^{+} + mB \longrightarrow Na^{+}A^{-}\text{—}(A)_{n+1}\text{—}(B)_{m-1}\text{-}B^{-}\text{—}Na^{+} \text{ itd.}$$

The grafted copolymers are branched structures. Their main chain is composed of groups of monomers A and the side chains of monomers B:

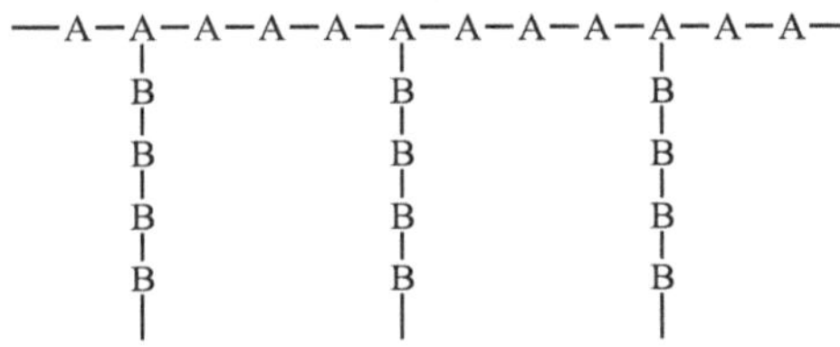

The grafted copolymers are obtained from macromolecules of polymer of A, in which are formed active centers that can then initiate polymerization of another monomer B. The active centers in polymer A may be formed by the reaction with free radicals arising from the decomposition of reaction initiators, and also during the irradiation, oxidation and under influence of ultrasounds, etc.

The copolymerization differs in some particular aspects from the homopolymerization as the initiation of free radical copolymerization of two monomers A and B may run as follows:

$$R\cdot + A \longrightarrow R\text{—}H + A\cdot$$

$$R\cdot + B \longrightarrow R\text{—}H + B\cdot$$

The resulting radicals A • B • may react in the following way:

$$\sim\sim A\cdot + A \xrightarrow{k_{1,1}} \sim\sim AA\cdot$$

$$\sim\sim A\cdot + B \xrightarrow{k_{1,2}} \sim\sim AB\cdot$$

$$\sim\sim B\cdot + A \xrightarrow{k_{2,1}} \sim\sim BA\cdot$$

$$\sim\sim B\cdot + B \xrightarrow{k_{2,2}} \sim\sim BB\cdot$$

The coefficients $k_{1,1}$, and $k_{2,2}$ denote the reaction rate constants of attachment of monomers A and B to their radicals. The rate constants $k_{1,2}$ and $k_{2,1}$, relate to attachment of these monomers to foreign radicals.

The steady-state concentration of radicals A• and B• does not change. Therefore, under these conditions the rate of the process of the radical transformation A• into B• is equal to the rate of the inverse radical transformation: B• into A•:

$$k_{1,2}[A^\bullet][B] = k_{2,1}[B^\bullet][A]$$

The rate equations of the copolymerization process have the following form:

$$-\frac{d[A]}{dt} = k_{1,1}[A^\bullet][A] + k_{2,1}[B^\bullet][A]$$

$$-\frac{d[B]}{dt} = k_{1,2}[A^\bullet][B] + k_{2,2}[B^\bullet][B]$$

Dividing these equations side by side one obtains:

$$\frac{d[A]}{d[B]} = \frac{[A]}{[B]}\left(\frac{k_{1,1}[A^\bullet] + k_{2,1}[B^\bullet]}{k_{1,2}[A^\bullet] + k_{2,2}[B^\bullet]}\right)$$

Since in the steady state:

$$[A] = \frac{k_{1,2}[A^\bullet][B]}{k_{1,2}[B^\bullet]}$$

then

$$\frac{d[A]}{d[B]} = \frac{[A]}{[B]}\left(\frac{\frac{k_{1,1}}{k_{1,2}}[A] + [B]}{[A] + \frac{k_{2,2}}{k_{2,1}}[B]}\right)$$

This is the copolymerization equation, whose validity was confirmed experimentally.

The occurring in the equation ratios of the reaction rate constants are defined as the monomer reactivity ratios r_1 and r_2:

$$r_1 = \frac{k_{1,1}}{k_{1,2}} \quad r_2 = \frac{k_{2,2}}{k_{2,1}}$$

These ratios are derived from the copolymerization kinetic equations and they determine, respectively:

r_1 – the ratio of the rate constant of reaction of monomer A with radical A to the rate constant of reaction of radical A with monomer B,

r_2 – the ratio of the rate constant of reaction of radical B with monomer B to the rate constant of the reaction of radical B with monomer A.

Depending on the values of parameters r_1 and r_2 the following copolymerization options may take place:

a) When $r_1 < 1$ and $r_2 < 1$ – monomer A reacts easier with radical B• than with the own radical A•. Similarly, monomer B reacts more easily with radical A• than with radical B•. Both monomers compile to form a copolymer with no long segments corresponding to only one of the comonomers.
b) When $r_1 > 1$ and $r_2 < 1$, facilitated is the reaction of both radicals A• and B• with monomer A. As result the formed polymer is enriched in monomer A in comparison to the initial monomer mixture.
c) When $r_1 > 1$ and $r_2 > 1$, a mixture of homopolymers is formed.
d) When $r_1 \times r_2 = 1$ then the composition of the resulting copolymer is directly proportional to the composition of the starting monomer mixture. Such a system is called a perfect one.
e) When $r_1 = r_2 \approx 0$, alternating copolymers are formed.
f) When $r_1 = r_2 = 1$, the copolymer composition is identical with that of the initial comonomer mixture.

The selected reactivity ratios r_1 and r_2 for different monomers are given in Table 8.18.

There are various methods of determining the coefficients r_1 and r_2. As described above, the copolymerization equation assumes that:

$$a = \frac{[A]}{[B]} \quad i \quad b = \frac{d[A]}{d[B]} = \frac{-[A]-}{-[B]-}$$

where – [A] – and – [B] – indicate the contents of mers A and B in copolymer, respectively.

TABLE 8.18 Reactivity Coefficients for Selected Monomer Systems Used in the Synthesis of Copolymers

No.	Monomer 1	Monomer 2	r_1	r_2
1	acrylamide	acrylic acid	0.48	1.73
2	acrylamide	acrylonitrile	1.04	0.94
3	acrylamide	vinyl chloride	19.6	0
4	acrylamide	methyl methacrylate	0.44	2.6
5	acrylamide	Styrene	0.3	1.44
6	methyl acrylate	vinylidene chloride	0.8	0.5
7	methyl acrylate	2-vinylpirydine	0.19	0.23
8	methyl acrylate	Styrene	0.17	0.77
9	1,3-butadiene	butyl acrylate	0.99	0.08
10	1,3-butadiene	acrylonitrile	0.18	0.03
11	1,3-butadiene	vinylidene chloride	1.9	0.05
12	1,3-butadiene	2-vinylpirydine	0.75	0.85
13	1,3-butadiene	styrene	1.4	0.5
14	ethylene	butyl acrylate	0.2	11
15	ethylene	vinyl chloride	0.6	1.85
16	ethylene	vinyl acetale	0.7	3.7
17	N-phenylmaleimide	vinyl chloride	4.37	0.03
18	N-phenylmaleimide	methyl methacrylate	0.183	1.022
19	N-phenylmaleimide	vinyl acetale	1.269	0
20	N-phenylmaleimide	styrene	0.047	0.012
21	metmethyl acrylate	styrene	0.35	0.35
22	metmethyl acrylate	2-hydroxyethyl methacrylate	0.296	1.054
23	vinyl acetate	vinyl chloride	0.23	1.68
24	vinyl acetate	vinylidene chloride	0	3.6
25	vinyl acetate	styrene	0.01	55
26	styrene	maleic anhydride	0.04	0.015
27	styrene	N-(4-chlorophenyl)maleimide	0.22	0.03
28	styrene	2-vinylpirydine	0.56	0.9
29	styrene	vinylidene chloride	2.1	0.45
30	styrene	N-vinylcarbazol	5.6	0.062

After R.Z. Greenley: J. Macromol. Sci. Chem. A14, 445 (1980).

After introducing the coefficients r_1 and r_2 the copolymerization equation takes the form:

$$b = a\left(\frac{r_1 a + 1}{r_2 + a}\right) = \frac{r_1 + 1}{\frac{r_2}{a} + 1}$$

A graphic image of this equation is a straight line.

The most frequently used methods of determining the reactivity ratios are:

- Mayo-Lewis method, and
- Finneman-Ross method.

The Mayo-Lewis method consists on determining the value of b in a series of successive copolimerization tests carried out at different contents of copolymerizable monomers used: A and B. The obtained values are entered into the copolymerization equation, which is solved in order to get r_1 and r_2.

The solution of the equation for r_2 is:

$$r_2 = a\frac{r_1 a + 1}{b} - a = r_1\frac{a^2}{b} + \frac{a}{b} - a$$

After substituting into the equation the *a* and *b* values obtained from measurements, one gets a series of straight lines corresponding to different dependencies $r_2 = f(r_1)$. Substituting the rising value of r_1 (e.g., from 0 to 1) and calculating r_2 one obtains a family of straight lines intersecting at one point (Figure 8.6). At this intersection point all obtained values of *a* and *b* correspond to the values of coefficients r_1 and r_2.

The Finnemana-Ross method consists on the solving the copolymerization equation with respect to a and b parameters:

$$a - \frac{a}{b} = r_1\frac{a^2}{b} - r_2$$

Then a graph, showing the dependence:

$$a - \frac{a}{b} = f\left(\frac{a^2}{b}\right)$$

is drawn.

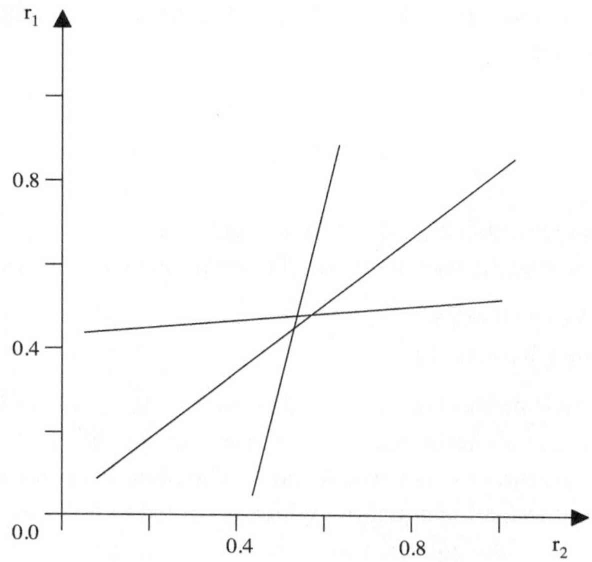

FIGURE 8.6 Graphical determination of reactivity ratios r_1 and r_2 by the Mayo-Lewis method.

The value of r_2 is the abscissa on (a–a/b) axis and r_1 is the slope of the curve. The Finneman-Ross method is simpler than the Mayo-Lewis method, because it requires obtaining of only one curve, describing the above dependence, only. But it is not always accurate (cf. Figure 8.7).

In order to simplify the way used to determine the reactivity ratios Alfrey and Price proposed a semiempirical relationship in which these factors is expressed as a fixed characteristic of the monomer, but independent of the comonomer. In this way it is not necessary to determine experimentally the relationship for each pair of individual monomers in the copolymerization reaction.

This relationship is called Alfrey-Price's Q-e system:

$$r_1 = \left(\frac{Q_A}{Q_B}\right) e^{-e_A(e_A - e_B)}$$

$$r_2 = \left(\frac{Q_B}{Q_A}\right) e^{-e_B(e_B - e_A)}$$

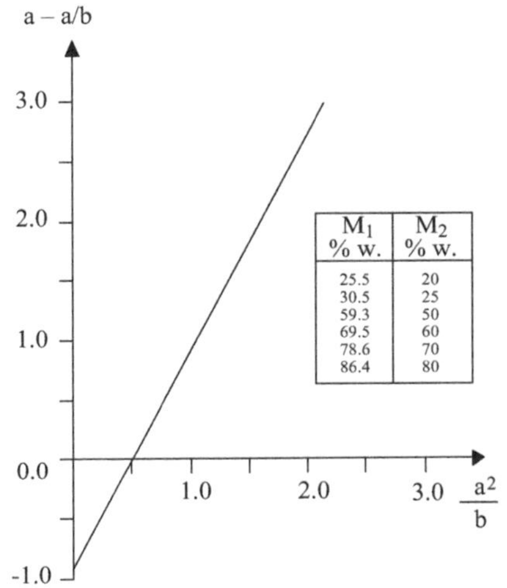

FIGURE 8.7 Graphical determination of the reactivity ratios r_1 and r_2 by the Finneman-Ross method for copolymerization of acrylic acid with acrylonitrile.

where Q_A and Q_B are the measures of the reactivity of monomers A, B, respectively and are connected by the resonance stabilization of monomer. Constants e_A and e_B are the measures of the polarity of monomers.

Styrene monomer was chosen as standard with assigned values of $Q = 1.00$ and $e_A = -0.80$.

The Q values increase with the increasing resonance stabilization, while the values of e becomes to be less negative when the groups attached to atoms forming a double bond attract stronger electrons.

Q and e values for selected monomers are given in Table 8.19.

It should be noted that the values given in Table 8.19 are only approximate and in the best case they provide only semi-quantitative informations. The latter are, however, very useful, as they help to predict the value of the reactivity coefficients in the situation when no other experimental data are available.

The reactivity coefficients are affected by the sum of steric, resonance and polar effects.

Resonance in the monomer molecule affects the reactivity of the intermediate radical. A large resonance stabilization of the monomer reduces its reactivity in the chain growth reaction. For example, it is difficult to copolymerize styrene with vinyl acetate. This happens because the resonance

TABLE 8.19 Q and e values for Selected Monomers

No.	Monomer	Q	e
1	acrylaldehyde (propenal)	0.8	1.31
2	acrylamide	0.23	0.54
3	acrylonitrile (vinyl cyanide)	0.48 ± 0.07	1.23 ± 0.08
4	ethyl acrylate	0.41	0.55
5	octyl acrylate	0.63	2.01
6	acrylic acid	0.83 ± 0.47	0.88 ± 0.23
7	1,3-butadiene	1.7	–0.5
8	vinyl chloride	0.056	0.16
9	vinylidene chloride	0.31	0.34
10	ethylene	0.016	0.05
11	1-hexene	0.035	0.92
12	indene	0.13	–0.71
13	diethyl maleate	0.053	1.08
14	maleic anhydride	0.86	3.69
15	maleimide	0.94	2.86
16	N-phenylmaleimide	2.81	3.24
17	butyl methacrylate	0.88	3.7
18	ethyl methacrylate	0.76	0.17
19	methyl methacrylate	0.78 ± 0.06	0.40 ± 0.08
20	propylene	0.009	–1.69
21	styrene	1	–0.8
22	N-vinylcarbazole	0.26	–1.29
23	2-vinylpirydine	1.41	–0.42
24	4-vinylpirydine	2.47	0.84

After R.Z. Greenley: J. Macromol. Sci. Chem. A14, 427 (1980).

stabilized sterile radical shows no tendency of addition to vinyl acetate, since the resulting radical would not be stabilized.

On the contrary, styrene copolymerizes readily with methyl methacrylate, which forms, like styrene, a resonance stabilized radical and exhibits the acceptor-donor effect.

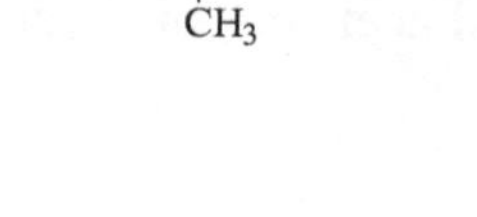

In order to clarify the cases in which there exists a strong tendency to form alternating copolymers (r_1 and $r_2 \approx 0$) two theories were proposed. The first one takes into account the polar effects. The second explore the mutual interactions connected with the charge transfer.

Effect of polar effects can be seen on the example of the styrene copolymerization with maleic anhydride. Attachment of the maleic radical to styrene leads to a transition state, which can be stabilized by the resonance structure formed as a result of the electron transfer:

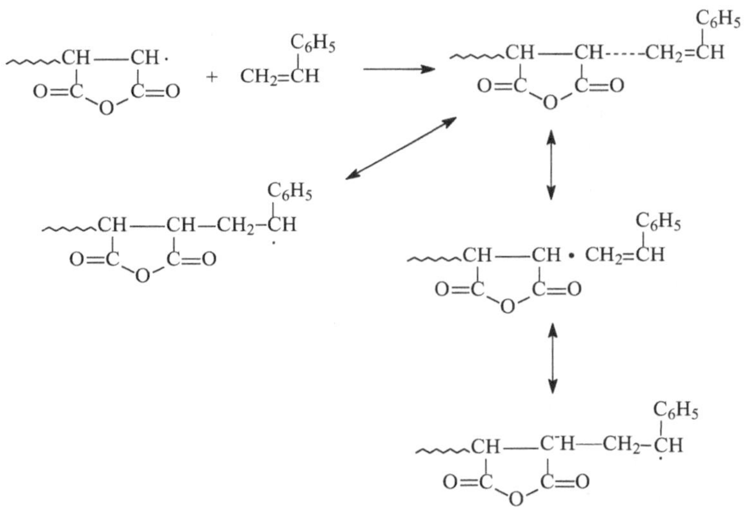

This type of structure may arise when a monomer includes numerous electronegative substituents, as it is the case of maleic anhydride.

The attachment of the styrile radical to the maleic anhydride gives rise to a transition state, which is stabilized in a similar way. In sum, these interactions take such an effect that the styrile radicals prefer to join the maleic anhydride and the maleic radicals prefer to combine with styrene.

A newer theory of the formation of alternating copolymers assumes the formation of complexes between comonomer molecules, associated with a charge transfer.

According to this theory the alternating copolymerization reaction is in fact a homolymerization or an alternate copolymerization of formed complex with the charge transfer, which reacts favorably with other charge transfer complexes. This is confirmed by the observed increasing tendency to form alternate copolymers when the reactants increasing the acceptor properties of one of the monomers are introduced to the system. This phenomenon is observed, for example, when the ethyl aluminum sesquichloride is added to the styrene maleic anhydride or the zinc chloride to the butadiene-acrylonitrile system. This is also confirmed by the fact, according to which after the introduction of other highly reactive monomers the latter do not react. It was also found that the molecular weight regulators have only an insignificant influence on the synthesis of alternating copolymers.

Other examples of compounds, which copolymerize with vinyl monomers are: carbon monoxide, forming polyketones and sulfur dioxide which reacts forming a polysulfone. It should be noted that sulfur dioxide forms complex compounds with alkenes in the 1:1 ratio.

KEYWORDS

- **elastomers**
- **macromolecules**
- **plastic materials (plastics)**
- **polymers**
- **thermoplastic polymers**

REFERENCES AND FURTHER READING

1. Abusleme, J., Giannetti, E., Emulsion Polymeryzation, Macromolecules 1991, 24.
2. Adesamya, I., Classification of Ionic Polymers, Polym. Eng. Sci., 1988, 28, 1473.
3. Akbulut, U., Toppare, L., Usanmaz, A., Onal, A., Electroinitiated Cationic Polymeryzation of Some Epoksides, Makromol. Chem. Rapid. Commun., 1983, 4, 259.
4. Andreas, F., Grobe, K., Chemia propylenu, WNT 1974, Warszawa .
5. Barg, E. I., Technologia tworzyw sztucznych, PWT 1957, Warszawa.
6. Billmeyer, F. W., Textbook of Polymer Science, J. Wiley & Sons, New York 1984.
7. Borsig, E., Lazar, M., Chemical Modyfication of Polyalkanes, Chem. Listy 1991, 85, 30.
8. Braun, D., Czerwiński, W., Disselhoff, G., Tudos, F., Analysis of the Linear Methods for Determining Copolymeryzation Reactivity Ratios, Angew. Makromol. Chem. 1984, 125, 161.

9. Burnett, G. M., Cameron, G. G., Gordon, G., Joiner, S. N., Solvent Effects on the Free Radical Polymeryzation of Styrene, J. Chem. Soc. Faraday Trans., 1, 1973, 69, 322.
10. Challa, R. R., Drew, J. H. Stannet, V. T., Stahel, E. P., Radiationindueed Emulsion Polymeryzation of Vinyl Acetate, J. Appl. Polym. Sci., 1985, 30, 4261.
11. Candau, F., Mechamism and Kinetics of the Radical Polymeryzation of Acrylamide In Inverse Micelles, Makromol. Chem., Makromol Symp., 1990, 31, 27.
12. Ceresa, R. J., Kopolimery blokowe i szczepione, WNT 1965, Warszawa.
13. Charlesby, A., Chemia radiacyjna polimerów, WNT 1962, Warszawa.
14. Chern, C.-S., Poehlein, G. W., Kinetics of Cross-linking Vinyl Polymerization, Polym. Plast. Technol. Eng., 1990, 29, 577.
15. Chiang, W.-Y., Hu, C.-M., Studies of Reactions with Polymers, J. Appl. Polym. Sci., 1985, 30, 3895 i 4045.
16. Chien, J. C., Salajka, Z., Syndiospecific Polymeryzation of Styrene, J. Polym. Sci. A, 1991, 29, 1243.
17. Chojnowski, J., Stańczyk, W. E., Procesy chlorowania polietylenu, Polimery, 1972, 17, 597.
18. Collomb, J., Morin, B., Gandini, A., Cheradame, H., Cationic Polymeryzation Iduced by Metal Salts, Europ. Polym. J., 1980, 16, 1135.
19. Czerwiński, W. K., Solvent Effects on Free - Radical Polymerization, Makromol. Chem., 1991, 192, 1285 i 1297.
20. Dainton, F. S., Reakcje łańcuchowe, PWN, 1963 Warszawa.
21. Ding, J., Price, C., Booth, C., Use of Crown Ether in the Anionic Polymeryzation of Propylene Oxide, Eur. Polym. J., 1991, 27, 895 i 901.
22. Dogatkin, B. A., Chemia elastomerów, WNT 1976, Warszawa.
23. Duda, A., Penczek, S., Polimeryzacja jonowa i pseudojonowa, Polimery 1989, 34, 429.
24. Dumitrin, S., Shaikk, A. S., Comanita, E., Simionescu, C., Bifunetional Initiators, Eur. Polym. J., 1983, 19, 263.
25. Dunn, A. S., Problems of Emulsion Polymeryzation, Surf. Coat. Int., 1991, 74, 50.
26. Eliseeva, W. I., Asłamazova, T. R., Emulsionnaja polimeryzacja v otsustivie emulgatora, Uspiechi Chimii, 1991, 60, 398.
27. Erusalimski, B. L., Mechanism dezaktivaci rastuszczich ciepiej v procesach ionnoj polimeryzacji vinylovych polimerov, Vysokomol. Soed. A., 1985, 27, 1571.
28. Fedtke, M., Chimiczieskije reakcji polimerrov, Chimica, Moskwa 1990.
29. Fernandez-Moreno, D., Fernandez-Sanchez, C., Rodriguez, J. G., Alberola, A., Polymeryzation of Styrene and 2-vinylpyridyne Using AlEt-VCl Heteroaromatic Bases, Eur. Polym. J., 1984, 20, 109.
30. Fieser, L. F., Fieser, M., Chemia organiczna, PWN 1962, Warszawa.
31. Flory, P. J., Nauka o makrocząsteczkach, Polimery 1987, 32, 346.
32. Frejdlina, R. Ch., Karapetian, A. S., Telomeryzacja, WNT, Warszawa 1962.
33. Funt, B. L., Tan, S. R., The Fotoelectrochemical Initiation of Polymeryzation of Styrene, J. Polym. Sci. A, 1984, 22, 605.
34. Gerner, F. J., Hocker, H., Muller, A. H., Shulz, G. V., On the Termination Reaction In The Anionic Polymeryzation of Methyl Methacrylate in Polar Solvents, Eur. Polym. J. ,1984, 20, 349.

35. Ghose, A., Bhadani, S. N., Elektrochemical Polymeryzation of Trioxane in Chlorinated Solvents, Indian, J. Technol., 1974, 12, 443.
36. Giannetti, E., Nicoletti, G. M., Mazzocchi, R., Homogenous Ziegler-Natta Catalysis, J. Polym. Sci. A, 1985, 23, 2117.
37. Goplan, A., Paulrajan, S., Venkatarao, K., Subbaratnam, N. R., Polymeryzation of N,N'- methylene Bisacrylamide Initiated by Two New Redox Systems Involving Acidic Permanganate, Eur. Polym. J., 1983, 19, 817.
38. Gózdz, A. S., Rapak, A., Phase Transfer Catalyzed Reactions on Polymers, Makromol. Chem. Rapid Commun., 1981, 2, 359.
39. Greenley, R. Z., An Expanded Listing of Revized Q and e Values, J. Macromol. Sci. Chem. A, 1980, 14, 427.
40. Greenley, R. Z., Recalculation of Some Reactivity Ratios, J. Macromol. Sci. Chem. A, 1980, 14, 445.
41. Guhaniyogi, S. C., Sharma, Y. N., Free-radical Modyfication of Polipropylene, J. Macromol. Sci. Chem. A, 1985, 22, 1601.
42. Guo, J. S., El Aasser, M. S., Vanderhoff, J. W., Microemulsion polymerization of Styrene, J. Polym. Sci. A, 1989, 27, 691.
43. Gurruchaga, M., Goni, I., Vasguez, M. B., Valero, M., Guzman, G. M., An Approach to the Knowledge of The Graft Polymerization of Acrylic Monomers Onto Polysaccharides Using Ce (IV) as Initiator, J. Polym. Sci. Part C, 1989, 27, 149.
44. Hasenbein, N., Bandermann, F., Polymeryzation of Isobutene With VOCl3 in Heptane, Makromol. Chem., 1987, 188, 83.
45. Hatakeyama, H. et all., Biodegradable Polyurethanes from Plant Components, J. Macromol. Sci. A., 1995, 32, 743.
46. Higashimura, T., Law, Y.-M., Sawamoto, M., Living Cationic Polymerization, Polym. J., 1984, 16, 401.
47. Hirota, M., Fukuda, H., Polymerization of Tetrahydrofuran, Nagoya-ski Kogyo Kenkynsho Kentyn Hokoku 1984, 51.
48. Hodge, P., Sherrington, D. C., Polymer-supported Reactions in Organic Syntheses, J. Wiley & Sons, New York 1980, tłum. na rosyjski Mir, Moskwa 1983.
49. Inoue, S., Novel Zine Carboxylates as Catalysts for the Copolymeryzation of CO2 With Epoxides, Makromol. Chem. Rapid Commun., 1980, 1, 775.
50. Janović, Z., Polimerizacije I polimery, Izdanja Kemije Ind., Zagrzeb 1997.
51. Jedliński, Z. J., Współczesne kierunki w dziedzinie syntezy polimerów, Wiadomości Chem., 1977, 31, 607.
52. Jenkins, A. D., Ledwith, A., ed., Reactivity, Mechanism and Structure in Polymer Chemistry, J. Wiley and sons, London 1974, tłum. na rosyjski, Mir, Moskwa 1977.
53. Joshi, S. G., Natu, A. A., Chlorocarboxylation of Polyethylene, Angew. Makromol. Chem. 1986, 140, 99 i 115.
54. Kaczmarek, H., polimery, a środowisko, Polimery 1997, 42, 521.
55. Kang, E. T., Neoh, K. G., Halogen Induced Polymerization of Furan, Eur. Polym. J. 1987, 23, 719.
56. Kang, E. T., Neoh, K. G., Tan, T. C., Ti, H. C., Iodine Induced Polymerization and Oxidation of Pryridazine, Mol. Cryst. Lig. Cryst., 1987, 147, 199.
57. Karnojitzki, V., Organiczieskije pierekisy, I. I. L., 1961 Moskwa.

58. Karpiński, K., Prot, T., Inicjatory fotopolimeryzacji i fotosieciowania, Polimery 1987, 32, 129.
59. Keii, T., Propene Polymerization with MgCl2 - Supported Ziegler Catalysts, Makromol. Chem. 1984, 185, 1537.
60. Khanna, S. N., Levy, M., Szwarc, M., Complexes formed by anthracene with 'living' polystyrene, *Dormant Polymers. Trans Faraday Soc,* 1962, 58, 747–761.
61. Koinuma, H., Naito, K., Hirai, H., Anionic Polymerization of Oxiranes., Makromol. Chem., 1982, 183, 1383.
62. Korniejev, N. N, Popov, A. F, Krencel, B. A., Komplieksnyje metalloorganiczeskije katalizatory, Chimia, Leningrad 1969.
63. Korszak, W. W., Niekotoryje problemy polikondensacji, Vysokomol. Soed. A, 1979, 21, 3.
64. Korszak, W. W., Technologia tworzyw sztucznych, WNT, Warszawa 1981.
65. KowalskaE., Pełka, J., Modyfikacja tworzyw termoplastycznych włóknami celulozowymi, Polimery 2001, 46, 268.
66. Kozakiewicz, J., Penczek, P., Polimery jonowe, Wiadomości Chem. 1976, 30, 477.
67. Kubisa, P., Polimeryzacja żyjąca, Polimery 1990, 35, 61.
68. Kubisa, P., Neeld, K., Starr, J., Vogl, O., Polymerization of Higher Aldehydes, Polymer 1980, 21, 1433.
69. Kucharski, M., Nitrowe i aminowe pochodne polistyrenu, Polimery 1966, 11, 253.
70. Kunicka, M. Choć, H. J., Hydrolitic degradation and mechanical properties of hydrogels, Polym. Degrad. Stab. 1998, 59, 33.
71. Lebduska, J., Dlaskova, M., Roztokova kopolymerace, Chemicky Prumysl 1990, 40, 419 i 480.
72. Leclerc, M., Guay, J., Dao, L. W., Synthesis and Characterization of Poly(alkylanil ines),Macromolecules 1989, 22, 649.
73. Lee, D. P., Dreyfuss, P., Triphenylphesphine Termination of Tetrahydrofuran Polymerization, J. Polym. Sci., Polym. Chem. Ed. 1980, 18, 1627.
74. Leza, M. L., Casinos, I., Guzman, G. M., Bello, A., Graft Polymerization of 4-vinylpyridine Onto Cellulosic Fibres by the Ceric on Method, Angew. Makromol. Chem. 1990, 178, 109 i 119.
75. Lopez, F., Calgano, M. P., Contrieras, J. M., Torrellas, Z., Felisola, K., Polymerization of Some Oxiranes, Polymer International 1991, 24, 105.
76. Lopyrev, W. A., Miaczina, G. F., Szevaleevskii, O. I., Hidekel, M. L., Poliacetylen, Vysokomol. Soed. 1988, 30, 2019.
77. Mano, E. B., Calafate, B. A. L., Electrolytically Initiated Polymerization of N-vinylcarbazole, J. Polym. Sci., Polym. Chem. Ed. 1983, 21, 829.
78. Mark, H. F., Encyclopedia of Polymer Science and Technology, Concise 3rd ed., Wiley-Interscience, Hoboken, New York 2007.
79. Mark, H., Tobolsky, A. V., Chemia fizyczna polimerów, PWN 1957, Warszawa .
80. Matyjaszewski, K., Davis, T. P., Handbook of Radical Polymerization, Wiley-Interscience, New York 2002.
81. Matyjaszewski, K., Mülle, A. H. E., 50 years of living polymerization, Progress in Polymer Science, 2006, 31, 1039-1040.
82. Mehrotra, R., Ranby, B., Graft Copolymerization Onto Starch, J. Appl. Polym. Sci. 1977, 21, 1647, 3407 i 1978, 22, 2991.

83. Morrison, R. T., Boyd, N., Chemia organiczna, PWN 1985, Warszawa.
84. Morton, M., Anionic Polymerization, Academic Press, New York 1983.
85. Munk, P., Introduction to Macromolecular Science, J. Wiley & Sons, New York, 1989.
86. Naraniecki, B., Ciecze mikroemulsyjne, Przem. Chem. 1999, 78(4), 127.
87. Neoh, K. G., Kang, E. T., Tan, K. L., Chemical Copolymerization of Aniline with Halogen-Substituted Anilines, Eur. Polym. J. 1990, 26, 403.
88. Nowakowska, M., Pacha, J., Polimeryzacja olefin wobec kompleksów metaloorganicznych, Polimery 1978, 23, 290.
89. Ogata, N., Sanui, K., Tan, S., Synthesis of Alifatic Polyamides by Direct Polycondensation with Triphenylphosphine, Polymer, J. 1984, 16, 569.
90. Onen, A., Yagci, Y., Bifunctional Initiators, J. Macromol. Sci-Chem. A, 1990, 27, 743, Angew. Makromol. Chem. 1990, 181, 191.
91. Osada, Y., Plazmiennaja polimerizacia i plazmiennaja obrabotka polimerov, Vysokomol. soed. A, 1988, 30, 1815.
92. Pasika, W. M., Cationic Copolymerization of Epichlorohydrin With Styrene Oxide And Cyclohexene Oxide, J. Macromol. Sci-Chem. A, 1991, 28, 545.
93. Pasika, W. M., Copolymerization of Styrene Oxide And Cyklohexene Oxide, J. Polym. Sci. Part A, 1991, 29, 1475.
94. Pielichowski, J., Puszyński, A., Preparatyka monomerów, W. P. Kr. Kraków.
95. Pielichowski, J., Puszyński, A., Preparatyka polimerów, TEZA WNT 2005 Kraków.
96. Pielichowski, J., Puszyński, A., Technologia tworzyw sztucznych, WNT 2003, Warszawa.
97. Pielichowski, J., Puszyński, A., Wybrane działy z technologii chemicznej organicznej, W. P. Kr. Kraków.
98. Pistola, G., Bagnarelli, O., Electrochemical Bulk Polymeryzation of Methyl Methacrylate in the Presence of Nitric Acid, J. Polym. Sci., Polym. Chem. Ed. 1979, 17, 1002.
99. Pistola, G., Bagnarelli, O., Maiocco, M., Evaluation of Factors Affecting The Radical Electropolymerization of Methylmethacrylate in the Presence of HNO_3, J. Appl. Electrochem. 1979, 9, 343.
100 Połowoński, S., Techniki pomiarowo-badawcze w chemii fizycznej polimerów, W. P. L., Łódź 1975.
101 Porejko, S., Fejgin, J., Zakrzewski, L., Chemia związków wielkocząsteczkowych, WNT, Warszawa 1974.
102 Praca zbiorowa: Chemia plazmy niskotemperaturowej, WNT, Warszawa 1983.
103 Puszyński, A., Chlorination of Polyethylene in Suspension, Pol. J. Chem. 1981, 55, 2143.
104 Puszyński, A., Godniak, E., Chlorination of Polyethylene in Suspension in the Presence of Heavy Metal Salts, Macromol. Chem. Rapid Commun, 1980, 1, 617.
105 Puszyński, A., Dwornicka, J., Chlorination of Polypropylene in Suspension, Angew. Macromol. Chem. 1986, 139, 123.
106 Puszyński, J. A., Miao, S., Kinetic Study of Synthesis of SiC Powders and Whiskers In the Presence KClO3 and Teflon, Int. J. SHS 1999, 8(8), 265.
107 Rabagliati, F. M., Bradley, C. A., Epoxide Polymerization, Eur. Polym. J. 1984, 20, 571.
108 Rabek, J. F., Podstawy fizykochemii polimerów, W. P. Wr., Wrocław 1977.

109 Rabek, T. I., Teoretyczne podstawy syntezy polielektrolitów i wymieniaczy jonowych, PWN, Warszawa 1962.

110 Regas, F. P., Papadoyannis, C. J., Suspension Cross-linking of Polystyrene With Friedel-Crafts Catalysts, Polym. Bull. 1980, 3, 279.

111 Rodriguez, M., Figueruelo, J. E., On the Anionic Polymeryzation Mechanism of Epoxides, Makromol. Chem. 1975, 176, 3107.

112 Roudet, J., Gandini, A., Cationic Polymerization Induced by Arydiazonium Salts, Makromol. Chem. Rapid Commun. 1989, 10, 277.

113 Sahu, U. S., Bhadam, S. N., Triphenylphosphine Catalyzed Polimerization of Maleic Anhydride, Makromol. Chem. 1982, 183, 1653.

114 Schidknecht, C. E., Polimery winylowe, PWT, Warszawa 1956.

115 Szwarc, M., Levy, M. Milkovich, R., Polymerization initiated by electron transfer to monomer. A new method of formation of block copolymers, *J Am Chem Soc*, 1956, 78, 2656–2657. .

116 Szwarc, M., 'Living' polymers, *Nature*, 1956, 176, 1168–1169.

117 Sen, S., Kisakurek, D., Turker, L., Toppare, L., Akbulut, U., Elektroinitiated Polymerization of 4-Chloro-2,6-Dibromofenol, New Polymeric Material, 1989, 1, 177.

118 Sikorski, R. T., Chemiczna modyfikacja polimerów, Prace Nauk. Inst. Technol. Org. Tw. Szt. P. Wr. 1974, 16, 33.

119 Sikorski, R. T., Podstawy chemii i technologii polimerów, PWN, Warszawa 1984.

120 Sikorski, R. T., Rykowski, Z., Puszyński, A., Elektronenempfindliche Derivate von Poly-methylmethacrylat, Plaste und Kautschuk 1984, 31, 250. .

121 Simionescu, C. I., Geta, D., Grigoras, M., Ring-Opening Isomerization Polymeryzation of 2-Methyl-2-Oxazoline Initiated by Charge Transfer Complexes, Eur. Polym. J. 1987, 23, 689.

122 Sheinina, L. S., Vengerovskaya Sh. G., Khramova, T. S., Filipowich, A. Y., Reakcji piridinov i ich czietvierticznych soliej s epoksidnymi soedineniami, Ukr. Khim. Zh. 1990, 56, 642.

123 Soga, K., Toshida, Y., Hosoda, S., Ikeda, S., A Convenient Synthesis of a Polycarbonate, Makromol. Chem. 1977, 178, 2747.

124 Soga, K., Hosoda, S., Ikeda, S., A New Synthetic Route to Polycarbonate, J. Polym. Sci., Polym. Lett. Ed. 1977, 15, 611.

125 Soga, K., Uozumi, T., Yanagihara, H., Siono, T., Polymeryzation of Styrene with Heterogeneous Ziegler-Natta Catalysts, Makromol. Chem. Rapid. Commun., 1990, 11, 229.

126 Soga, K., Shiono, T., Wu, Y., Ishii, K., Nogami, A., Doi, Y., Preparation of a Stable Supported Catalyst for Propene Polymeryzation, Makromol. Chem. Rapid Commun. 1985, 6, 537.

127 Sokołov, L. B., Logunova, W. I., Otnositielnaja reakcionnosposobnost monomerov i prognozirovanic polikondensacji v emulsionnych sistemach, Vysokomol. Soed. A., 1979, 21, 1075.

128 Soler, H., Cadiz, V., Serra, A., Oxirane Ring Opening with Imide Acids, Angew. Makromol. Chem. 1987, 152, 55.

129 Spasówka, E., Rudnik, E., Możliwości wykorzystania węglowodanów w produkcji biodegradowalnych tworzyw sztucznych, Przemysł Chem. 1999, 78, 243.

130 Stevens, M. P., Wprowadzenie do chemii polimerów, PWN, Warszawa 1983.
131 Stępniak, I., Lewandowski, A., Elektrolity polimerowe, Przemysł Chem. 2001, 80(9), 395.
132 Strohriegl, P., Heitz, W., Weber, G., Polycondensation using silicon tetrachloride, Makromol. Chem. Rapid Commun. 1985, 6, 111.
133 Szur, A. M., Vysokomolekularnyje soedinenia, Vysszaja Szkoła, Moskwa 1966.
134 Tagle, L. H., Diaz, F. R., Riveros, P. E., Polymeryzation by Phase Transfer Catalysis, Polym. J. 1986, 18, 501.
135 Takagi, A., Ikada, E., Watanabe, T., Dielectric Properties of Chlorinated Polyethylene and Reduced Polyvinylchloride, Memoirs of the Faculty of Eng., Kobe Univ. 1979, 25, 239.
136 Tani, H., Stereospecific Polymeryzation of Aldehydes and Epoxides, Advances in Polym. Sci. 1973, 11, 57.
137 Vandenberg, E. J., Coordynation Copolymeryzation of Tetrahydrofuran and Oxepane with Oxetanes and Epoxides, J. Polym. Sci. Part A, 1991, 29, 1421.
138 Veruovic, B., Navody pro laboratorni cviceni z chemie polymeru, SNTL, Praha 1977.
139 Vogdanis, L., Heitz, W., Carbon Dioxide as a Mmonomer, Makromol. Chem. Rapid Commun. 1986, 7, 543.
140 Vollmert, B., Grundriss der Makromolekularen Chemic, Springer-Verlag, Berlin 1962.
141 Wei, Y., Tang, X., Sun, Y., A study of the Mechanism of Aniline Polymeryzation, J. Polym. Sci. Part A, 1989, 27, 2385.
142 Wei, Y., Sun, Y., Jang, G.-W., Tang, X., Effects P-Aminodiphenylamine on Electrochemical Polymeization of Aniline, J. Polym. Sci., Part C, 1990, 28, 81.
143 Wei, Y., Jang, G.-W., Polymerization of Aniline and Alkyl Ringsubstituted Anilines in the Presence of Aromatic Additives, J. Phys. Chem. 1990, 94, 7716.
144 Wiles, D. M., Photooxidative Reactions of Polymers, J. Appl. Polym. Sci., Appl. Polym. Symp. 1979, 35, 235.
145 Wilk, K. A., Burczyk, B., Wilk, T., Oil in Water Microemulsions as Reaction Media, 7th International Conference Surface and Colloid Science, Compiegne 1991.
146 Winogradova, C. B., Novoe v obłasti polikondensacji, Vysokomol. Soed. A, 1985, 27, 2243.
147 Wirpsza, Z., Poliuretany, WNT, Warszawa 1991.
148 Wojtala, A., Wpływ właściwości oraz otoczenia poliolefin na przebieg ich fotodegradacji, Polimery 2001, 46, 120.
149 Xue, G., Polymerization of Styrene Oxide with Pyridine, Makromol. Chem. Rapid Commun. 1986, 7, 37.
150 Yamazaki, N., Imai, Y., Phase Transfer Catalyzed Polycondensation of Bishalomethyl Aromatic compounds, Polym. J. 1983, 15, 905.
151 Yasuda, H., Plasma Polymerization, Academic Press, Inc. Orlando 1985.
152 Yuan, H. G., Kalfas, G., Ray, W. H., Suspension polymerization, J. Macromol. Sci. Rev. Macromol. Chem. -Phys., 1991, C 31, 215.

153 Zubakowa, L. B., Tevlina, A. C., Davankov, A. B., Sinteticzeskije ionoobmiennyje materiały, Chimia, Moskwa 1978.
154 Zubanov, B. A., Viedienie v chimiu polikondensacionnych procesov, Nauka, Ałma Ata 1974.
155 Żuchowska, D., Polimery konstrukcyjne, WNT, Warszawa 2000.
156 Żuchowska, D., Struktura i właściwości polimerów jako materiałów konstrukcyjnych, W. P. Wr., Wrocław 1986.
157 Żuchowska, D., Steller, R., Meisser, W., Structure and properties of degradable polyolefin – starch blends, Polym. Degrad. Stab. 1998, 60, 471.

CHAPTER 9

UPDATES ON PARTICULATE-FILLED POLYMER NANOCOMPOSITES

G. E. ZAIKOV,[1] G. V. KOZLOV,[2] A. K. MIKITAEV,[2] and A. K. HAGHI[3]

[1]*N.M. Emanuel Institute of Biochemical Physics of Russian Academy of Sciences, Kosygin St., 4, Moscow-119334, Russian Federation*

[2]*Kh.M. Berbekov Kabardino-Balkarian State University, Chernyshevsky St., 173, Nalchik, 360004, Russian Federation*

[3]*University of Guilan, Rasht, Iran*

CONTENTS

ABSTRACT

A number of the main mechanical characteristics (yield stress, impact toughness, microhardness) of particulate-filled polymer nanocomposites were described quantitatively within the framework of general conception – fractal analysis. Such approach allows to study the main specific features of the indicated nanomaterials mechanical behavior. The influence of both nanofiller initial particles size and their aggregation degree on nanocomposites mechanical properties has been shown.

9.1 INTRODUCTION

Mechanical properties represent a very important part of polymer materials characteristics, particularly if the talk is about their application as engineering materials. Nevertheless, even if the indicated materials have another functional assignment, mechanical properties remain always-practical application important factor in this case as well. Particulate-filled polymer nanocomposites mechanical properties have a specific features number, which will be considered below.

As it is well-known [1], the yield stress of polymeric materials is an important operating characteristic, restricting the range of their application as engineering materials from above. Therefore, the theoretical treatment of yielding process was always paid special attention, to that resulted in the development of a large number of theoretical models, describing this process [2]. For particulate-filled polymer nanocomposites the specific feature of the dependence of yield stress σ_Y on nanofiller contents is observed [3, 4]: unlike microcomposites of the same class [5], the value σ_Y is not increase to some extent perceptibly at nanofiller contents growth and even can be reduced. It is obvious, that the indicated effect is a negative one from the point of view of these polymeric materials exploitation, since it restricts their using possibilities as engineering materials.

The authors of papers [6, 7] found out, that the introduction of particulate nanofiller (calcium carbonate ($CaCO_3$)) into high density polyethylene (HDPE) results in nanocomposites HDPE/$CaCO_3$ impact toughness A_p in comparison with the initial polymer by about 20%. The authors [6, 7] performed this effect detailed fractographic analysis and explained the observed A_p increase by nanocomposites HDPE/$CaCO_3$ plastic deformation mechanism change in comparison with the initial HDPE. Without going into details of the indicated analysis, one should note some reasons for doubts in its correctness. In Figure 9.1 the schematic diagrams load-time

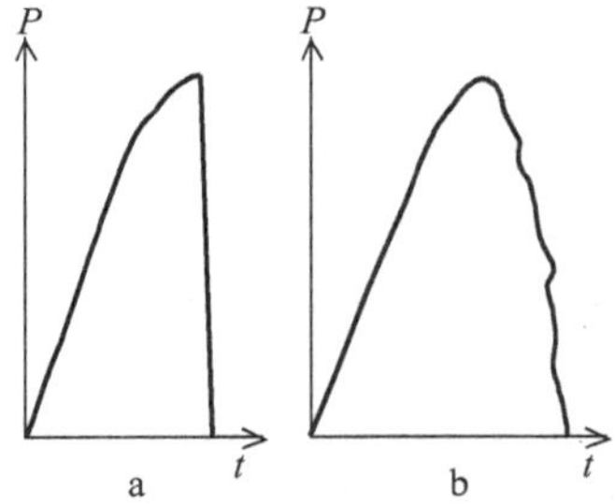

FIGURE 9.1 The schematic diagrams load-time (*P–t*) in instrumented impact tests. Failure by instable (a) and stable (b) crack.

(*P-t*) for two cases of polymeric materials samples failure in impact tests are adduced: by instable (a) and stable (b) cracks. As it is known [8], the value A_p is characterized by the area under *P-t* diagram, which gives mechanical energy expended on samples failure. The polymeric materials macroscopic fracture process, defined by main crack propagation, begins at the greatest load *P*. From the schematic diagrams *P-t* it follows, that fracture process as a matter of fact practically does not influence on A_p value in case of crack instable propagation and influences only partly in case of stable crack. Although the authors [6, 7] performed impact tests on instrumented apparatus, allowing to obtain diagrams *P-t*, these diagrams were not adduced. Moreover, the structural aspect of fracture process has been considered in works [6, 7] with the usage of secondary structures (crazes, shear zones and so on). Their interconnection with the initial undeformed material structure is purely speculative. It is obvious, that it does not occur possible to obtain quantitative relationships structure properties at such method of analysis.

At present it is known [10–12], that microhardness H_v is the property sensitive to morphological and structural changes in polymeric materials. For composite materials the existence of the filler, whose microhardness exceeds by far polymer matrix corresponding characteristic, is an additional powerful factor [13]. The introduction of sharpened indentors in the form of a cone or a pyramid in polymeric material a stressed state is localized in small enough microvolume and it is supposed, that in such tests polymeric materials real structure is found [14]. In connection with the fact, that polymer nanocomposites are complex enough [15], the question arises, which structure component reacts on indentor forcing and how far this reaction alters with particulate nanofiller introduction.

The interconnection of microhardness, determined according to the results of the tests in a very localized microvolume, with such macroscopic properties of polymeric materials as elasticity modulus E and yield stress σ_Y is another problem aspect. At present a large enough number of derived theoretically and obtained empirically relationships between H_v, E and σ_Y exists [16].

Proceeding from the said above, the purpose of the present work is the indicated mechanical properties of particulate-filled polymer nanocomposites treatment within the framework of general structural approach, namely, the fractal analysis.

9.2 EXPERIMENTAL PART

Polypropylene (PP) "Kaplen" of mark 01030 with average weight molecular mass of ~ $(2–3)\times10^3$ and polydispersity index 4.5 was used as a matrix polymer. Nanodimensional calcium carbonate ($CaCO_3$) in a compound form of mark Nano-Cal P-1014 (production of China) with particles size of 80 nm and mass contents of 1–7 mass % and globular nanocarbon (GNC) (production of corporations group "United Systems," Moscow, Russian Federation) with particles size of 5–6 nm, specific surface of 1400 m^2/g and mass contents of 0.25–3.0 mass % were applied as nanofiller.

Nanocomposites PP/$CaCO_3$ and PP/GNC were prepared by components mixing in melt on twin-screw extruder Thermo Haake, model Reomex RTW 25/42, production of German Federal Republic. Mixing was performed at temperature 463–503 K and screw speed of 50 rpm during 5 min. Testing samples were prepared by casting under pressure method on a casting machine Test Sample Molding Apparate RR/TS MP of firm Ray-Ran (Taiwan) at temperature 483 K and pressure 43 MPa.

Uniaxial tension mechanical tests have been performed on the samples in the shape of a two-sided spade with the sizes according to GOST 112 62–80. The tests have been conducted on a universal testing apparatus Gotech Testing Machine CT-TCS 2000, production of German Federal Republic, at temperature 293 K and strain rate ~ 2×10^{-3} s^{-1}.

The impact tests have been conducted by Sharpy method on samples by sizes of 80×10×4 mm. Samples have V-like notch with length of 0.8 mm. Tests have been performed on pendulum apparate model Gotech Testing

Machine GT-7045-MD, production of Taiwan, with the energy dial of 1 J so that no less than 10% and no more than 80% of energy reserve was consumed on sample failure, with distance between supports (span) of 60 mm. No less than 5 samples were used for each test.

The microhardness H_v measurements by Shore (scale D) were performed according to Gost 24 621–91 on Scleroscope HD-3000, model 05-2 of form "Hildebrand," production of German Federal Republic. The samples have cylindrical shape with diameter of 40 mm and height of 3 mm.

9.3 RESULTS AND DISCUSSION

9.3.1 YIELD STRESS

For the dependence of yield stress σ_Y on particulate nanofiller contents theoretical analysis the dispersive theory of strength was used, where nanocomposite yield stress at shear τ_n is determined as follows [17]:

$$\tau_n = \tau_m^{'} + \frac{G_n b_B}{\lambda} \tag{9.1}$$

where $\tau_m^{'}$ is shear yield stress of polymer matrix, G_n is shear modulus of nanocomposite, b_B is Burgers vector, λ is distance between nanofiller initial particles in nanocomposite.

In case of nanofiller particles aggregation the Eq. (9.1) has the look [17]:

$$\tau_n = \tau_m^{'} + \frac{G_n b_B}{k(r)\lambda} \tag{9.2}$$

where $k(r)$ is aggregation parameter.

It is easy to see, that the Eq. (9.2) describes the initial nanoparticles aggregation influence on nanocomposite yield stress. This effect is important from both theoretical and practical points of view in virtue of well-known nanoparticles tendency to aggregation, which is expressed by the following relationship [15]:

$$k(r) = 7.5 \times 10^{-3} S_u \tag{9.3}$$

where S_u is nanofiller specific surface, which is determined as follows [18]:

$$S_u = \frac{6}{\rho_n D_p} \tag{9.4}$$

where ρ_n is nanofiller density, D_p is its particles diameter.

From the Eqs. (9.3) and (9.4) it follows, that the nanofiller particles size decreasing results in S_u enhancement, that intensifies nanofiller initial particles tendency to aggregation.

Let us consider determination methods of the parameters, included in the Eq. (9.2). The general relationship between normal stress σ and shear stress τ has the look [19]:

$$\tau = \frac{\sigma}{\sqrt{3}} \tag{9.5}$$

The stress $\tau_m^{'}$ is determined according to the equation [17]:

$$\tau_m^{'} = \tau_m \left(1 - \varphi_n^{2/3}\right) \tag{9.6}$$

where τ_m is shear yield stress of matrix polymer, φ_n is nanofiller volume content, determined according to the well-known formula [1]:

$$\varphi_n = \frac{W_n}{\rho_n} \tag{9.7}$$

where W_n is nanofiller mass content and the value ρ_n for nanoparticles is determined as follows [15]:

$$\rho_n = 188\left(D_p\right)^{1/3}, \text{kg/m}^3 \tag{9.8}$$

where D_p is given in nm.

The shear modulus G_n is connected with Young's modulus E_n by the following simple relationship [20]:

$$G_n = \frac{E_n}{d_f} \tag{9.9}$$

where d_f is fractal dimension of nanocomposite structure, which is determined according to the equation [20]:

$$d_f = (d-1)(1+\nu) \quad (9.10)$$

where d is dimension of Euclidean space, in which a fractal is considered (it is obvious, that in our case d=3), ν is Poisson's ratio, estimated according to the mechanical test results with the aid of the relationship [14]:

$$\frac{\sigma_Y}{E_n} = \frac{1-2\nu}{6(1+\nu)} \quad (9.11)$$

where E_n is nanocomposite elasticity modulus.

The value of Burgers vector b_B for polymeric materials is determined according to the equation [2]:

$$b_B = \left(\frac{60.5}{C_\infty}\right)^{1/2}, \text{ Å} \quad (9.12)$$

where C_∞ is characteristic ratio, connected with dimension d_f by the equation [2]:

$$C_\infty = \frac{2d_f}{d(d-1)(d-d_f)} + \frac{4}{3} \quad (9.13)$$

It is obvious, that for the value τ_n theoretical estimation according to the Eq. (9.2) an independent method of parameter $k(r)\lambda$ determination is necessary. The following equation gives such method [3]:

$$k(r)\lambda = 2.09\times10^{-2} D_p (S_u/\phi_n)^{1/2} \quad (9.14)$$

where D_p is given in nm, S_u – in m^2/g.

The value S_u estimation according to the Eq. (9.4) gave the following results: S_u=3280 and 93 m^2/g for GNC and $CaCO_3$, respectively.

In Figure 9.2, the comparison of the received experimentally σ_Y and calculated according to the described above method σ_Y^T yield stress values for nanocomposites PP/GNC is adduced. As one can see, the theory and experiment good correspondence is observed (the average discrepancy

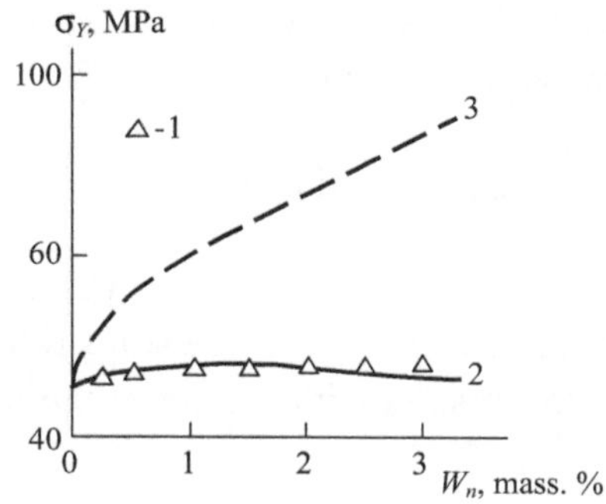

FIGURE 9.2 The dependences of yield stress σ_Y on nanofiller mass contents W_n for nanocomposites PP/GNC. 1 – experimental data; 2, 3 – calculation according to the Eqs. (9.2) and (9.1), respectively.

between σ_Y and σ_Y^T makes up 5.5%). Besides, the value σ_Y for nanocomposites does not differ to some extent significantly from the corresponding parameter for matrix PP: for nanocomposites σ_Y=36.0–32.9 MPa, for PP σ_Y=31.5 MPa, i.e., σ_Y enhancement at nanofiller introduction does not exceed 15%. The causes of such effect can be elucidated by the Eq. (9.1) using, where the value λ is calculated as follows [17]:

$$\lambda = \left[\left(\frac{4\pi}{3\varphi_n}\right)^{1/3} - 2\right]\frac{D_p}{2} \quad (9.15)$$

The Eq. (9.1) supposes nanofiller initial particles aggregation absence ($k(r)$=1.0) and the dependence σ_Y^T (W_n), calculated according to the indicated equation, is also adduced in Figure 9.2. The absence of GNC initial particles aggregation results in nanocomposites PP/GNC yield stress strong increasing within the range of W_n=0.25–3.0 mass % – from 44 up to 86 MPa.

In Figure 9.3, the similar dependences of yield stress on W_n for nanocomposites PP/$CaCO_3$ are adduced. As one can see, σ_Y^T estimation according to the Eq. (9.2) gives an excellent correspondence to the experiment, the average discrepancy between σ_Y and σ_Y^T makes up 0.7% only. Besides, for nanocomposites PP/$CaCO_3$ σ_Y reduction within the range of W_n=1–7 mass % is observed. And at last, $CaCO_3$ initial nanoparticles aggregation suppression does not give positive effect for these nanocomposites. It is obvious, that the cause of the indicated σ_Y reduction for non-aggregated

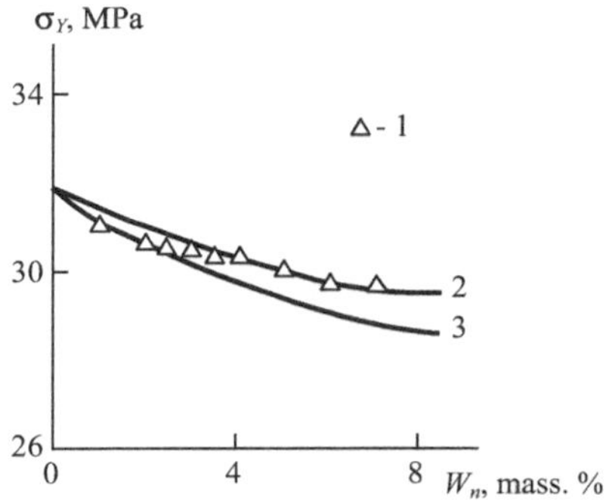

FIGURE 9.3 The dependences of yield stress σ_Y on nanofiller mass contents W_n for nanocomposites PP/$CaCO_3$. 1 – experimental data; 2, 3 – calculation according to the Eqs. (9.2) and (9.1), respectively.

$CaCO_3$ is a relatively large diameter of its initial nanoparticles, approaching to upper limit of nanoparticles dimensional range, which is equal to ~ 100 nm [21]. Owing to that λ value for nanocomposites PP/$CaCO_3$ varies within the limits of 200–66 nm within the range of W_n=1–7 mass %, whereas for nanocomposites PP/GNC the value λ is essentially smaller: 17–4 nm within the range of W_n=0.25–3.0 mass %. Thus, in particulate-filled polymer nanocomposites yield stress value definition two competeting factors played critical role: nanofiller initial particles size and their aggregation level. It is important to note, that weak dependence of yield stress on nanofiller contents is typical not only for particulate-filled polymer nanocomposites, but also for other classes of these nanomaterials: polymer/organoclay [22] and polymer/carbon nanotubes [23].

Let us consider alternative, specific for nanocomposites with semicrystalline matrix, treatment of yield stress change. The Eqs. (9.10) and (9.11) combination at the condition d=3 allows to obtain the following dependence of ratio E_n/σ_Y on the main structural characteristic d_f:

$$\frac{E_n}{\sigma_Y} = \frac{3d_f}{\left(3-d_f\right)} \tag{9.16}$$

In Figure 9.4, the dependence of ratio E_n/σ_Y on dimension d_f is adduced, which demonstrates strong nonlinear growth of the indicated ratio at d_f increasing, specifically at $d_f \geq 2.7$. Thus, the postulated in work [24] σ_Y and E_n proportionality is true in special case only, namely, in case of polymeric

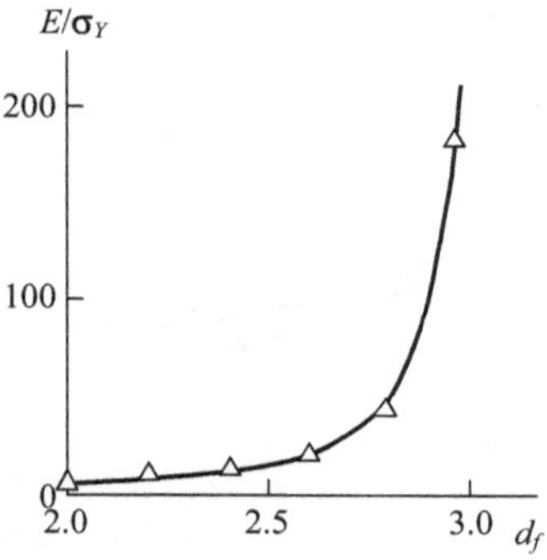

FIGURE 9.4 The dependence of elasticity modulus and yield stress ratio E/σ_Y on structure fractal dimension d_f for polymeric materials.

material structure invariability, that is, in case d_f =const. This rule is fulfilled for particulate-filled polymer nanocomposites with amorphous matrix: phenylone/β-sialone and phenylone/oxynitride silicium-yttrium [15]. For the indicated nanocomposites d_f =const=2.416 and then E_n/σ_Y=12.4, that is confirmed experimentally [15].

Let us consider further yield stress σ_Y behavior as a function of nanofiller mass contents W_n for the considered nanocomposites. The value d_f for them can be estimated by an independent mode, using the following equation [2]:

$$d_f = 3 - 6\left(\frac{\varphi_{cl}}{SC_\infty}\right)^{1/2} \tag{9.17}$$

where φ_{cl} is local order domains (nanoclusters) relative fraction, S is macromolecule cross-sectional area, which is equal to 27.2 Å^2 for PP [2].

The value φ_{cl} can be estimated according to the following percolation relationship [2]:

$$\varphi_{cl} = 0.03(1-K)(T_m - T)^{0.55} \tag{9.18}$$

where K is crystallinity degree, T_m and T are melting and testing temperatures, respectively. For the considered nanocomposites the value K according to DSC data varies within the limits of 0.637–0.694 for PP/GNC and 0.637–0.668 for PP/$CaCO_3$ and the value T_m for PP was accepted equal to 445 K [25].

Since the value φ_{cl} is considered for nanocomposites, where nanoclusters are concentrated in polymer phase only, then one should use its reduced value φ_{cl}^{red}, which is equal to [15]:

$$\varphi_{cl}^{red} = \varphi_{cl}\left(1-\varphi_n\right) \tag{9.19}$$

In Figure 9.5 the comparison of the received experimentally and calculated according to the described above method dependences of yield stress σ_Y on nanofiller mass contents W_n for the considered nanocomposites is adduced. As one can see, a good correspondence of theory and experiment is obtained (their mean discrepancy makes up 2.5%). This circumstance allows to explain the cause of insignificant increasing and in case of nanocomposites PP/$CaCO_3$ even reduction of yield stress at nanofiller contents growth. As it is known [15], the fractal dimension d_f of crystallizing polymeric materials structure depends on their crystallinity degree K as follows:

$$d_f = 2 + K + \varphi_{if} \tag{9.20}$$

where φ_{if} is a relative fraction of interfacial (crystallizing also) regions.

Therefore the values d_f for the considered nanocomposites are within the range of ~ 2.75–2.80, that is, within the range, where the ratio E_n/σ_Y strong increase begins (see Figure 9.4). So, for nanocomposites PP/GNC the ratio E_n/σ_Y value varies within the range of 31.0–41.8 and for PP/$CaCO_3$ – within the limits of 31.0–37.2. This increase corresponds by absolute value to

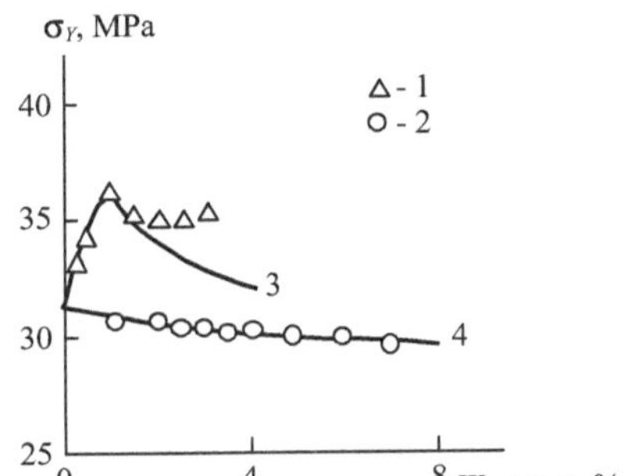

FIGURE 9.5 The dependences of yield stress σ_Y on nanofiller mass contents W_n for nanocomposites PP/GNC (1, 3) and PP/$CaCO_3$ (2, 4). 1, 2 – the theoretical calculation; 3, 4 – experimental data.

nanocomposites elasticity modulus enhancement, which makes up ~ 40% for PP/GNC and ~ 13% for PP/$CaCO_3$. Hence, the ratio E_n/σ_Y increasing compensates E_n growth owing to nanofiller introduction and in case of nanocomposites PP/$CaCO_3$ E_n small increase results in σ_Y reduction. Let us note, that the proposed model is true for nanocomposites with amorphous matrix as well, having high enough d_f values (e.g., for rubbers) [4].

9.3.2 IMPACT TOUGHNESS

In Figure 9.6, the dependences of impact toughness A_p on nanofiller volume contents φ_n are adduced for the considered nanocomposites. As it follows from the data of this figure, for both nanocomposites the dependence $A_p(\varphi_n)$ has an extreme character, whose maximum is reached at $\varphi_n \approx 0.03$. A_p increasing for nanocomposites in comparison with the corresponding parameter for matrix polymer can be significant: so, for nanocomposite PP/$CaCO_3$ at $\varphi_n = 0.03$ A_p value exceeds impact toughness for PP in 1.5 times. Nanocomposites PP/GNC and PP/$CaCO_3$ mechanical properties study has shown that a similar extreme dependence of property on nanofiller contents has yield stress σ_Y only (see Figs. 9.2 and 9.4). As it has been shown above, such dependence $\sigma_Y(\varphi_n)$ shape is due to nanofiller initial particles aggregation, which is intensified at φ_n growth. This interconnection is not accidental: as it has been noted above, A_p value is proportional to area under P–t diagram or curve stress-strain (σ–ε). In its turn, for plastic polymeric materials, which are investigated nanocomposites, a stress is restricted from above by yield

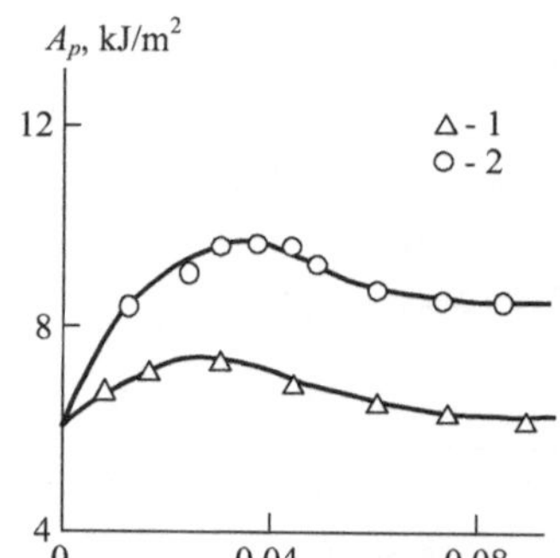

FIGURE 9.6 The dependences of impact toughness A_p on nanofiller volume contents φ_n for nanocomposites PP/GNC (1) and PP/$CaCO_3$ (2).

stress σ_Y and limiting strain is equal to failure strain ε_f. Therefore, it is to be expected that impact toughness A_p is proportional to product $\sigma_Y\varepsilon_f$. Within the framework of fractal analysis the value ε_f is determined as follows [2]:

$$\varepsilon_f = C_\infty^{D_{ch}-1} - 1 \tag{9.21}$$

where D_{ch} is fractal dimension of a polymer chain part between its fixation points (chemical cross-linking nodes, physical entanglements, nanoclusters, etc.).

The parameters C_∞ and D_{ch} characterize polymer chain statistical flexibility and molecular mobility level, respectively [2]. The dimension D_{ch} can be determined with the aid of the following equation [2]:

$$\frac{2}{\varphi_{cl}} = C_\infty^{D_{ch}} \tag{9.22}$$

In Figure 9.7 the dependence $A_p(\sigma_Y\varepsilon_f)$ for the considered nanocomposites is adduced, which proves to be linear and passing through coordinates origin, that allows to describe it analytically by the following empirical equation:

$$A_p = \left(0.4\times10^{-3}\,m\right)\sigma_Y\varepsilon_f \tag{9.23}$$

The adduced above analysis allows to elucidate the cause of higher values A_p for nanocomposites PP/$CaCO_3$ in comparison with PP/GNC. This cause is higher D_{ch} values: for PP/$CaCO_3$ D_{ch}=1.33–1.34, for PP/GNC D_{ch}=1.13–1.29, that is, higher molecular mobility level for nanocomposites PP/$CaCO_3$, although σ_Y values are somewhat higher for PP/GNC.

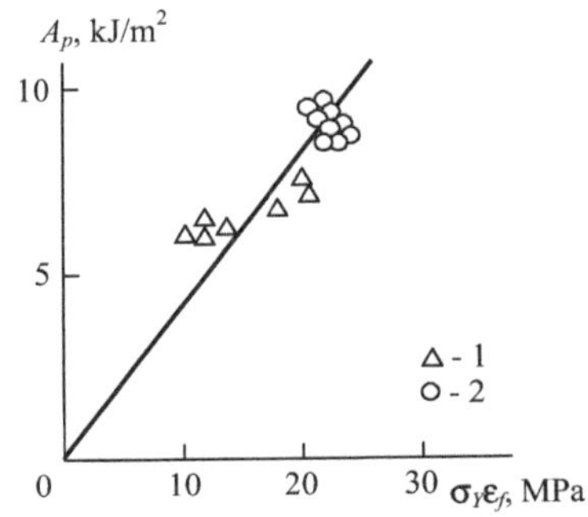

FIGURE 9.7 The dependence of impact toughness A_p on parameter $\sigma_Y\varepsilon_f$ for nanocomposites PP/GNC (1) and PP/$CaCO_3$ (2).

9.3.3 MICROHARDNESS

Let us consider the interconnection of microhardness H_v and other mechanical characteristics, among their number yield stress σ_Y for nanocomposites PP/GNC and PP/$CaCO_3$. The following relationship was received by Tabor [26] for metals, which were considered as rigid perfectly plastic solids, between H_v and σ_Y:

$$\frac{H_v}{\sigma_Y} \approx c \tag{9.24}$$

where c is constant, which is approximately equal to 3.

The relationship (9.24) implies, that the exerted in microhardness tests pressure under indentor is higher than yield stress in quasistatic tests owing to restriction, imposed by undeformed polymer, surrounding indentor. However, in works [12, 16, 22, 27, 28] it has been shown that the value c can differ essentially from 3 and varied in wide enough limits: ~ 1.5–30. In the work [28] it has been found out, that for the composites HDPE/$CaCO_3$ depending on strain rate $\dot{\varepsilon}$ and type of quasistatic tests, in which the value σ_Y was determined (tensile or compression) c magnitude varies within the limits of 1.80–5.83. To c=3 the ratio H_v/σ_Y approaches only at minimum value $\dot{\varepsilon}$ and at using σ_Y values, received by compression tests. Therefore, in the work [28] the conclusion has been obtained, that the value c=3 can be received only at comparable strain rates in microhardness and quasistatic tests and at interfacial boundaries polymer-filler failure absence.

An elasticity role in indentation process was proposed to consider for the analysis spreading on a wider interval of solids. For the solid, having elasticity modulus E and Poisson's ratio ν, Hill has obtained the following equation [16]:

$$H_v = \frac{2}{3}\left[1 + \ln\frac{E}{3(1-\nu)\sigma_Y}\right]\sigma_Y \tag{9.25}$$

and empirical Marsh equation has the look [16]:

$$H_v = \left(0.07 + 0.6\ln\frac{E}{\sigma_Y}\right)\sigma_Y \tag{9.26}$$

The Eqs. (9.25) and (9.26) allow the microhardness H_v theoretical estimation for particulate-filled polymer nanocomposites at the known E and σ_Y condition and the value ν can be calculated according to the Eq. (9.11).

In Table 9.1, the comparison of experimental H_v and calculated according to the Eq. (9.26) H_v^T microhardness values for the considered nanocomposites is adduced. The Eq. (9.26) was chosen according to a simple reason that it gives better correspondence to experiment than the Eq. (9.25) for all classes of nanocomposites [4, 22, 29]. As it follows from the data of this table, a good enough correspondence of theory and experiment is obtained (the mean discrepancy between H_v and H_v^T makes up ~ 8%). This correspondence indicates, that H_v value for the considered nanocomposites is controlled by their macroscopic mechanical properties to the same extent, as for other materials.

TABLE 9.1 The Comparison of Obtained Experimentally H_v and Calculated According to the Eq. (26) H_v^T Microhardness Values for Nanocomposites PP/GNC and $PP/CaCO_3$

Nanocomposite	Nanofiller mass content, mass %	H_v, MPa	H_v^T, MPa	Δ, %
PP/GNC	0	68	66.3	2.5
	0.25	75	71.7	4.4
	0.50	73	75.8	9.4
	1.0	74	78.7	6.4
	1.50	72	76.4	6.1
	2.0	72	75.5	4.9
	2.50	72	75.2	4.4
	3.0	72	75.9	4.8
$PP/CaCO_3$	1.0	72	67.0	10.7
	2.0	75	66.8	10.9
	2.5	76	66.8	12.1
	3.0	75	66.5	10.9
	3.5	75	66.7	11.3
	4.0	75	66.4	11.1
	5.0	75	66.7	11.5
	6.0	75	66.2	11.7
	7.0	75	66.5	11.3

Δ is relative discrepancy between parameter H_v and H_v^T.

Let us consider the physical nature of the ratio H_V/σ_Y deviation from the constant $c\approx 3$ in the Eq. (9.24), using for this purpose the relationships (9.25) and (9.26). The value d_f can be determined according to the Eq. (9.10) and then the relationships combination allows to obtain the following equations [4, 22]:

$$\frac{H_v}{\sigma_Y}=\frac{2}{3}\left\{1+\ln\left[\frac{2d_f}{\left(4-d_f\right)\left(3-d_f\right)}\right]\right\} \tag{9.27}$$

and

$$\frac{H_v}{\sigma_Y}=\left[0.07+0.6\ln\left(\frac{3d_f}{3-d_f}\right)\right] \tag{9.28}$$

for case d=3.

From the Eqs. (9.27) and (9.28) it follows, that the ratio H_V/σ_Y is defined by structural state of nanocomposite (matrix polymer) only, which is characterized by its fractal dimension d_f. In Figure 9.8, the dependences of the ratio H_V/σ_Y on d_f is adduced, calculated according to the Eqs. (9.27) and (9.28), which found out complete similarity, but absolute values H_V/σ_Y, calculated by the Eq. (9.27), proved to be on about 15% higher than the analogous magnitudes, calculated according to the Eq. (9.28). The identical results for extrudates of polymerization-filled compositions on the basis of ultra-high-molecular

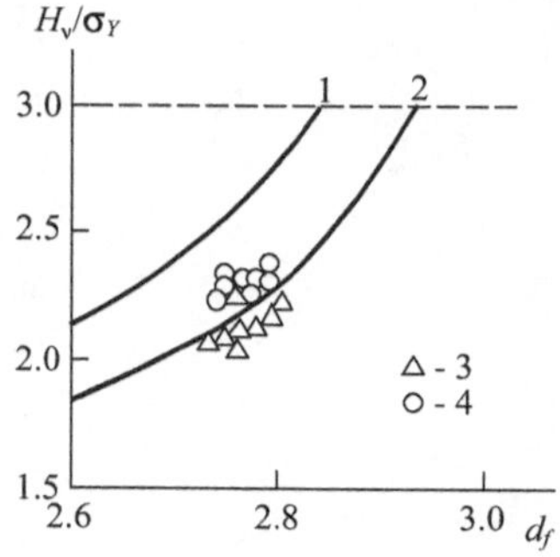

FIGURE 9.8 The dependences of ratio H_V/σ_Y on structure fractal dimension d_f. 1, 2 – calculation according to the Eqs. (9.27) (1) and (9.28) (2); 3, 4 – experimental data for nanocomposites PP/GNC (3) and PP/$CaCO_3$ (4). The horizontal stroked line indicates Tabor criterion H_V/σ_Y=c=3.

polyethylene were obtained in work [12] and for polymer nanocomposites of different classes – in works [4, 22, 29]. In Figure 9.8, the condition $H_V/\sigma_Y = c = 3$ according to the Eq. (24) is given by a horizontal stroke line. As it follows from this figure data, the indicated condition is reached at $d_f \approx 2.85$ according to the Eq. (9.27) and $d_f \approx 2.93$ according to the Eq. (9.28). As it is known [20], for real solids the limiting greatest value d_f is equal to 2.95. Thus, these d_f values at $c=3$ indicate again, that the empirical Marsh equation (the Eq. (9.28)) gives more precise estimation of ratio H_V/σ_Y, than more strictly derived Hill relationship (the Eq. (9.27)). Hence, the adduced above results show, that the ratio H_V/σ_Y value is defined by polymer nanocomposites structural state only and Tabor criterion $c=3$ is realized for Euclidean solids only.

In Figure 9.8, the dependence of the obtained experimentally values H_V/σ_Y on nanocomposites structure fractal dimension d_f is shown by points. As one can see, the obtained experimentally dependence $H_V/\sigma_Y(d_f)$ corresponds well to the theoretical curve, calculated according to the Eq. (9.28) (the mean discrepancy between theory and experiment makes up 8.5%), whereas calculation according to the Eq. (9.27) gives overstated absolute values of the indicated ratio.

9.4 CONCLUSIONS

The yield stress value of particulate-filled polymer nanocomposites is controlled by two competeting factors: diameter of nanofiller initial particles and their aggregation degree. The cause of weak increasing (and even reduction) of the indicated nanocomposites yield stress at nanofiller contents growth is the initial nanoparticles strong aggregation. The application of nanoparticles disaggregation artificial methods might be worth-while only for nanocomposites with a small diameter of the initial nanoparticles. High enough values of nanocomposites structure fractal dimension are an additional factor, influencing on this effect.

The impact toughness of particulate-filled polymer nanocomposites is defined by a number of factors on various structural levels: molecular, topological and suprasegmental ones. The indicated levels characteristics are interconnected and changed at nanofiller introduction. The molecular mobility level is the main parameter, defining the considered nanocomposites impact toughness.

The ratio of microhardness and yield stress for particulate-filled polymer nanocomposites is defined by this polymer nanomaterials structural state only. Tabor criterion is correct for Euclidean (or close to them) solids only.

KEYWORDS

- **fractal analysis**
- **mechanical properties**
- **nanocomposite**
- **structure**
- **ultrafine particles**

REFERENCES

1. Narisawa I. Strength of Polymeric Materials. Moscow, Chemistry, 1987, 400 p.
2. Kozlov G.V., Zaikov G.E. Structure of the Polymer Amorphous State. Utrecht, Boston, Brill Academic Publishers, 2004, 465 p.
3. Kozlov G.V., Zaikov G.E. Structure and Properties of Particulate-Filled Polymer Nanocomposites. Saarbrücken, Lambert Academic Publishing, 2012, 112 p.
4. Kozlov G.V., Yanovskii Yu.G., Zaikov G.E. Particulate-Filled Polymer Nanocomposites. Structure, Properties, Perspectives. New York, Nova Science Publishers, Inc. 2014, 273 p.
5. Kozlov G.V., Yanovskii Yu.G., Zaikov G.E. Structure and Properties of Particulate-Filled Polymer Composites: The Farctal Analysis. New York, Nova Science Publishers, Inc. 2010, 282 p.
6. Tanniru M., Misra R.D.K. Mater. Sci. Engng., 2005, v. A405, № 1, p. 178–193.
7. Deshmane C., Yuan Q., Misra R.D.K. Mater. Sci. Engng., 2007, v. A452–453, № 3, p. 592–601.
8. Bucknall C.B. Toughened Plastics. London, Applied Science, 1977, 315 p.
9. Bartenev G.M., Frenkel S.Ya. Physics of Polymers. Leningrad, Chemistry, 1990, 432 p.
10. Balta-Calleja F.J., Kilian H.G. Colloid Polymer Sci., 1988, v. 266, № 1, p. 29–34.
11. Balta-Calleja F.J., Santa Cruz C., Bayer R.K., Kilian H.G. Colloid Polymer Sci., 1990, v. 268, № 5, p. 440–446.
12. Aloev V.Z., Kozlov G.V. The Physics of Orientational Effects in Polymeric Materials. Nalchik, Polygraphservice and T, 2002, 288 p.
13. Perry A.J., Rowcliffe D.J. J. Mater. Sci. Lett., 1973, v. 8, № 6, p. 904–907.
14. Kozlov G.V., Sanditov D.S. Anharmonic Effects and Physical-Mechanical Properties of Polymers. Novosibirsk, Science, 1994, 261 p.

15. Mikitaev A.K., Kozlov G.V., Zaikov G.E. Polymer Nanocomposites: Variety of Structural Forms and Applications. New York, Nova Science Publishers, Inc. 2008, 319 p.
16. Kohlstedt D.L. J. Mater. Sci., 1973, v. 8, № 6, p. 777–786.
17. Sumita M., Tsukumo Y., Miyasaka K., Ishikawa K. J. Mater. Sci., 1983, v. 18, № 5, p. 1758–1764.
18. Bobryshev A.N., Kozomazov V.N., Babin L.O., Solomatov V.I. Synergetic of Composite Materials. Lipetsk, NPO ORIUS, 1994, 154 p.
19. Honeycombe R.W.K. The Plastic Deformation of Metals. Cambridge, Edwards Arnold Publishers, 1968, 402 p.
20. Balankin A.S. Synergetic of Deformable Body. Moscow, Publishers of Ministry Defense SSSR, 1991, 404 p.
21. Buchachenko A.L. Achievements of Chemistry, 2003, v. 72, № 5, p. 419–437.
22. Kozlov G.V., Mikitaev A.K. Structure and Properties of Nanocomposites Polymer/Organoclay. Saarbrücken, LAP LAMBERT Academic Publishing GmbH and Comp., 2013, 318 p.
23. Yanovsky Yu.G., Kozlov G.V., Zhirikova Z.M., Aloev V.Z., Karnet Yu.N. Intern. J. Nanomechanics Science and Technology, 2012, v. 3, № 2, p. 99–124.
24. Brown N. Mater. Sci. Engng., 1971, v. 8, № 1, p. 69–73.
25. Kalinchev E.L., Sakovtseva M.B. Properties and Processing of Thermoplastics. Leningrad, Chemistry, 1983, 288 p.
26. Tabor D. The Hardness of Metals. New York, Oxford University Press, 1951, 329 p.
27. Aphashagova Z.Kh., Kozlov G.V., Burya A.I., Zaikov G.E. Theoretical Fundamentals of Chemical Technology, 2007, v. 41, № 6, p. 699–702.
28. Suwanprateeb J. Composites, Part A, 2000, v. 31, № 3, p. 353–359.
29. Zhirikova Z.M., Kozlov G.V., Aloev V.Z. Main Problems of Modern Materials Science, 2012, v. 9, № 1, p. 82–85.

CHAPTER 10

THE REINFORCEMENT OF PARTICULATE-FILLED POLYMER NANOCOMPOSITES

G. E. ZAIKOV,[1] G. V. KOZLOV,[2] A. K. MIKITAEV,[2] and A. K. HAGHI[3]

[1]*N.M. Emanuel Institute of Biochemical Physics of Russian Academy of Sciences, Kosygin St., 4, Moscow-119334, Russian Federation*

[2]*Kh.M. Berbekov Kabardino-Balkarian State University, Chernyshevsky St., 173, Nalchik, 360004, Russian Federation*

[3]*University of Guilan, Rasht, Iran*

CONTENTS

ABSTRACT

The applicability of irreversible aggregation model for theoretical description of nanofiller particles aggregation process in polymer nanocomposites has been shown. The main factors, influencing on nanoparticles aggregation process, were revealed. It has been shown that strongly expressed particulate nanofiller particles aggregation results in sharp (in about 4 times) formed fractal aggregates real elasticity modulus reduction. Nanofiller particles aggregation is realized by cluster-cluster mechanism and results in the formed fractal aggregates density essential reduction, that is the cause of their elasticity modulus decreasing. As distinct from microcomposites nanocomposites require consideration of interfacial effects for elasticity modulus correct description in virtue of a well-known large fraction of phases division surfaces for them.

10.1 INTRODUCTION

In the course of technological process of particulate-filled polymer composites in general [1] and nanocomposites [2–4] in particular preparation of the initial filler powder particles aggregation in more or less large particles aggregates always occurs. The aggregation process exercises essential influence on composites (nanocomposites) macroscopic properties [1–5]. For nanocomposites the aggregation process gains special significance, since its intensity can be such, that nanofiller particles aggregates size exceeds 100 nm – the value, which is assumed (although conditionally enough [6]) as upper dimensional limit for a nanoparticle. In other words, the aggregation process can be resulted in the situation, when initially supposed nanocomposite ceases to be as such. Therefore, at present a methods number exists, allowing to suppress nanoparticles aggregation process [4, 7].

Analytically this process is treated as follows. The authors [5] obtain the relationship:

$$k(r) = 7.5 \times 10^{-3} S_u \tag{10.1}$$

where $k(r)$ is aggregation parameter, S_u is specific surface of nanofiller initial particles, which is given in m^2/g.

In its turn, the value S_u is determined as follows [8]:

$$S_u = \frac{6}{\rho_n D_p} \tag{10.2}$$

where ρ_n is nanofiller density, D_p is diameter of its initial particles.

From the Eqs. (10.1) and (10.2) it follows, that D_p reduction results in S_u growth, that in its turn reflects in the aggregation intensification, characterized by the parameter $k(r)$ increasing. Therefore, in polymer nanocomposites strengthening (reinforcing) element are not nanofiller initial particles themselves, but their aggregates [9]. This result in essential changes of nanofiller elasticity modulus, the value of which is determined with the aid of the equation [9]:

$$E_{agr} = E_{nan}\left(\frac{a}{R_{agr}}\right)^{3+d_l} \tag{10.3}$$

where E_{agr} is nanofiller particles aggregate elasticity modulus, E_{nan} is elasticity modulus of material, from which the nanofiller was obtained, a is an initial nanoparticles size, R_{agr} is a nanoparticles aggregate radius, d_l is chemical dimension of the indicated aggregate, which is equal to ~ 1.1 [9].

As it follows from the Eq. (10.3), the initial nanoparticles aggregation degree enhancement, expressed by R_{agr} growth, results in E_{agr} decrease (the rest of parameters in the Eq. (10.3) are constant) and, as consequence, in nanocomposite elasticity modulus reduction.

Very often the elasticity modulus (or reinforcement degree) of polymer composites (nanocomposites) is described within the frameworks of numerous micromechanical models, which proceed from elasticity modulus of matrix polymer and filler (nanofiller) and the latter volume contents [10]. Additionally it is supposed, that the indicated above characteristics of a filler are approximately equal to the corresponding parameters of compact material, from which a filler is prepared. This practice is inapplicable absolutely in case of polymer nanocomposites with fine-grained nanofiller, since in this case a polymer is reinforced by nanofiller fractal aggregates, whose elasticity modulus and density differ essentially from compact material characteristics (see the Eq. (10.3)) [5, 9]. Therefore, the microcomposites models application, as a rule, gives a large error at polymer composites elasticity modulus evaluation, that in its turn results in the appearance of an indicated models modifications large number [10].

Proceeding from the said above, the present work purpose is the theoretical treatment of particulate nanofiller aggregation process and elasticity modulus (reinforcement degree) particulate-filled polymer nanocomposites with due regard for the indicated effect within the framework of irreversible aggregation models and fractal analysis.

10.2 EXPERIMENTAL PART

Polypropylene (PP) “Kaplen” of mark 01030 with average weight molecular mass of ~ $(2–3)\times10^3$ and polydispersity index 4.5 was used as matrix polymer. Nanodimensional calcium carbonate ($CaCO_3$) in compound form of mark Nano-Cal P-1014 (production of China) with particles size of 80 nm and mass contents of 1–7 mass % and globular nanocarbon (GNC) (production of corporations group “United Systems,” Moscow, Russian Federation) with particles size of 5–6 nm, specific surface of 1400 m^2/g and mass contents of 0.25–3.0 mass % were applied as nanofiller.

Nanocomposites PP/$CaCO_3$ and PP/GNC were prepared by components mixing in melt on a twin screw extruder Thermo Haake, model Reomex RTW 25/42, production of German Federal Republic. Mixing was performed at temperature 463–503 K and screw speed of 50 rpm during 5 min. Testing samples were prepared by casting under pressure method on a casting machine Test Sample Molding Apparatus RR/TS MP of firm Ray-Ran (Taiwan) at temperature 483 K and pressure 43 MPa.

The nanocomposites melt viscosity was characterized by a melt flow index (MFI). MFI measurements were performed on an extrusion-type plastometer Noselab ATS A-MeP (production of Italy) with capillary diameter of 2.095±0.005 mm at temperature 513 K and load of 2.16 kg. The sample was maintained at the indicated temperature during 4.5±0.5 min.

Uniaxial tension mechanical tests have been performed on the samples in the shape of a two-sided spade with sizes according to GOST 112 62–80. The tests have been conducted on a universal testing apparatus Gotech Testing Machine CT-TCS 2000, production of German Federal Republic, at temperature 293 K and strain rate ~ 2×10^{-3} s^{-1}.

10.3 RESULTS AND DISCUSSION

The particulate nanofiller aggregation degree can be evaluated and aggregates diameter D_{agr} quantitative estimation can be performed within the framework of strength dispersive theory [11], where shear yield stress of nanocomposite τ_n is determined as follows:

$$\tau_n = \tau_m^{'} + \frac{G_n b_B}{\lambda} \tag{10.4}$$

where τ_m is shear yield stress of polymer matrix, b_B is Burgers vector, G_n is nanocomposite shear modulus, λ is distance between nanofiller particles.

In case of nanofiller particles aggregation the Eq. (10.4) has the look [11]:

$$\tau_n = \tau_m^{'} + \frac{G_n b_B}{k(r)\lambda} \tag{10.5}$$

where $k(r)$ is aggregation parameter.

The parameters, included in the Eqs. (10.4) and (10.5) are determined as follows. The general relationship between normal stress σ and shear stress τ has the look [12]:

$$\tau = \frac{\sigma}{\sqrt{3}} \tag{10.6}$$

The intercommunication of matrix polymer τ_m and nanocomposite polymer matrix $\tau_m^{'}$ shear yield stresses is given as follows [5]:

$$\tau_m^{'} = \tau_m \left(1 - \varphi_n^{2/3}\right) \tag{10.7}$$

where φ_n is nanofiller volume content, which can be determined according to the well-known formula [5]:

$$\varphi_n = \frac{W_n}{\rho_n} \tag{10.8}$$

where W_n is nanofiller mass contents, ρ_n is its density, which for nanoparticles is determined according to the equation [5]:

$$\rho_n = 188\left(D_p\right)^{1/3}, \text{ kg/m}^3 \tag{10.9}$$

where D_p is given in nm.

The value of Burgers vector b_B for polymeric materials is determined as follows [13]:

$$b_B = \left(\frac{60.5}{C_\infty} \right)^{1/2}, \text{ Å} \tag{10.10}$$

where C_∞ is characteristic ratio, connected with nanocomposite structure dimension d_f by the equation [13]:

$$C_\infty = \frac{2d_f}{d(d-1)(d-d_f)} + \frac{4}{3} \tag{10.11}$$

where d is dimension of Euclidean space, in which a fractal is considered (it is obvious, that in our case d=3).

The value d_f can be calculated according to the equation [14]:

$$d_f = (d-1)(1+\nu) \tag{10.12}$$

where ν is Poisson's ratio, estimated according to the mechanical tests results with the aid of the relationship [15]:

$$\frac{\sigma_Y}{E_n} = \frac{1-2\nu}{6(1+\nu)} \tag{10.13}$$

where σ_Y and E_n are yield stress and elasticity modulus of nanocomposite, respectively.

Nanocomposite moduli E_n and G_n are connected between themselves by the relationship [14]:

$$G_n = \frac{E_n}{d_f} \tag{10.14}$$

And at last, the distance λ between nanofiller nonaggregated particles is determined according to the equation [11]:

$$\lambda = \left[\left(\frac{4\pi}{3\varphi_n} \right)^{1/3} - 2 \right] \frac{D_p}{2} \tag{10.15}$$

From the Eqs. (10.5) and (10.15) k(r) growth from 5.65 up to 43.70 within the range of W_n=0.25–3.0 mass % for nanocomposites PP/GNC and from 1.0 up to 2.87 within the range of W_n=1–7 mass % for nanocomposites PP/$CaCO_3$ follows. Let us note, that the indicated variation *k*(*r*) for the considered nanocomposites corresponds completely to the Eqs. (10.1) and (10.2). Let us consider, how such *k*(*r*) growth is reflected on nanofiller particles aggregates diameter D_{agr}. The Eqs. (10.8), (10.9) and (10.15) combination gives the following relationship:

$$k(r)\lambda = \left[\left(\frac{0.251\pi D_{agr}^{1/3}}{W_n}\right)^{1/3} - 2\right]\frac{D_{agr}}{2} \tag{10.16}$$

allowing at D_p replacement on D_{agr} to determine real, that is, with accounting of nanofiller particles aggregation, nanoparticles aggregates diameter of the used nanofiller. Calculation according to the Eq. (10.16) shows D_{agr} increasing (corresponding to *k*(*r*) growth) from 25 up to 125 nm within the range of W_n=0.25–3.0 mass % for GNC and from 80 up to 190 nm within the range of 1–7 mass % for $CaCO_3$. Further nanofiller particles aggregates density can be calculated according to the Eq. (10.9) at the condition of D_p replacement by D_{agr}.

Within the framework of irreversible aggregation model D_{agr} value is given by the following relationship [16]:

$$D_{agr} \sim \left(\frac{4c_0 kT}{3\eta m_0}\right)^{1/d_f^{agr}} t^{1/d_f^{agr}} \tag{10.17}$$

where c_0 is nanoparticles initial concentration, *k* is Boltzmann constant, *T* is temperature, η is medium viscosity, m_0 is mass of initial nanoparticle, d_f^{agr} is fractal dimension of particles aggregate, *t* is aggregation process duration.

Let us consider estimation methods of the parameters, included in the relationship (10.17). In the simplest case it can be accepted that all particles of nanofiller initial powder have the same size and mass. In this case $c_0 \approx \varphi_n$, where φ_n value is determined according to the Eq. (10.8) with using nanofiller particles aggregates diameter D_{agr}. η value is accepted equal to reciprocal of MFI value and m_0 magnitude was calculated as follows.

In supposition of nanofiller initial particles spherical shape the nanoparticle volume was calculated according to the known values of their diameter D_p and then, using ρ_n value, calculated according to the Eq. (10.8), their mass m_0 can be estimated. T value is accepted as constant and equal to nanocomposites processing duration, that is, 300 s.

The fractal dimension of nanofiller particles aggregates structure d_f^{agr} was calculated with the aid of the equation [17]:

$$\rho_n = \rho_{dens}\left(\frac{D_{agr}}{2a}\right)^{d_f^{agr}-d} \qquad (10.18)$$

where ρ_{dens} is density of compact material of nanofiller particles, a is self-similarity (fractality) lower scale of nanofiller particles aggregates.

ρ_{dens} value for carbon is accepted equal to 2700 kg/m², for $CaCO_3$ – 2000 kg/m² [5] and a value is accepted equal to the initial GNC particle radius, that is, 2.5 nm. d_f^{agr} values, calculated according to the Eq. (10.18), are equal to 2.09–2.67 and 2.47–2.75 for GNC and $CaCO_3$ nanoparticles aggregates, respectively.

In Figure 10.1, the dependences $D_{agr}(W_n)$, plotted according to the Eqs. (10.16) and (10.17), comparison is adduced. As one can see, the good enough correspondence of estimations according to both indicated methods was obtained (the average discrepancy of D_{agr} values, calculated with the usage of these relationships, makes up ~ 16%). This circumstance

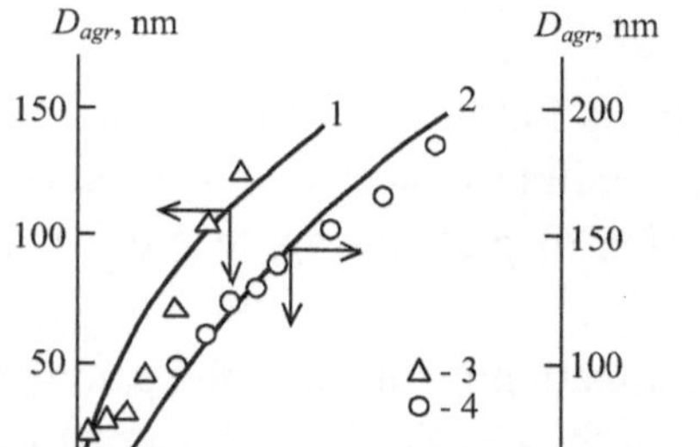

FIGURE 10.1 The dependences of nanofiller particles aggregates diameter D_{agr} on nanofiller mass contents W_n for nanocomposites PP/GNC (1, 3) and PP/$CaCO_3$ (2, 4). 1, 2 – calculation according to the Eq. (10.16); 3, 4 – calculation according to the relationship (10.17).

indicates, that irreversible aggregation models can be used for the theoretical description of particulate nanofiller particles aggregation processes. Besides, the Eq. (10.17) analysis demonstrates various factors influence on nanofiller particles aggregates size (or their aggregation degree). So, c_0, T and t increasing results in aggregation processes intensification and η, m_0 and d_f^{agr} enhancement – to their weakening.

Let us note in conclusion, that proportionality coefficient in the relationship (10.17) for GNC and $CaCO_3$ (c_{GNC} and c_{CaCO_3}, respectively) can be approximated by the following relationship:

$$\frac{c_{CaCO_3}}{c_{GNC}} = \left(\frac{m_0^{CaCO_3}}{m_0^{GNC}} \right)^{1/d_f^{av}} \tag{10.19}$$

where $m_0^{CaCO_3}$ and m_0^{GNC} are masses of the initial particles of $CaCO_3$ and GNC, respectively, d_f^{av} is average fractal dimension of the indicated nanoparticles aggregates.

Further elasticity modulus E_{agr} of nanofiller particles aggregates according to the Eq. (10.3) can be determined. Let us consider the concrete conditions of this equation usage in reference to nanocomposites PP/GNC. Two possible variants exist at parameter a choice in the indicated equation. The first from them supposes, that the value a is equal to GNC initial particles diameter [9], that is, 5.5 nm. Such supposition means, that GNC nanoparticles aggregates are formed by particle-cluster (P-Cl) mechanism, that is, by separate particles GNC joining to a growing aggregate [18]. However, such supposition gives unreal high E_{agr} values of order of 5×10^5 GPa. The other variant assumes, that nanofiller aggregation is realized by a cluster-cluster (Cl-Cl) mechanism, that is, small clusters association in larger ones [18]. In such model aggregate radius R_{agr}^{i-1} on the previous (i–1)th aggregation stage is accepted as a and then the Eq. (10.3) can be rewritten as follows:

$$E_n = E_{agr} \left(\frac{R_{agr}^{i-1}}{R_{agr}^{i}} \right)^{4.1} \tag{10.20}$$

The elasticity modulus E_{agr} real values within the range of 21.3–5.0 GPa were obtained at such calculation method. Further the simplest microcomposite models can be used for nanocomposite elasticity modulus E_n estimation. For the case of uniform strain in nanocomposite phases the theoretical value E_n $\left(E_n^T\right)$ is given by a parallel model [10]:

$$E_n^T = E_{agr}\varphi_n + E_m\left(1-\varphi_n\right) \tag{10.21}$$

where E_m is elasticity modulus of matrix polymer.

For the case of uniform stress in nanocomposite phases the lower theoretical boundary E_n^T is determined according to the serial model [10]:

$$E_n^T = \frac{E_{agr}E_m}{E_{agr}\left(1-\varphi_n\right)+E_m\varphi_n} \tag{10.22}$$

In Figure 10.2 the comparison of the received experimentally E_n and calculated according to the Eqs. (10.21) and (10.22) E_n^T elasticity modulus values of the considered nanocomposites PP/GNC is adduced. As one can see, the experimental data correspond better to the determined according to the Eq. (10.21) E_n^T upper boundary (in this case average discrepancy of E_n and E_n^T makes up ~ 8 %). The indicated discrepancy is due to objective causes. As it is known [10], at the Eqs. (10.21) and (10.22) derivation the equality of Poisson's ratio for nanocomposite both phases was supposed. In practice this condition non-fulfillment defines discrepancy between experimental and theoretical data.

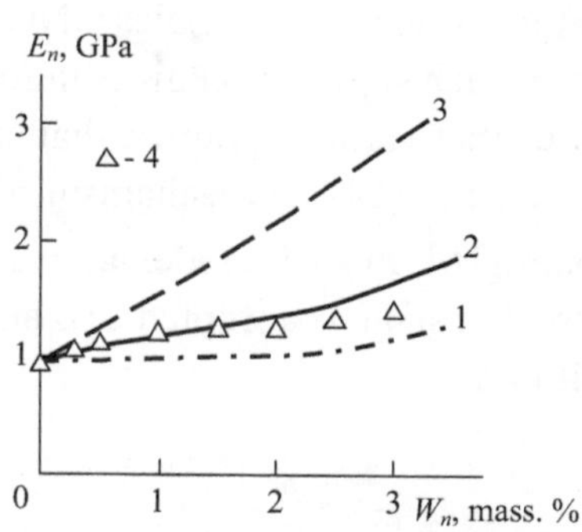

FIGURE 10.2 The dependences of elasticity modulus E_n on nanofiller mass contents W_n for nanocomposites PP/GNC. 1 – calculation according to the Eq. (10.22); 2 – according to the Eq. (10.21) at E_{agr}=variant; 3 – according to the Eq. (10.21) at E_{agr}=const=21.3 GPa; 4 – experimental data.

In Figure 10.2 the dependence E_n^T (W_n), calculated according to the Eq. (10.21) in supposition E_{agr}=const=21.3 GPa, is also adduced. As one can see, in this case the theoretical values of elasticity modulus E_n^T exceed essentially experimentally received ones E_n. Hence, the good correspondence of experiment and calculation according to the Eq. (10.21) is due to real values E_{agr} usage only.

It is obvious, that nanoparticles aggregates elasticity modulus reduction is due to the indicated aggregates diameter growth and, as consequence, their density ρ_n reduction, which can be calculated according to the Eq. (10.18). In Figure 10.3 the dependence $E_{agr}(\rho_n)$ is adduced, which, as was expected, proves to be linear, passing through coordinates origin and is described analytically by the following empirical equation:

$$E_{agr} = 12.6 \times 10^{-3} \rho_n, \qquad \text{GPa} \tag{10.23}$$

where ρ_n is given in kg/m^3.

The limiting magnitude ρ_n=ρ_{dens} allows to obtain the greatest value E_{agr}≈34 GPa for GNC aggregates, that is the real value of this parameter [1].

The authors [19] proposed to use for nanocomposites elasticity modulus E_n determination a modified mixtures rule, which in original variant gives upper limiting value of composites elasticity modulus [10]:

$$E_n = E_m\left(1 - \varphi_n\right) + bE_{nan}\varphi_n \tag{10.24}$$

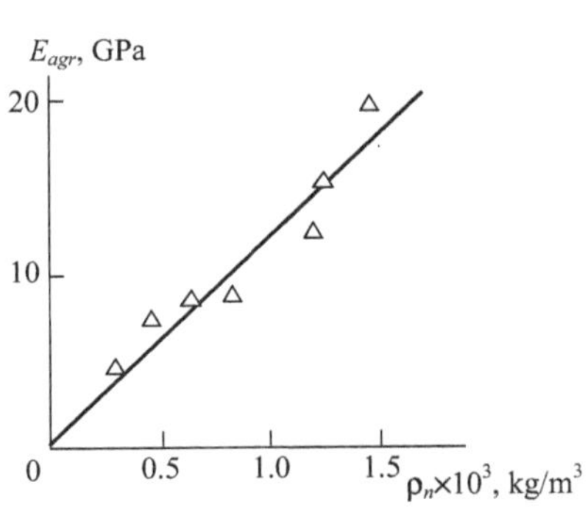

FIGURE 10.3 The dependence of GNC nanoparticles fractal aggregates elasticity modulus E_{agr} on their density ρ_n for nanocomposites PP/GNC.

where $b<1$ is coefficient, reflected nanofiller properties realization degree in polymer nanocomposite. In the present work context the parameter bE_{nan} as a matter of fact presents nanofiller effective modulus or, more precisely, its aggregates modulus E_{agr} (compare with the Eq. (10.21)).

In Figure 10.4, the dependence of parameter b in the Eq. (10.24) on nanofiller particles aggregates diameter D_{agr}, calculated according to the Eq. (10.16), for the studied nanocomposites is adduced. As one can see, this dependence disintegrates on two linear parts: at small D_{agr} fast decay of b at D_{agr} growth is observed and at large enough D_{agr} the value $b \approx const \approx 0.175$. Let us note, that dimensional interval of the indicated transition, showed in Figure 10.4 by a shaded area, makes up $D_{agr} \approx 70–100$ nm, that is, it coincides approximately with upper dimensional boundary of nanoparticles interval (although and conditional enough [6]), which is equal to about 100 nm. As a matter of fact, the indicated dimensional interval defines the transition from nanocomposites to microcomposites, the dependence $b(D_{agr})$ for which differs actually qualitatively. The adduced in Figure 10.4 dependence $b(D_{agr})$ can be described analytically by the following integrated equation:

$$b = 0.67 - 6.7 \times 10^{-3} D_{agr}, \qquad \text{for } D_{agr}\ 70 \text{ nm}$$

$$b = \text{const} = 0.175, \qquad \text{for } D_{agr} > 70 \text{ nm} \qquad (10.25)$$

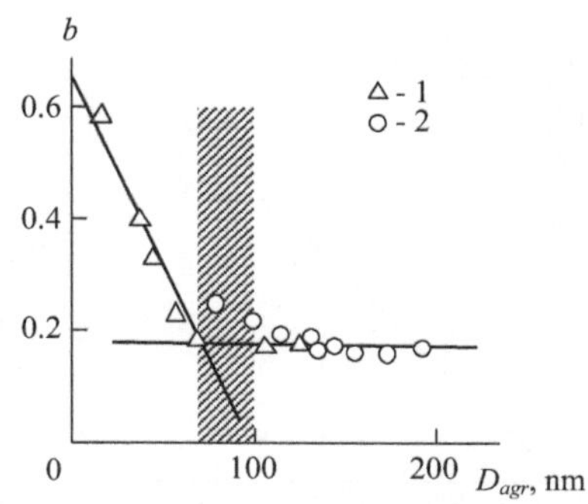

FIGURE 10.4 The dependence of parameter b on nanofiller particles aggregates diameter D_{agr} for nanocomposites PP/GNC (1) and PP/$CaCO_3$ (2). The shaded region indicates transition of nanofiller particles aggregates from nano- to microbehavior.

In Figure 10.5, the comparison of experimentally obtained and calculated according to the Eq. (10.25) dependences $E_n(\varphi_n)$ is adduced for the studied nanocomposites. In this case the parameter b value was estimated according to the Eq. (10.25) and values E_{nan} were accepted equal to 30 GPa for GNC and 15 GPa for $CaCO_3$. As one can see, the good correspondence of theory and experiment is obtained (their mean discrepancy makes up 3%, that approximately equal to the experimental error of E_n determination). Higher values E_n for nanocomposites PP/GNC in comparison with PP/$CaCO_3$ even at D_{agr}>100 nm are due to two factors: the initial nanoparticles smaller size, that gives higher values φ_n at the same W_n values (see the Eqs. (10.8) and (10.9)) and higher value E_{nan}. It is important to note close values E_{agr} for nanocomposites PP/GNC, determined according to the Eq. (10.20) and as bE_{nan}.

The authors [8] proposed the following percolation relationship for polymer microcomposites reinforcement degree E_c/E_m description:

$$\frac{E_c}{E_m} = 1 + 11(\varphi_n)^{1.7} \tag{10.26}$$

where E_c is elasticity modulus of microcomposite.

Later the relationship (10.26) was modified in reference to the polymer nanocomposites case [20]:

$$\frac{E_n}{E_m} = 1 + 11(\varphi_n + \varphi_{if})^{1.7} \tag{10.27}$$

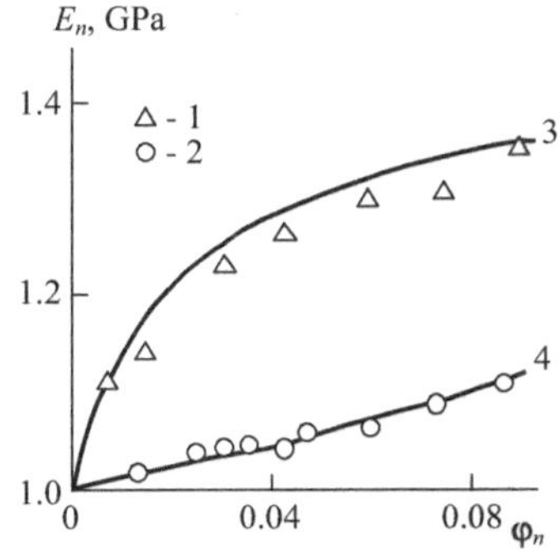

FIGURE 10.5 The comparison of experimentally received (1, 2) and calculated according to the Eqs. (10.24) and (25) (3, 4) dependences of elasticity modulus E_n on nanofiller volume contents φ_n for nanocomposites PP/GNC (1, 3) and PP/$CaCO_3$ (2, 4).

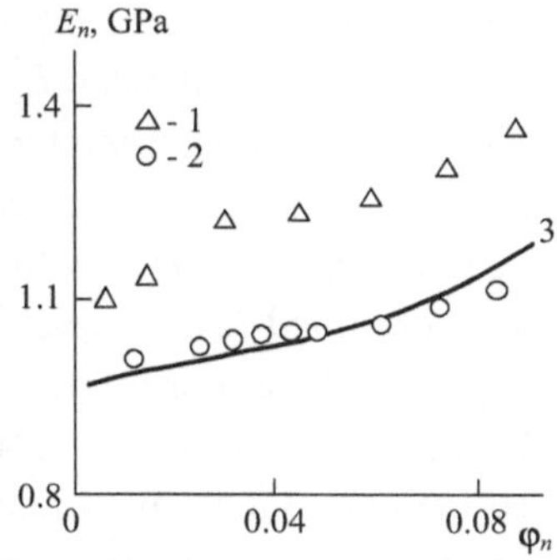

FIGURE 10.6 The comparison of experimentally received (1, 2) and calculated according to the Eq. (10.26) (3) dependences of elasticity modulus E_n on nanofiller volume contents φ_n for nanocomposites PP/GNC (1, 3) and PP/$CaCO_3$ (2, 4).

where φ_{if} is relative fraction of interfacial regions.

It is easy to see, that the modified relationship (27) takes into consideration a factor of sharp increase of division surfaces polymer matrix-nanofiller [21]. In Figure 10.6, the comparison of experimentally obtained and calculated according to the Eq. (10.26) dependences $E_n(\varphi_n)$ for the considered nanocomposites is adduced. As it follows from this figure data, the Eq. (10.26) describes well the experimental data for nanocomposites PP/$CaCO_3$, but the corresponding data for nanocomposites PP/GNC set essentially higher than theoretical curve. This discrepancy cause is obvious from the Eqs. (10.26) and (10.27) comparison – for nanocomposites PP/GNC interfacial effects accounting is necessary, that is, parameter φ_{if} accounting. Hence, in the considered case only compositions PP/GNC are true nanocomposites.

10.4 CONCLUSIONS

The applicability of irreversible aggregation models for theoretical desc ription of particulate nanofiller particles aggregation processes in polymer nanocomposites has been shown. Analysis within the framework of the indicated models allows to reveal either factors influence on aggregation degree.

Strongly expressed aggregation of particulate nanofiller particles results in sharp (in about 4 times) formed fractal aggregates real elasticity modulus reduction. In its turn, this process defines nanocomposites as the whole elasticity modulus reduction. Nanofiller particles aggregation is realized by a cluster-cluster mechanism and results in the formed fractal aggregates density essential reduction, that is the cause of their elasticity modulus decreasing.

A nanofiller elastic properties realization degree is defined by the aggregation of its initial particles level. Unlike microcomposites nanocomposites require interfacial effects accounting for elasticity modulus correct description in virtue of well-known large fraction of phases division surfaces for them.

KEYWORDS

- **aggregation**
- **calcium carbonate**
- **globular nanocarbon**
- **interfacial effects**
- **nanocomposite**
- **reinforcement**

REFERENCES

1. Kozlov, G. V., Yanovskii Yu. ., Zaikov, G. E. Structure and Properties of Particulate-Filled Polymer Composites: The Farctal Analysis. New York, Nova Science Publishers, Inc. 2010, 282 p.
2. Kozlov, G. V., Zaikov, G. E. Structure and Properties of Particulate-Filled Polymer Nanocomposites. Saarbrücken, Lambert Academic Publishing, 2012, 112 p.
3. Kozlov, G. V., Yanovskii Yu. G., Zaikov, G. E. Particulate-Filled Polymer Nanocomposites. Structure, Properties, Perspectives. New York, Nova Science Publishers, Inc. 2014, 273 p.
4. Edwards, D. C. J. Mater. Sci., 1990, v. 25, № 12, 4175–4185.
5. Mikitaev, A. K., Kozlov, G. V., Zaikov, G. E. Polymer Nanocomposites: Variety of Structural Forms and Applications. New York, Nova Science Publishers, Inc. 2008, 319 p.
6. Buchachenko, A. L. Achievements of Chemistry, 2003, v. 72, № 5, 419–437.

7. Kozlov, G. V., Yanovsky Yu. G., Zaikov, G. E. Synergetics and Fractal Analysis of Polymer Composites Filled with Short Fibers. New York, Nova Science Publishers, Inc. 2011, 223 p.
8. Bobryshev, A. N., Kozomazov, V. N., Babin, L. O., Solomatov, V. I. Synergetics of Composite Materials. Lipetsk, NPO ORIUS, 1994, 154 p.
9. Witten, T. A., Rubinstein, M., Colby, R. H. J. Phys. II France, 1993, v. 3, № 3, 367–383.
10. Ahmed, S., Jones, F. R. J. Mater. Sci., 1990, v. 25, № 12, 4933–4942.
11. Sumita, M., Tsukumo, Y., Miyasaka, K., Ishikawa, K. J. Mater. Sci., 1983, v. 18, № 5, 1758–1764.
12. Honeycombe, R. W. K. The Plastic Deformation of Metals. Cambridge, Edwards Arnold Publishers, 1968, 402 p.
13. Kozlov, G. V., Zaikov, G. E. Structure of the Polymer Amorphous State. Utrecht, Boston, Brill Academic Publishers, 2004, 465 p.
14. Balankin, A. S. Synergetics of Deformable Body. Moscow, Publishers of Ministry Defence SSSR, 1991, 404 p.
15. Kozlov, G. V., Sanditov, D. S. Anharmonic Effects and Physical-Mechanical Properties of Polymers. Novosibirsk, Science, 1994, 261 p.
16. Weitz, D. A., Huang, J. S., Lin, M. Y., Sung, J. Phys. Rev. Lett., 1984, v. 53, № 17, 1657–1660.
17. Brady, L. M., Ball, R. C. Nature, 1984, v. 309, № 5965, 225–229.
18. Shogenov, V. N., Kozlov, G. V. Fractal Clusters in Physics-Chemistry of Polymers. Nalchik, Polygraphservice and T, 2002, 268 p.
19. Komarov, B. A., Dzhavadyan, E. A., Irzhak, V. I., Ryabenko, A. G., Lesnichaya V. A., Zvereva, G. I., Krestinin, A. V. Polymer Science, Series A, 2001, v. 53, № 6, 897–905.
20. Malamatov, A.Kh., Kozlov, G. V., Mikitaev, M. A. Reinforcement Mechanisms of Polymer Nanocomposites. Moscow, Publishers of, D. I. Mendeleev RKhTU, 2006, 240 p.
21. Andrievsky, R. A. Russian Chemical Journal, 2002, v. 46, № 5, 50–56.

CHAPTER 11

A STUDY ON ELECTROSPUN NANOFIBER MATS

A. L. IORDANSKII,[1] S. G. KARPOVA,[2] A. A. OLKHOV,[1,4] O. V. STAROVEROVA,[1] A. V. KHVATOV,[2] A. GRUMEZESCU,[3] G. E. ZAIKOV,[2] and A. A. BERLIN[1]

[1]*N. Semenov Institute of Chemical Physics, RAS, 4 Kosygin Str., Moscow, 119991, Russian Federation; E-mail: aljordan08@gmail.com*

[2]*N. Emanuel Institute of Biochemical Physics, RAS, 4 Kosygin Str., Moscow, 119334, Russian Federation; E-mail: gezaikov@yahoo.com*

[3]*Department of Science and Engineering of Oxide Materials and Nanomaterials, Faculty of Applied Chemistry and Materials Science, University Politehnica of Bucharest, 1–7 Polizu Street, 011061 Bucharest, Romania*

[4]*G. Plekhanov Russian University of Economics, 9 Stremyannoy per. Moscow, 117997, Russian Federation*

CONTENTS

ABSTRACT

The chapter is focused on the study of segmental dynamics in electrospun PHB fibers and the same fibers subjected to cold rolling as well as in PHB films for the comparison of spin probe characteristics obtained by ESR technique. It was disclosed the presence of two TEMPO probe populations with different correlation times indicating the heterogeneous structure of intercrystalline areas in the films, fiber mats and the cold rolled fiber mats. The ESR data are in agree with the 2-mode model of amorphous state in semicrystalline polymers. The difference in peak intensities shows that effective correlation time in the electrospun fibers ($3.5 \pm 0.6\times10^{-9}$ s) exceeds the same characteristic in the film ($1.4 \pm 0.3\times10^{-9}$ s) that also indicates the slower molecular mobility in the low-dense amorphous fraction of PHB fibers as compared to the film. Taking into account the ESR technique data, the analysis of temperature, water and ozone impacts upon probe mobility enable us to suggest that fiber electrospinning and especially the cold-rolling procedure for the fiber mats lead to denser field formation in the intercrystalline area of PHB that furthers the fiber stabilization against the exterior aggressive factors.

11.1 INTRODUCTION

The ultrathin fibrillar structures (fibers, medical threads, meshes, mats and scaffolds) are of utmost interest as the modern functional materials with specific properties such as the high surface/volume ratio of a single filament, effective surface modification, special physical-chemical behavior, and anomalous diffusion [1–5]. At the present time ultrathin fibers and articles on their base are efficiently applicable in biomedicine, tissue engineering, in filtration and separation processes, for composite design, in electronics, analytical supplies, sensor-based diagnostics, and in other innovative applications [6–9].

For conventional fiber technologies there are significant challenges in arranging fibers and threads with the diameter in submicron range [10].

Currently the alternative manufacture of ultrathin fibers is the unconventional technology based on electrospinning of polymer solutions and melts that enables one to produce fibers and nonwoven materials (mats) in the nano-sized range. The electrospinning presents technologically uncomplicated but multivariate process for ultrathin fibers' formation that promotes their wide applications [11]. The process is based on the combination of mechanical and electrodynamic forces affecting a polymer solution (or melt), which is aligned in electric field [12]. Therefore, a number of technological parameters such as electric potential, distance between collector and electrode, solution conductivity, and the others influence on morphology, surface properties, functionality, porosity and fiber diameter [13].

Recently, for the example of poly(3-hydroxybutyrate) (PHB) and a number of its composites [14–16] we have studied physical-chemical, dynamic and transport characteristics of macroscopic biodegradable matrices and microparticles of PHB which were designed for controlled drug release [16, 17]. High biocompatibility, controlled biodegradation and appropriate mechanical properties allow one to consider this biopolymer as one of the most promising biomedical polymers. Besides therapeutical aims, PHB is widely used as bone implants, nervous conduits, matrices in cell engineering, filters and membranes, in cardiology and in the other areas [14, 18, 19].

It is currently known a large number of methods for polymer structure orientation, such as extrusion, blow molding, hot shaping, drawing, nozzle extrusion etc. All of them dominate in polymer technology but unfortunately the stretching processes of this type are accompanied by the undesirable effect of microcavitation development [20]. The fibers obtained by electrospinning in electric field are less prone to defect formation. To arrange an additional orientation of macromolecules in fibers, we used the cold rolling belonging to the grope of compression methods that likewise minimizes the probability of defect cavities forming in the fibrillar mates [21]. In spite of a number of limitations, mechanical rolling does not demand high-energy expenses therefore it remains one of the most economical processes in metal- and polymer-technologies.

The transition from films to ultrathin fibers and unwoven fiber materials is accompanied by the changes of physical-chemical and transport characteristics, which related with spatial confinements modulating dynamics of polymer molecules [22]. In this relation, at the molecular level the comparison of structure-dynamic characteristics will be performed for PHB matrices, ultrathin fiber mates and the same mates after rolling process to elucidate the specific behavior of diffusion and chemical degradation proceeding under spatial confinement conditions. Study of local segmental dynamics in ultrathin fibers plays a significant role in understanding structure-property relationships shown by biodegradable polymers. Using probe ESR spectroscopy, it will be shown the impact of specific aggressive media (water and ozone) on molecular dynamics of PHB in the mates with comparison of the matrices [23] that enables segmental mobility to be considered at earlier stages of interactions between aggressive components and polymer molecules.

11.2 MATERIALS AND METHODS

The PHB was kindly presented by Biomer Co (Krailing, Germany) as lot 16F. The initial polymer was in the form of white powder with particle size 5–20 μm, MW = 2.06×10^5 Da. Ultrathin fibers were obtained by electrospinning from 5 to 9% PHB chloroform solutions. The details of this procedure were described earlier in Ref. [24]. Physical modifications of PHB structure were produced by cold rolling at 22±1°C at rolling rate 30 s^{-1} and compression 200 kg/cm^2 [25=22]. As result of compression the film thickness decreased from 120±15 to 95±10 μm.

Segmental mobility was studied by probe ESR method with Radiopan spectrometer. The probe was 2,2,6,6-tetramethylpyperidin-1-oxil (TEMPO), that was incorporated from vapor phase at 25°C. Correlation times (τ) were calculated from ESR spectra [26] using the formula

$$\tau = 6.65\times10^{-10}\,\Delta H^{+}[(I^{+}/I^{-})^{1/2}-1],\ [c],$$

where ΔH^{+} – the weak-field peak on the half height, $I+/I^{-}$ is the ratio of peaks in weak and stark magnetic field areas. Standard deviation of measuring was 7%.

Ozone oxidation was performed in oxygen-ozone atmosphere at 5×10^{-5} mole/l. The oxidation degree was controlled by FTIR spectrum on the spectrometer "Bruker IFS 48" (Germany) in area 1650–1700 cm^{-1}. DSC thermograms were obtained by DSC calorimeter DSC 204 F1 (Netzsch) in inert atmosphere of Ar at heating rate 10°C/min. The averaged standard deviation of polymer fusion was 2%.

11.3 RESULTS AND DISCUSSION

Under electrospinning a number of polymer solution characteristics affect both morphology of ultrathin fibers and such important indicator as their diameter. In Figure 11.1 we have illustratively shown that the relatively small increase of spinning solution from 5 to 9 wt.% changes the diameter distribution in the PHB fibers. It is reasonable to note that when developing medical sensors and chemical detectors, a challenge issue to decrease a fiber diameter in the nanometer range but in the case of biomedical applications (e.g., scaffold performance in tissue engineering) such reduction results in poor cell adhesion and dysregulation of tissue growth [27]. The possibility of control in ultrathin fiber geometry by the variation of solution characteristics was described in the recent publication [28]. In this chapter, we have chosen a 7% PHB solution in chloroform as a basic option owing to the good reproducibility of geometry and quality of the fibers.

The segmental mobility measurements, performed by the ESR technique, which will show beneath, were combined with thermophysical measurements obtained by the DSC method. The melting endotherms for the PHB mats before and after cold rolling reveal similar melting

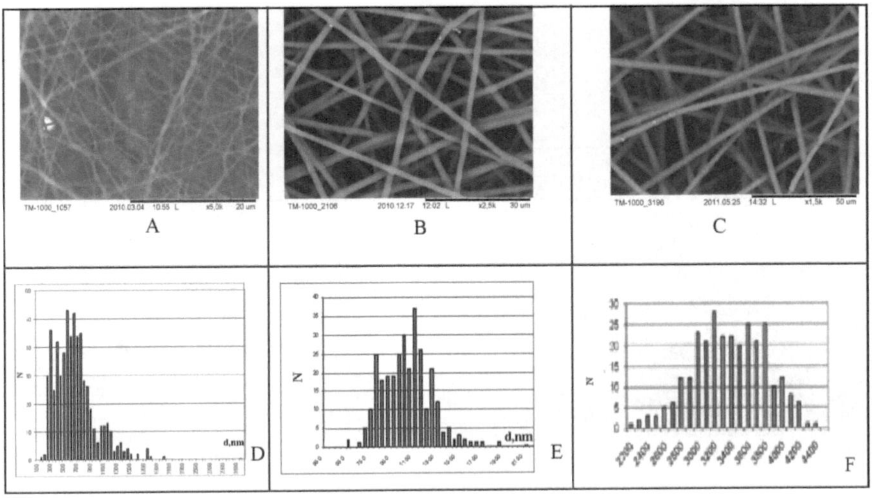

FIGURE 11.1 SEM images of spun PHB fibers obtained from solution at concentrations 5% (A), 7% (B), and 9% (C) and corresponding fiber diameter distributions with the averaged values 620 nm (D), 1100 nm (E), and 3350 nm (F).

behavior but their phase transitions are located at the different temperatures. In spite of the single peaks for first melting (1st run) of samples before and after cold rolling are very close each other: 448 K (175°C) and 449 K (176°C) respectively, the positions in melting temperatures (bimodal peaks) after 2nd run are displaced quite significantly. First, for both samples the single peak transformed to the bimodal peaks located respectively at 434.6 K (161.6°C) and 422.3 K (149.3°C) for initial mats and 439 K (166.5°C) and 430.5 (157.5°C) for mats after cold rolling. Second, as the result of repeated heating the high-temperature peaks are displaced to area of lower temperatures on decrement ~10°C, that attests nonperfect organization of crystalline structure PHB after its controlled cooling (10°/min) to room temperature. Additionally, by comparison the initial and mechanically treated mats the cold-rolled mats are characterized by two melting peaks at higher temperatures. The main conclusions that could be obtained from the analysis of thermograms (Figure 11.2) are: (a) mechanical treatment of the mats leads to bimodal

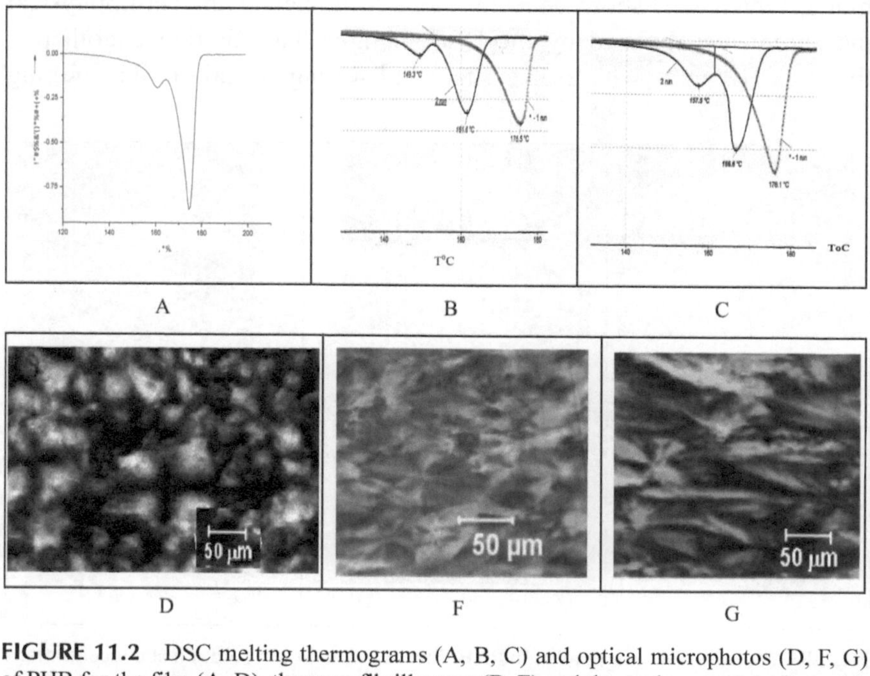

FIGURE 11.2 DSC melting thermograms (A, B, C) and optical microphotos (D, F, G) of PHB for the film (A, D), the spun fibrillar mat (B, F) and the analogous mat after cold-rolling (C, G). The microphotos were obtained after the first DSC scan.

distribution of crystalline entities including more perfect and less perfect spherulites; (b) if compare the peak temperature positions for the bulk film and the fiber mate unrolled or cold-rolled, the temperature shift for PHB increased in the set film > unrolled mate > cold-rolled mate under approximately constant degree crystallization (~79%); (c) the two-modal crystalline structure would be most likely conjugated with two modes of amorphous phase. To confirm the last suggestion we have used the probe ESR method that enables to estimate not only dynamic characteristics of segment mobility (τ) but the difference of amorphous densities in the framework of modern dual-mode model of amorphous phase in semicrystalline polymers, considering the intercrystalline space as combination of denser and less dense fields [29].

In Figure 11.3, it is shown the typical ESR spectra of the TEMPO probe encapsulated in the PHB film and the PHB mat respectively. The spectral characteristics of the mats and the films such as the ratio of intensities for low-field and high-field components of spectra and effective time correlations are essentially distinguished. Irrespective of the producing method (film casting, electrospinning or cold rolling) all PHB spectra represent superposition of two single spectra pertaining to individual radicals with different correlation times (τ_1 and τ_2). The inherent correlation time τ_1 determines the state of the probe in denser amorphous fields with slow rotation and τ_2 does the fast probe rotation in less dense amorphous fields of the polymer. The presence of two TEMPO populations in amorphous

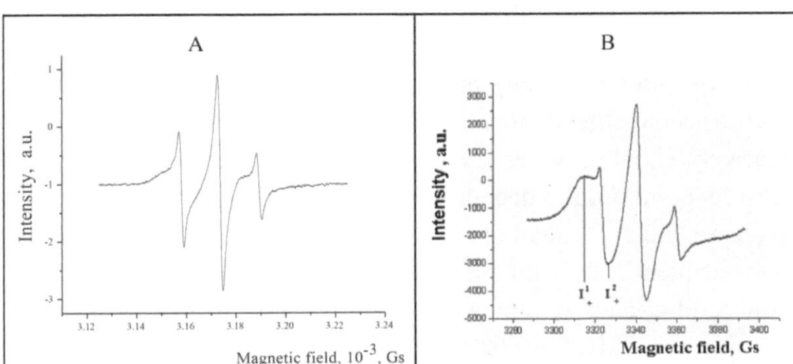

FIGURE 11.3 The ESR spectra of the probe TEMPO in the PHB film (A) and the PHB fiber mat (B) at 25°C. The spectrum B shows the features of superposition for two rotating populations: slow (I^1_+) and fast (I^2_+) modes.

phase of PHB with distinct rotation frequencies indicates heterogeneous structure of intercrystalline areas which could be approximated by 2-mode model of amorphous state in semicrystalline polymers such as PHB, polylactides, and poly(ethylene terephthalate) proposed earlier [29, 30].

The quality analysis of ESR spectra was pursued by using the ratio of intensities of two first low-field peaks, I^1_+ and I^2_+, belonging to slow and fast rotation, respectively (see Figure 11.3B). Additionally the spectrum characteristic was selected as the distance between the first peak and the third one (L) in single triplet that identifies mobility in denser areas of amorphous phase. In ESR spectra comparison the value L for the bulk film of PHB (50 Gs) was slightly less than for the fibers (62 Gs) that allows one to suggest slower rotation of the probe in denser field of the fiber mat than in the same field of the PHB film. The distinction between behaviors of two probe populations in the film are more noticeable than in the fiber mates. In fact the ratio I_+^1/I_+^2 in the rolled mats has higher value (0.52) as compared with the initial fiber mats (0.37). The difference in peak intensities shows that effective correlation time in the fibers (3.5×10^{-9} s) exceeds the same characteristic in the film (1.36×10^{-9} s) that also indicates the slow molecular mobility in the low-dense amorphous fraction of PHB rolled fibers as compared to the film.

Additionally to dynamic measurements the static sorption experiments performed by integration of ESR spectra were carried out to determine the TEMPO concentration in equal mass quantities of PHB prepared as the film, the fiber mat, and the fiber mat after cold rolling. Integration spectra calculations shown that the radical contents in the PHB items are 4.6×10^{18} (the film), 2.9×10^{18} (the fiber mat), and 2.3×10^{18} spin/cm^3 (the cold-rolled mat). Taking into account the identical temperature (25°C), time of sorption (24 h) and relatively equal polymer crystallinity for all types of samples (~79%), these results point out the decrease in specific volume of PHB being available to absorb the probe, namely, the decrease in the denser amorphous fraction in the sequence: the film > the mat after electrospinning > the same mat after cold rolling. Thus, comparative study of ESR spectra in differential and integration forms indicates that all three samples are estimably different in amorphous phase organization. This effect leads to the change of effective segmental mobility that influence rate of radical rotation in the samples of different origin (different conditions of stretching).

Using the measurements of ESR spectra at different temperatures, the effective time correlation (τ) dependences on temperature are presented to calculate activation energy of radical rotation in the samples after different stretching treatment, namely, after electrospinning and cold rolling (see Figure 11.4). Corresponding calculations reflected in semilogarithm coordinates gave the activation energies 27 for films, 42 for the electrospun mats and 55 kJ/mol for the same mats after cold rolling. In passing from the conventional films to the cold-rolled mats the progressive increasing in the activation energy testifies the increase in activation energy barrier of TEMPO rotation that is related with rendering segmental mobility in dense fields of amorphous phase. The tendency to mobility decreasing is manifested in the same sequence as the radical concentration sequence: in the film > in the fiber mat > in the cold-rolled fiber mats.

The semilogarithmic relation (lg(τ) – 1/T) for the cold-rolled mats has shown the anomalous behavior that was observed at elevated temperatures (~70°C) as deviation from the function linearity (Figure 11.4B). This phenomenon could be interpreted as the result in bimodal distribution of the probe in the oriented samples with two individual mobilities for each population. While the temperature decreases below 70°C, the dense amorphous fraction increases (see also Ref. [31]) that leads to the decline of effective mobility in intercrystalline area of cold-rolled PHB. For such samples the beginning of dense fraction formation precedes cold crystallization of PHB observed by DSC method at about 60°C and slightly higher [29]. Under dynamic measurements by ESR technique the segmental ordering

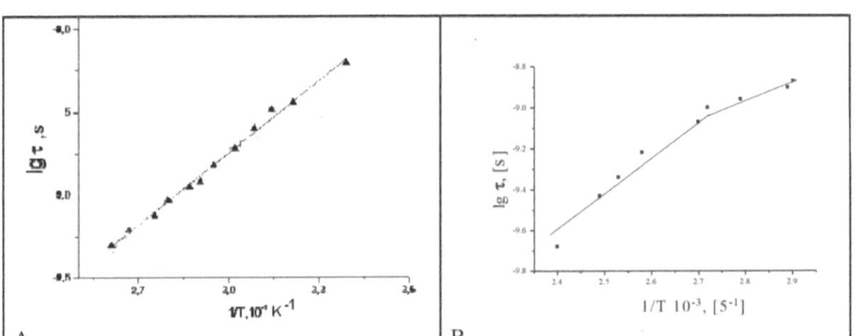

FIGURE 11.4 Time correlation dependence on the reciprocal temperature for the PHB mat (A) and the same mat after cold-rolling (B). The solid lines reflect approximation for energy activation calculations.

in PHB as the process preceding cold crystallization is detected at slightly higher temperature ~70°C, when the probe correlation time begins to deviate from a lg(τ)–1/T function linearity as it is shown in Figure 11.4B.

For the same reason the polymer exposure in liquid water that is the moderate PHB plasticizer [32] does not enhance molecular dynamics but leads to opposite effect, namely, to some reduction in segmental mobility. In Figure 11.5 the dependences of effective time correlation on the water contact duration for the PHB mats at 40°C, that is, in the vicinity of physiological temperature are shown. From the figure it is seen that with increasing in contact duration the correlation time is increased as well. This trend attests decreasing in segmental mobility due to segmental redistribution between denser fields and less dense fields in amorphous phase similarly as it has occurred at temperature impact (see the previous paragraph) when the concentration of polymer segments with low mobility is raised. The additional orientation of polymer molecules after cold-drawing produces the concentration rise in the dense amorphous fields and, hence, leads to the increase of effective time correlation (τ) as compared with characteristic of the initial mats that is demonstrated for all times of water contact in Figure 11.5.

Special experiments were carried out for the estimation of ultrathin fibers' behavior in ozone atmosphere as in effective oxide-active medium.

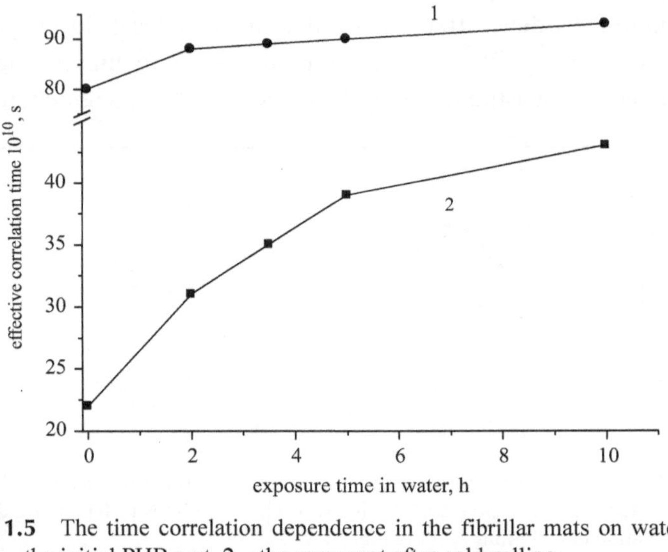

FIGURE 11.5 The time correlation dependence in the fibrillar mats on water contact duration. 1 – the initial PHB mat, 2 – the same mat after cold-rolling.

In Figure 11.6, the dependences of effective correlation time for initial and cold-rolled mats on ozone oxidation time are represented. The figure shows that for the first hour there is a sharp drop of the τ values and then for next 3 hours the decrease of τ is very small, that is, the dynamic characteristic is stabilized. After 4 hours of oxidation, the final stage shows the initiation of correlation time decrease. Note that the initial PHB mats and the cold-rolled mats demonstrate the symbatic probe rotation when in accordance with temperature- and water-influence data the mobility of TEMPO in the cold-rolled oxidized mat is decreased relative to the initial mat after ozonolysis. Taking into account the previous characteristics of crystalline structure and ESR data in amorphous area of the PHB and PHBV films after ozone exposition [23], it can be assumed that at the first stage of oxidation the partial destruction of macromolecules occurs that leads to the increase of probe mobility. On this stage only more accessible and defect molecules take part in reaction with ozone which are situated in less dense fields of PHB. After their concentration depletion the PHB-ozone interaction is stabilized for the next 3 hours that could be treated as induction period (the plateau in the Figure 11.6) and than the oxidation

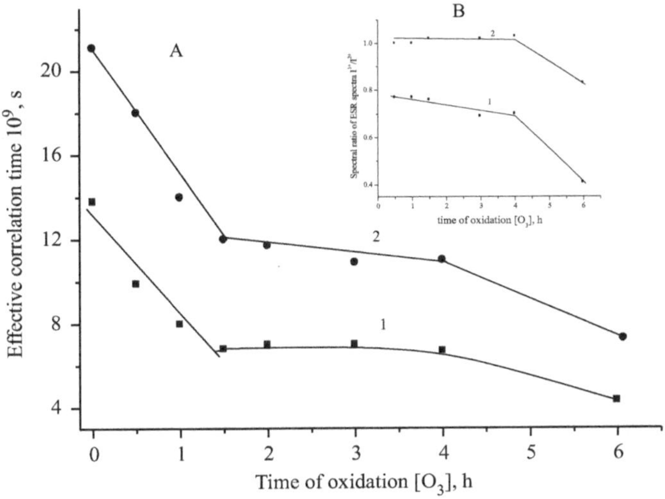

FIGURE 11.6 A: The dependence of effective time correlation of TEMPO probe on time of PHB mat exposition in ozone atmosphere. 1 – the initial fiber mat; 2 – the same mat after cold-rolling treatment. B: The ratio of fast and slow components in ESR probe spectra as function of time oxidation.

process is enhanced to involve not only the low-dense fields (see above) but the denser fields in intercrystalline space of PHB.

This mechanism is supported by ESR spectra analysis of oxidated mats based on calculations of the ratio for two first peaks I_+^1 and I_+^2 (Figure 11.6B) belonging to slow and fast-rotating fraction of the probe accumulated in denser and less dense amorphous areas respectively. The figure shows the constant ratio of the peaks for the 4–5 hours that testifies induction period as it was described above.

The further exposure of the polymer to ozone leads to the probe mobility increase related with oxidative destruction of PHB. It is worth to note that the drop in rotation mobility for the mats after cold-rolling is less pronounced than for the initial mats. Taking into account the previous ESR data this effect enables one to suggest that the cold-rolling leads to denser field formation in the intercrystalline area of PHB furthering its stabilization against attack of temperature, water and oxidative agent.

11.4 CONCLUSIONS

In summary, a coherent study on ESR dynamics for PHB prepared in the forms of matrices and ultrathin fiber mats reveals the presence of two spin probe populations with different rotation mobilities. The presence of two TEMPO probe populations indicates heterogeneous structure of intercrystalline areas in the films, fiber mats and the cold rolled fiber mats, which could be approximated by 2-mode model of amorphous state in semicrystalline polymers. The difference in peak intensities shows that effective correlation time in the fibers ($3.5 \pm 0.6 \times 10^{-9}$ s) exceeds the same characteristic in the film ($1.4 \pm 0.3 \times 10^{-9}$ s) that also indicates the slower molecular mobility in the low-dense amorphous fraction of PHB rolled fibers as compared to the film. The comparative study of ESR spectra after differential and integration treatments revealed that all three samples (cast films, electrospun fibers and the same mats after cold rolling) are estimably different in amorphous phase organization. This effect leads to the change of effective segmental mobility that influence rate of ESR probe rotation in the samples with different origin. Taking into account the ESR technique data, the analysis of temperature, water and ozone impacts upon probe mobility enables one to suggest that the cold-rolling procedure for the fiber mats leads to denser field formation in the intercrystalline area of PHB furthering its stabilization against the attack of the exterior aggressive factors.

ACKNOWLEDGMENTS

This work was supported by the Russian Foundation for Basic Research (Project No 14-03-01086/14a) and by the Division of Chemistry and Materials Sciences, Russian Academy of Sciences, under the academic research program Creation of Macromolecular Structures of New Generations (03/OKhM-14). Authors appreciate S. M. Lomakin, PhD, and N. G. Shilkina, PhD, for the essential impact on the DSC thermogram performance.

KEYWORDS

- **aggressive media**
- **electrospun fibers**
- **mats**
- **poly(3-hydroxybutyrate)**
- **spin probe ESR technique**
- **TEMPO**

REFERENCES

1. Natural and Synthetic Biomedical Polymers. (Edited by S. Kumbar, C. Laurencin, M. Deng) Elsevier, Hardbound. 2014. 420 pp. ISBN: 978-0-12-396983-5.
2. Kim, I. -D. *Macromol. Mater. Eng.* 2013, 298, 473–474. Advances in Electrospun Functional Nanofibers.
3. Agarwal, S., Greiner, A., Wendorff, J. H. Functional materials by electrospinning of polymers. *Progress in Polymer Science* 2013, 38 (6), 963–991.
4. Reneker, D. H., Yarin, A. L., Zussman, E., Xu, H. Electrospinning of Nanofibers from Polymer Solutions and Melts. *Advances in Applied Mechanics*, 2007. V.41, 43–195, 345–346.

5. Fernandes, J. G., Correia, D. M., Botelho, G., Padrão, J., Dourado, F., Ribeiro, C., Lanceros-Méndez, S., Sencadas, V. PHB-PEO electrospun fiber membranes containing chlorhexidine for drug delivery applications. *Polymer Testing*, 2014, V.34(1), 64–71.
6. Gunn, J., Zhang, M. Polyblend nanofibers for biomedical applications: perspectives and challenges. *Trends in Biotechnology*, 2010. 28 (4), 189–197.
7. Dvir, T., Timko, B. P., Kohane, D. S., Langer, R. Nanotechnological strategies for engineering complex tissues. *Nature Nanotechnology* 2011, 6 (1), 13–22.
8. Persano, L., Camposeo, A., Pisignano, D. *Progress in Polymer Science, In Press, Available online 13 October 2014.* http://dx.doi.org/doi:10.1016/j.progpolymsci.2014.10.001. Active polymer nanofibers for photonics, electronics, energy generation and micromechanics.
9. Sundarrajan, S., Tan, K. L., . Lim, S. H, Ramakrishna, S. Electrospun nanofibers for air filtration applications. *Procedia Engineering*, 2014. V. 75, 159–163.
10. Bhardwaj, N., Kundu, S. C. Biotechnology Advances. Electrospinning: A fascinating fiber fabrication technique. 2010. 28 (2), 325–347.
11. Baji, A., Mai Y-W., Wong S-C., Abtahi, M., Chen, P. Electrospinning of polymer nanofibers: Effects on oriented morphology, structures and tensile properties. *Composites Science and Technology.* 2010, 70, 703–718.
12. Raghavan, P., Lim D-H., Ahn J-H., Nah Ch., Sherrington, D. C., Ryu H-S, Ahn H-J, Electrospun polymer nanofibers: The booming cutting edge technology. *Reactive and Functional Polymers* 2012, 72, 915–930.
13. Palangetic, L., Reddy, N. K., Srinivasan, S., Cohen, R. E., McKinley, G. H., Clasen, C. Dispersity and spinnability: Why highly polydisperse polymer solutions are desirable for electrospinning? Polymer. 2014. 55, 4920–4931.
14. Iordanskii, A. L., Bonartseva, G. A., Pankova Yu. N., Rogovina, S. Z, Gumargalieva, K. Z., Zaikov, G. E., and Berlin, A. A. Current Status and Biomedical Application Spectrum of Poly(3- Hydroxybutyrate as a Bacterial Biodegradable Polymer Volume, I. Current State-of-the-Art on Novel Materials. (Editors: Devrim Balköse, Daniel Horak, Ladislav Šoltés). Ch. 12. 2014. Apple Academic Press. New York. 450pp. ISBN: 978-1-926895-79-6. Cat# N11007.
15. Shchegolikhin, A. N., Iordanskii, A. L., Filatova, A. V., Gumargalieva, K. Z., Fomin, S. V., Potapov, E. E. Water transport, FTIR, and morphology characterizations of novel biodegradable blends based on poly(3-hydroxybutyrate). J. Polym Eng. 2011. 31, 283–288.
16. Bonartsev, A. P., Livshits, V. A., Makhina, T. A., Myshkina, V. L., Bonartseva, G. A., Iordanskii, A. L. Controlled release profiles of dipyridamole from biodegradable microspheres on the base of poly(3-hydroxybutyrate). *eXPRESS Polymer Letters,* 2007,1 (12), 797–803.
17. Ivantsova, E. L., Iordanskii, A. L., . Kosenko, R.Yu, Rogovina, S. Z., Prut, E. V. A novel biodegradable composition PHB-chitosan for controlled release of bioactive components. Chemical-pharmaceutical, J. 2011. 45 (1), 39–44.
18. Corre Y-M., Bruzaud, S., Audic, J.-L., Grohens, Y. Morphology and functional properties of commercial polyhydroxyalkanoates: A comprehensive and comparative study. Polymer Testing. 2012. 31, 226–235.
19. Chanprateep, S. Current trends in biodegradable polyhydroxyalkanoates. Journal of Bioscience and Bioengineering. 2010. 110 (6), 621–632.

20. Galeski, A. Strength and toughness of crystalline polymer systems. Progress in Polymer Science. 2003. 28 (12), 1643–1699.
21. Pluta, M., Bartczak, Z., Galeski, A. Changes in the morphology and orientation of bulk spherulitic polypropylene due to plane-strain compression. Polymer. 2000. 41, 2271–2288.
22. Ma, Q., Mao, B., Cebe, P. Chain confinement in electrospun nanocomposites: using thermal analysis to investigate polymer-filler interactions. *Polymer* 2011, 52, 3190–3200. doi:10.1016/j.polymer.2011.05.015
23. Karpova, S. G., Iordanskii, A. L., Popov, A. A., Shilkina, N. G., Lomakin, S. M., Shcherbina, M. A., Chvalun, S. N., Berlin, A. A., Effect of External Influences on the Structural and Dynamic Parameters of Polyhydroxybutyrate–hydroxyvalerate Based Biocomposites. Russian Journal of Physical Chemistry B, 2012, 6 (1), 72–80.
24. Filatov, Y., Budyka, A., Kirichenko, V. Electrospinning of Micro- and Nanofibers: Fundamentals in Separation and Filtration Processes. New York, Begell House Inc., 2007. 404p.
25. Vlasov, S. V., Olkhov, A. A. Rolling influence on properties of isotropic and oriented films. Plastic Masses 1996, # 6, 40–42.
26. Vasserman, A. M., Buchachenko, A. L., Kovarskii, A. L., Neiman, M. B. Study of molecular motion in polymers by the paramagnetic probe method. Polymer Science, U.S.S.R. 1968. 10(8), 2238–2246.
27. Staroverova, O. V., Shushkevich, A. M., Kuzmicheva, G. M., Olkhov, A. A., Voinova, V. V., Zharkova, I. I., Shaitan, K. V., Sklyanchuk, E. D., Guryev, E. D. Fibrillar matrixes for tissue engineering constructions from poly(3-hydroxybutyrate) and its composites (In Rus). Technology of living systems. 2013, 10(8), 74–79.
28. Rutledge, G. C., Fridrikh, S. V. Formation of fibers by electrospinning. Advanced Drug Delivery Reviews 2007, 59 (14), 1384–1391.
29. Kamaev, P. P., Aliev, I. I., Iordanskii, A. L., Wasserman, A. M. Molecular dynamics of the spin probes in dry and wet poly(3-hydroxybutyrate) films with different morphology. Polymer. 2000. V.42. №2. 515–520.
30. Di Lorenzo, M. L., Gazzano, M., Righetti, M. C. The Role of the Rigid Amorphous Fraction on Cold Crystallization of Poly(3-hydroxybutyrate). Macromolecules 2012, 45, 5684−5691. dx.doi.org/10.1021/ma3010907.
31. Di Lorenzo, M. L., Righetti, M. C. J. Therm. Anal. Calorim. online. DOI 10.1007/s10973–012–2734–3 Evolution of crystal and amorphous fractions of poly[(R)-3-hydroxybutyrate] upon storage.
32. Iordanskii, A. L., Kamaev, P. P., Zaikov, G. E. Water immobilization and transport in poly(3-hydroxybutyrate). Int. J. Polymer. Mater 1998, 41, 55.

INDEX

D

E

F

G

H

I

K

O

P